PHILLIP H. SMITH, Ph.D.

48.50

8/83

D1266771

STEREOLOGICAL METHODS

Vol. 1

Practical Methods for Biological Morphometry

STEREOLOGICAL METHODS

Vol. 1

Practical Methods for Biological Morphometry

Ewald R. Weibel

Department of Anatomy, University of Berne,
Berne, Switzerland

1979

ACADEMIC PRESS
London · New York · Toronto · Sydney · San Francisco
A Subsidiary of Harcourt Brace Jovanovich, Publishers

ACADEMIC PRESS INC. (LONDON) LTD.
24/28 Oval Road
London NW1

United States Edition published by
ACADEMIC PRESS INC.
111 Fifth Avenue
New York, New York 10003

Copyright © 1979 by
ACADEMIC PRESS INC. (LONDON) LTD.

All Rights Reserved
No part of this book may be reproduced in any form by photostat, microfilm, or any other
means, with written permission from the publishers

Weibel, Ewald R.
  Stereological methods
  Vol. 1: Practical methods for Biological Morphometry
  1. Morphology—Mathematics
  2. Stereology
  I. Title
  574.4'028   QH351   78-75269
  ISBN 0-12-742201-3

Printed in Great Britain by Page Bros (Norwich) Ltd,
Mile Cross Lane, Norwich

Contributors to Chapter 8

ROBERT P. BOLENDER, Department of Biological Structure, School of Medicine, University of Washington, Seattle, Washington 98195, U.S.A.

PETER H. BURRI, Anatomisches Institut, Bühlstrasse 26, CH 3000 Berne 9, Switzerland.

BRENDA EISENBERG, Department of Pathology, Rush Medical College, Chicago, Illinois 60612, U.S.A.

HERBERT HAUG, Anatomisches Institut, Medizinische Hochschule, Ratzeburger Allee 160, D24 Lübeck, West Germany.

HERMANN HECKER, Schweizerisches Tropeninstitut, Socinstrasse 57, CH 4051, Basel, Switzerland.

TERRY M. MAYHEW, University of Sheffield, Department of Human Biology and Anatomy, Sheffield S10 2TN, England.

ERNEST PAGE, Department of Medicine, University of Chicago, Illinois 60637, U.S.A.

HANSPETER ROHR, Pathologisches Institut, Schönbeinstrasse 40, CH 8056 Basel, Switzerland.

ROBERT K. SCHENK, Anatomisches Institut der Universität Bern, Abt. für Systematische Anatomie, Bühlstrasse 26, CH 3000 Bern 9, Switzerland.

HUBERT E. SCHROEDER, Department of Oral Structural Biology, Dental Institute, University of Zürich, Plattenstrasse 11, CH 8028, Zürich, Switzerland.

LÉON SIMAR, Laboratoire d'Anatomie Pathologique, Institut de Pathologie, Université de Liège au Sart Tilman, 4000 Liège, Belgium.

Preface

A GUIDE TO THIS BOOK

When setting out to write this book several years ago my intention was to produce guide-lines, drawn from practical experience, which should help biologists of varied denominations to "do good stereology", in other words, to diligently use the powerful methods of stereology in the framework of their morphometric endeavours. For this purpose, the text should be general and simple, but I found, in due time, that these two postulates could not be realized together: to keep the exposition simple would have entailed a great loss in generality because of the necessity to restrict the theoretical justification of stereological methods to some well-defined boundary conditions; to make it general requires a rigorous mathematical approach, and this would certainly shy away many biologists.

I therefore decided to split this book into two volumes which are, to a certain extent self-contained, but are also, in part, mutually dependent on each other. The aim of Volume 1 is to present the practical methods of stereology and to provide the user with that minimum of theoretical background which is required in order to prevent serious mistakes. Volume 2 presents the theoretical background for these methods in full. Stereology is actually a branch of applied mathematics. Practical stereological methods are based on usually very simple formulas, but these are so simple only because they were derived by often rather complex mathematical reasoning. A diligent use of these simple methods under sometimes tricky circumstances may make an occasional consideration of their theoretical roots unavoidable.

After a brief general introduction, Volume 1 starts out by an elementary introduction to stereological principles (Chapter 2). Requiring no further mathematical knowledge than that of a high school level it presents plausible justifications of the basic stereological methods, discussing first the fundamental operation of sectioning and its relation to geometrical probability. It then goes on to discuss the basic measurements (volume, surface, numerical density etc.) and the relations between the size of particles and that of their profiles seen on sections. This chapter uses very simple model structures, such as spherical particles; the conclusions drawn are compatible with the more general, and hence more realistic case, as presented in Volume 2, but the insight provided is evidently limited because it is based on special cases. Chapter 3 then deals with the important topic of sampling, important because obviously the study of a spatial structure on a section needs must involve sampling, and a difficult and critical type of sampling. Chapters 4–6 present the practical methods for applying stereological principles. Point counting methods are dealt with in Chapter 4, from the design of appropriate test systems to the calculation of stereological parameters and their corrections for inherent errors. Chapter 5 exposes the methods for obtaining estimates of particle size, whereas various additional methods are presented in Chapter 6. The systematic presentation ends with some guide-lines on how one should set about performing a stereological study (Chapter 7); the discussion of instrumentation is purposely kept very general, firstly because the technological developments are so fast that any specific presentation of instruments is bound to be outdated in a few years, and secondly, because instrumentation is of secondary importance in stereology: one can do good stereology with rather primitive equipment!

In Chapter 8 some special cases of application of stereological methods are presented. I have asked a number of biological morphometrists to explain, on a few pages, the way they have used stereological methods in order to solve some specific morphometric problems relating to various organs. This illustrates the use that can be made of the more general principles exposed in the preceding chapters. The Appendices finally collate some of the more pertinent bits of information that are frequently used in practical work.

Volume 2 is devoted to the theoretical foundations of stereology. It is unavoidable that this must make extensive use of mathematical reasoning so that some advanced knowledge of mathematics is required of the reader. However, an attempt was made to still keep this presentation accessible for non-mathematicians, at least on an intuitive basis; indeed full-fledged mathematicians may find the presentation primitive and naive on many accounts, but the book was not written primarily for their benefit, but rather for biologists. The introduction (Chapter 1) presents some background information, from some remarks on the history of stereology to the mathematical tools required. The exposition of the foundations of stereological principles (Chapters 2–10) retraces the sequence of topics presented in

Volume 1, Chapters 2–6. However, a number of topics that do not lend themselves easily to elementary explication are here dealt with in full, whereas they
had only been alluded to in Volume 1.

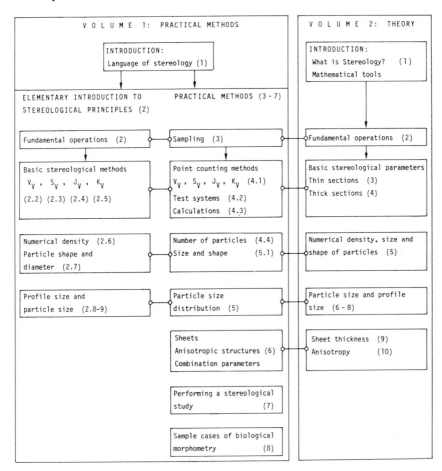

How should this book be used? The above diagram outlines the structure
of the book and some of the dependencies. I imagine that most readers from
biology will want to start with volume 1 and consult volume 2 when the need
for deeper understanding arises. A *newcomer to stereology* may first browse
or even read through Chapters 1 and 2 of Volume 1 and pick out those bits
of information that he finds necessary in order to find out what stereology is
about and what it might offer for his own work. He would then probably
jump to Chapter 4, reading section 4.1 to gain an overview and then proceed
to those sections of Chapters 4–6 which he may want to put to practical use

in his study. When setting out to do the work, he should consult Chapters 3 and 7 which can assist him in proper planning of the study. At one time or another it may be profitable to consult the relevant sections of Chapter 8 where a number of useful practical hints and some specific references to successful use of stereological methods are to be found.

The initiated stereologist may go directly to the presentation of practical methods in Chapters 4–6 and refer to the other chapters when the need comes up. He may familiarize himself with more advanced stereological methods by consulting Volume 2; this gives an adequate background if the development of new methods or substantial modifications of existing methods are required.

If this text is to be used in a course for *students* I would wish that Volume 2 be used to provide them with a sound theoretical basis, presenting the practical side of stereological methods as a result of the theoretical developments, rather than the theory as a justification of some practical methods. As the use of stereological methods becomes more and more sophisticated one cannot do without a sound theoretical basis.

Many a biologist will in time find it necessary to *consult a mathematician* or statistician in trying to solve some tricky problems. If this mathematician is not familiar with stereology – which will often be the case – Volume 2 could serve as a basis for discussion, rather than Volume 1. This would prevent the frequent event that one wastes a lot of time and energy at "reinventing the wheel". The mathematician will also find more appropriate references to the literature in Volume 2 than in Volume 1.

I hope that this book will help a variety of potential readers in doing good stereology.

August, 1978   Ewald R. Weibel

Acknowledgements

The writing of a book of this kind is a lonely task, but it cannot be done alone. I would therefore like to express my deep gratitude first to all those who have helped me acquire the notions presented in this book, and then to those who assisted me in the actual preparation of the text.

The seed was sown two decades ago when I learned to think in terms of mathematical models while working with Domingo Gomez at Columbia University and later on with Bruce W. Knight at Rockefeller University in New York. I was then fortunate to share the decisive moments in the early years of the International Society for Stereology with experienced "stereologists" from various disciplines, and it is particularly through the interaction with Hans Elias, Robert T. De Hoff, Ervin E. Underwood, Herbert Haug, Hans Giger, John E. Hilliard, August Hennig and many other colleagues that I could gradually get a grasp on the basic notions of stereology, that I could learn to correct some of the mistakes I had made and to avoid many of the pit-falls. A decisive event for me was the acquaintance and then friendship with Roger E. Miles; it will become obvious from the text, particularly in volume 2, how much I owe to him for having taught me what mathematical stereology truly is. I hope to have done him justice in my attempts to "translate" some of his thinking into the language of biology.

Among those who helped me in the process of assembling this text over a period of several years, my secretary Ms Gertrud Reber deserves to be mentioned and thanked in the first place; without her devoted and meticulous assistance this task could not have been accomplished. No lesser gratitude goes to Mr Karl Babl who not only did all the graphical work for this book, diligently assisted by Ms Marianne Rüfenacht, but also contributed importantly to all the technical developments done in our workshop. The contributions of my colleagues in the department are countless and I cannot list them all; I am deeply grateful for them particularly to Peter Burri, Robert Schenk, Robert Bolender, Dagmar Paumgartner, Hans-Rudolf Gnägi, Hansjörg Keller and Luis-Manuel Cruz-Orive who critically read the manuscript.

Finally I should thank the institutions which have generously supported our stereological work, particularly the Swiss National Science Foundation, the University of Berne, and the Roche Research Foundation. I am also grateful to the authors of the contributions to Chapter 8 for having generously consented to collaborate with me in this venture.

Contents

To VERENA
who cared and shared.

CHAPTER 1

Introduction

1.1. What is stereology?

The word "stereology" is still not to be found in contemporary dictionaries in 1979; it was coined in 1961 by a small group of scientists gathered on the Feldberg in Germany under the leadership of Hans Elias (see Elias, 1963). The purpose of this informal conference was to exchange views on the "spatial interpretation of sections" of solid tissues or materials. It was attended by a few biologists and a couple of mathematicians, but it drew immediate attention from some material scientists, metallurgists and mineralogists, who had been struggling with this kind of problem for a long time. It was no more than two years before the "First International Congress for Stereology" took place in Vienna. The title of this conference was certainly ambitious, for it was attended by no more than about thirty participants, but these represented a truly impressive range of disciplines (Haug, 1963). This Congress was the occasion where the International Society for Stereology was formally founded and registered by a handful of people.

In the intervening years three more congresses and several symposia were held on that topic and the membership of the Stereological Society has grown to several hundred. We have gradually also overcome the phase of groping for what "stereology" means; the definition of the term is as follows:

Stereology
is
(1) *a body of mathematical methods*
relating
(2) *three-dimensional parameters defining the structure*
to
(3) *two-dimensional measurements obtainable on sections*
of the structure.

1

Although the etymological root of the word "stereology" is rather broad, its connotative meaning has become quite specific: it is a methodology, perhaps a concept, by which we can learn to "spatially interpret sections", the daily problem of any microscopist, be he biologist or material scientist.

The problem encountered—often unconsciously—in interpreting microscopic images is that one is, in general, offered a flat image of a *section* through a spatial structure and is forced to visualize the spatial context, often without much objective guidance about what was really there before the section was cut. Our eyes are unused to looking at sections: the flat images we normally perceive are projections of three-dimensional objects, and we have a natural tendency to interpret any flat image as a projection. Thus, a circular "profile" impresses us, at first sight, as the image of a spherical object, but it could just as well have been derived by cutting a cylindrical rod. Many misrepresentations of tissue fine structure—still perpetuated in many textbooks—stem from this kind of misinterpretation of section images (Elias, 1972).

The problems of interpreting section images become very disturbing if we want to obtain from a study of sections information on the dimensions of spatial objects making up the structure. *How is the size of a profile related to the size of the object of interest*? Clearly, if a structure made of spherical objects is cut by a section plane most of the spheres will not be cut through their centers, so that, evidently, most profiles will have a diameter which is smaller than that of the spheres. Can nevertheless some information on the sphere profiles be gathered which allows some statements about the size of the spheres?

This is the field of stereology: as a theoretical branch of science it tries to develop relationships between some parameters of structure—e.g. the size of particles—and some appropriate measurements on the size of profiles that one finds on the sections. Mathematics enters into play because the appearance of profiles depends on the chance of cutting the objects in a certain way; stereology is therefore deeply statistical in nature.

The task of stereology as a science proper is to develop such relations by rigorous mathematical reasoning; but the aim of the exercise is to provide the microscopist with sound and useful methods by which he can obtain reliable quantitative data on structure, even though he may be restricted to the study of sections. The methods of stereology are particularly useful in biological morphometry.

Clearly then, there are two sides to stereology, and accordingly two populations of stereologists. The first group, made of mathematicians of varying denomination, deals with the theoretical foundations of stereology, of the relations between structure and sections; the second, made of biologists and material scientists, seeks to extract from these foundations, methods that help solve some practical problems of microscopic morphometry. What this book intends to do is to present in Volume 1 the solutions to problems, as they may be useful to biologists, and to expand in Volume 2 on their mathematical foundations.

1.2. The language and symbolism of stereology

As is the case with most scientific disciplines, stereology has developed its own language, particularly by giving some terms a restrictive connotative meaning. Some of these terms need to be explained.

A *structure* is generally defined as "something made up of interdependent parts in a definite pattern of organization" (Merriam–Webster). This definition holds for stereology as well: a structure is made up of at least two parts or components, and to clarify the pattern of organization—degree of interconnections between the parts, their relative size etc.—is the declared purpose of stereology.

The *components* of the structure can be described in various ways. Firstly, the different components must be clearly separated and identifiable; regions of transition from one to the other component (e.g. zones of diffusion) are difficult to handle. We call a *phase* the aggregate of all parts which are identical in nature, e.g. all mitochondria of a cell constitute its mitochondrial phase. If the component is made up of discrete elements that could be isolated as units we will call these *objects* or *particles*. These particles may have an arbitrary shape, but one specially important class of particles are *convex solids* which are defined as follows: the line segment connecting any two points of the solid must lie totally within the solid, in other words, a line traversing the solid forms only one chord or intercept and likewise a section hitting the solid forms only one profile.

We shall, in the following, often describe a structure quantitatively as the "*containing space*", i.e. the space containing the components of interest, often together with some "uninteresting" phases. A structure of this kind is shown in Fig. 1.1. The fundamental quantitative descriptor of the make-up of the structure is the *density* of the various components within the structure, i.e. within the containing space. A "density" is defined as "the quantity per unit volume, unit area or unit length" (Merriam–Webster). The basic quantities we shall be interested in are the volume of the component phases, their surface, their length, or the number of their particulate elements, so that the terms *volume density, surface density, length density, numerical density* describe these quantitative properties of the structure in relation to the structure space. The volume density is the volume of the phase in the unit volume of the structure, the surface density the surface area of the phase contained in the unit volume of the structure, etc.

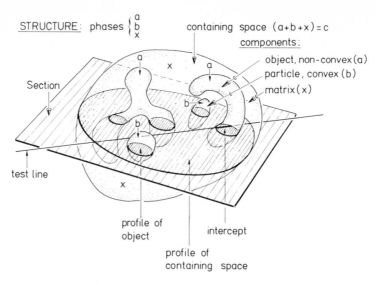

STRUCTURE: phases $\begin{cases} a \\ b \\ x \end{cases}$ containing space $(a+b+x)=c$

components:

object, non-convex (a)
particle, convex (b)
matrix (x)

Section

test line

profile of object

intercept

profile of containing space

Fɪɢ. 1.1.: Model structure and section.

The concept of density is particularly useful in stereology because of the very nature of the sampling process used: sectioning. A *section* is defined as a plane intersecting the structure; it hence also intersects some of the components of the structure. The "image" of any spatial component (and of the structure) on a section is called a *"profile"*, i.e. the sharply outlined flat trace on the section plane. It is intuitively understandable that a relationship must exist between the density of component profiles within the profile of the structure (the "quantity of profile per unit area") and the density of components within the structure. It is the basic goal of stereology to establish such relationships by mathematical reasoning.

It is found that the procedure of sectioning, i.e. the probing of a structure by means of a plane, can be extended to include other *geometric probes*, such as *test slices, test lines* (Fig. 1.1) and *test point sets*. In each case, the image of the structural composition on these probes again quantitatively reflects the make-up of the structure: the density of *intercept lengths* of some component on the (unit length of) test line is related to its volume density, and so is the "density of points" (per unit point set) falling into the component related to its volume density in the structure.

This narrative description of the nature of stereological procedures reveals one important point: all stereological measurements are in principle obtained as relative measurements, more precisely as a ratio of at least two joint measurements: one relating to the components, the other to the structure as a whole; the latter is called the "reference system". Stereological principles establish the precise relationship between such ratios measured on sections, and the corresponding ratios in the spatial structure.

This fundamental nature of stereological principles is reflected in the set of stereological symbols that have come into general usage over the past decade: the ratios are noted as a double symbol of two capital letters: the first capital letter defines the parameter, the second, usually written as a subscript, the parameter of the containing space or, more generally, of the reference system. For example, the numerical density of some particulate component (b) in the containing space (c) is written as

$$N_V = N(b)/V(c) \qquad (1.1)$$

or the areal density of the profiles of (b) in the profile of the structure (c) as

$$A_A = A(b)/A(c) \qquad (1.2)$$

The list of the symbols used in this text is given in Table 1.1, and is repeated in slightly expanded form in Appendix 1. There has been a tendency for unification of symbols in the International Society for Stereology, so as to facilitate reading of articles by various authors. Whilst this endeavour is certainly laudable we have found it impossible to restrict the symbols to a few selected letters because this can lead to ambiguities. For example, it is inadequate to use the letter P ("points") for both point hits by a point lattice *and* intersection points between a test line and a surface; we have therefor introduced the symbol I for "intersections" and have retained P for "point hits". Likewise the use of L for both test line length and length of a profile boundary leads to ambiguities, so that we have substituted B for "boundary length". The second column in Table 1.1 lists alternative symbols used by other authors.

A particular problem concerns the identification of the phases or components to which the parameter symbol relates. We have chosen to use lower-case roman letters for this purpose and to attach them to the parameter symbol either as a subscript or in parenthesis, such as

$$N_b = N(b) \qquad (1.3)$$

for the number of objects of phase b. The numerical density is then written as

$$N_{Vb} = N_V(b) = N(b)/V(c) \qquad (1.4)$$

If the containing space or reference system needs to be identified, its symbol is added following a comma:

$$N_{Vb,c} = N_V(b, c) = N(b)/V(c) \qquad (1.5)$$

Table 1.1
List of symbols.

Definition	Symbol	Alternative Symbols	Dimension
1. Parameters of structure			
Volume (of component or structure)	V		cm^3
Surface area (of components)	S		cm^2
Length of linear feature in space	J	L	cm
Mean curvature	K	M, H	cm^{-1}
Total (Gaussian) curvature	G		cm^{-2}
Number of objects	N		cm^0
Diameter (of spheres)	D		cm
Caliper diameter	H	D	cm
Sheet thickness	τ		cm
2. Parameters of profiles on section area			
Area of profiles	A		cm^2
Boundary length of profiles (perimeter)	B	L, P, C	cm
Transections of linear features	Q		cm^0
Tangents to profile boundary	T		cm^0
Number of profiles	N	n	cm^0
Diameter (caliper) of profiles	d, h		cm
Curvature of plane figure	C		
3. Parameter on test lines			
Length (total) of profiles on test line	L		cm
Intersections with surface or profile boundary	I	P, N, C	cm^0
Length of intercept or chord	l	L_2, L_3	cm
Number of intercepts	n	N	cm^0
4. Parameters on test point sets			
Point number on profiles or structure	P		cm^0

In such cases, one could evidently just as well write out the ratio in full; the main advantage of using the stereological double symbol, followed by phase identifiers, is that it reminds us that stereological estimates are never absolute but always relative estimates: we cannot obtain $N(b)$ by stereological methods, but only $N_V(b, c)$;

A last remark relates to the limitations imposed by the use of computers where the symbols have to be printed in a single line, and where no distinction between capital and lower-case letters is possible. By writing $NV(B, C)$ no ambiguity results: the first letter is the component parameter, the second the reference parameter, and the phase identifiers are in brackets.

Table 1.1 (continued)
List of symbols

Definition	Symbol	Alternative Symbols	Dimension
5. Parameters of test sets			
Test volume	V_T		cm^3
Test area	A_T		cm^2
Test line length	L_T		cm
Test point number	P_T		cm^0
Section (slice) thickness	t	T	cm
6. Densities in space			
Volume density	V_V		cm^0
Surface density	S_V		cm^{-1}
Length density in space	J_V	L_V	cm^{-2}
Mean curvature density	K_V	M_V	cm^{-4}
Total curvature density	G_V		cm^{-5}
Numerical density	N_V		cm^{-3}
7. Densities on test sets			
Areal density	A_A		cm^0
Boundary length density	B_A		cm^{-1}
Transection density on area	Q_A		cm^{-2}
Numerical density of profiles on area	N_A		cm^{-2}
Tangent density on area	T_A		cm^{-2}
Intercept length density on line	L_L		cm^0
Intersection density on line	I_L		cm^{-1}
Intercept number density	N_L		cm^{-1}
Point density	P_P		cm^0

1.3. Stereology is basically a mathematical approach

To construct an electron microscope is a difficult task, to be solved by physicists and engineers. The result is an instrument that can easily be used by biologists to produce good electron micrographs, requiring a minimum of technical knowledge of the underlying physical principles. Certainly, the number of biologists attempting to build their own electron microscope or even to modify it is rather small.

There is no difference with stereological methods. To develop them requires profound mathematical knowledge in such difficult fields as geometrical probability theory, integral geometry, or statistics. A group of highly qualified mathematicians have now taken hold of the field (see Miles and Serra, 1978), and improved stereological methods are rapidly developing. The increasing sophistication of the mathematical reasoning ensures a continuous improvement, particularly in defining boundary conditions for the use of stereological methods.

The astonishing fact is, however, that such complex mathematical reasoning in general leads to rather simple results: formulas that can be applied by microscopists without much knowledge about the process by which they were derived. In this first volume we therefore present these formulas as tools for practical morphometric work, and try to explain them in a rather elementary fashion. This is adequate for anybody who just wishes to do some measurements and obtain reliable results. However, the reader is reminded that the development of these formulas has often involved much more than is revealed by the elementary explanations given; he is therefore warned that he should remain modest and not try to modify the formulas or combine them without looking at their theoretical foundations, lest he is liable to make serious mistakes. In Volume 2, some of these foundations will be presented in such a way that a biologist with some background in advanced mathematics, or with the help of a mathematician, can get a better grasp of these methods.

1.4. Reading on stereology

For a long time stereological methods have been developed pragmatically in various fields. The need to measure on sections came up independently in the material sciences, mineralogy and metallurgy, and in the biological sciences. Accordingly, the references to stereological papers are wide-spread. Since biologists were late-comers in the field it is evident that much useful information can be obtained from consulting works from the material sciences (De Hoff and Rhines, 1968; Underwood, 1970). In the mathematical sciences explicit references to stereology have come about only in recent years; but much important ground-work, which is immediately relevant to stereology's mathematical foundations, has been done particularly in geometrical probability (see Kendall and Moran, 1963) and integral geometry (see Hadwiger, 1957, Santaló, 1976). The list of references given at the end of this volume presents a bibliography to some of the more important publications of a general nature; it also lists the journals in which papers on stereological methods regularly appear. A good overview on the development of stereology can also be obtained by consulting the proceedings of the various Congresses and Symposia organized or sponsored by the International Society for Stereology.

CHAPTER 2

Elementary introduction to stereological principles

2.1. Fundamental operations in stereology

2.1.1. SECTIONING

In morphological research one of the methods of choice is the microscopic study of sections across the structures under investigation. The great advantage of this approach is that the section "opens" the interior of the structure without disturbing the relative position of internal objects. But the picture offered by a section is in many ways different from that seen by dissection. In view of a quantitative stereological approach it is hence most important to examine, at the outset, the geometric aspects of sectioning.

The section exposes the internal objects in a very specific way; it will cut across some of the objects, but miss others; if it hits the object one will observe a trace of the object on the section plane, and the properties of this trace will depend on the properties of the object and on the way it has been cut.

Let us now try to arrive at a good and general definition of what a section is, so that we may derive some likewise general features of traces of structures on sections. Our common understanding of the term "section" implies a solid being cut by a plane, or, in other words, a two-dimensional space intersecting with a three-dimensional space. A "section" might thus be redefined as the intersection between two spaces.

With this general definition it is obvious that we may distinguish several degrees of sectioning. Looking at solid structures we may obtain three-dimensional sections or slices if the space intersecting the solid is itself three-dimensional. We may obtain a two-dimensional section—a true section— if a plane intersects the solid, or we may obtain a one-dimensional section if a line intersects the solid. In further reduction of this scheme we may even con-

9

sider a point or a set of points as forming a 0-dimensional section with the solid, if the points are introduced into the solid from outside.

To illustrate this somewhat abstract concept we must briefly look at the procedure by which we generate a section (Fig. 2.1). Following the common understanding we consider a section to be the trace left by the edge of a microtome knife pushed through the tissue along a straight line course perpendicular to the knife edge. The section thus is a plane, since the knife edge itself is a straight line too. To obtain a histological section a second identical operation is necessary: the knife edge is lowered by a certain small distance and again pushed through the tissue. If we would have the means to perform both cuts simultaneously the two parallel knife edges would be the edges of a narrow plane strip which is pushed through the tissue along a normal to this plane, thus generating a three-dimensional section or slice.

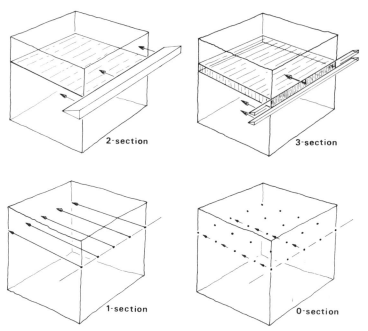

Fig. 2.1: Sectioning with a microtome knife can be used to illustrate the generalized concept of "sectioning" used in stereology. (Reproduced by permission from Weibel, 1967c.)

But what is the knife edge or line? Nothing else than an infinite number of points linearly arranged. If we now pick out a discrete number of these points and slide them through the tissue we obtain what we have defined as a one-dimensional section. And if, instead of continuously sliding them through the tissue, we let them jump by discrete intervals we deposit a set of points inside the tissue and obtain a 0-dimensional section.

What have we done? We have sent some *probes* into the tissue and these probes have picked out some *sample of the interior* on which we hope to gain insight into the internal construction of the tissue. And we may have chosen the dimension of our probe for convenience.

Let us now see how the "section" and the "structure" interact. We assume the structure to be a cube containing one simple object: a polyhedron

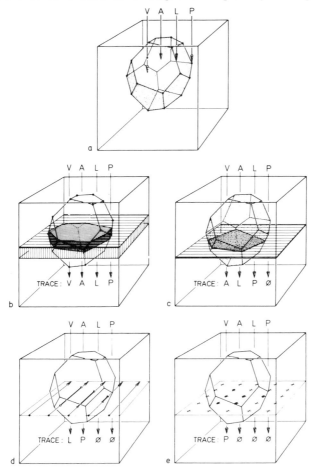

FIG. 2.2: (a) A polyhedron can serve as a general model object because it can be quantitatively defined by the four basic parameters: V = volume of the solid; A = area of the faces; L = length of the edges; P = number of corner points. (b)–(e) "Sectioning" with a slice (b), a plane (c), a set of lines (d), and a set of points (e) produces traces of different dimensions.

(Fig. 2.2a), which is quantitatively defined by its volume, V, the surface area of its faces, A, the length of its edges, L, and the number of its corner points, P. A three-dimensional section, or a slice, will cut out a slice of the polyhedron which contains a sample of all four parameters: volume, face area, edge length, and corner points (Fig. 2.2b). A plane two-dimensional section will show a flat polygonal trace of the polyhedron, the polygon being character- ized by area, A', a boundary edge length, L', and a number of corner points, P' (Fig. 2.2c).

But it is evident, that A' is related to the volume V of the polyhedron, L' to the surface area A, and P' to the edge length L; and furthermore, the corner points of the polyhedron are not represented on the section. Hence, the dimensions of the trace parameters are by 1 smaller than the dimensions of the object parameters from which they were derived. If we now probe with a line or one-dimensional section we observe that the volume is represented by a line segment (L'), the surface area by intersection points (P'), and that the polyhedral edge length is lost (Fig. 2.2d); the trace dimensions are re- duced by 2 as compared to the object dimensions. And finally if we probe with a point lattice or 0-section we will find some points (P') occupied by the object volume, all other parameters being lost (Fig. 2.2e); the dimensions of the trace parameters are reduced by 3.

This can be expressed in general form. Call the object parameter a, the section s, the trace t, and the respective dimensions $d(a)$, $d(s)$, $d(t)$, then we find that

$$d(t) = d(a) + d(s) - 3 \qquad (2.1)$$

You can easily verify this: if the object surface $(d(a) = 2)$ is cut with a line $(d(s) = 1)$ then the trace is a point $(d(t) = 0)$.

It should be noted that this formula has a limiting condition, namely

$$d(a) + d(s) \geqslant 3 \qquad (2.2)$$

which means that if the sum of the dimensions of object parameter and section

Table 2.1

Parameters and dimensions of object, section and trace (Modified after Weibel, 1967c)

	Parameters					Dimensions				
	TRACE	SECTION				$d(t)$	$d(s)$			
	↘	V	A	L	P	↘ 3	2	1	0	
OBJECT V	V	V	A	L	P	3	3	2	1	0
A	A	A	L	P	—	$d(a)$ 2	2	1	0	—
L	L	L	P	—	—	1	1	0	—	—
P	P	P	—	—	—	0	0	—	—	—

is smaller than 3, then no trace is formed: surface areas are not represented on a test point set.

These relationships are graphically shown in Table 2.1. It should particularly be noted that for every object parameter there exists a section dimension which produces a point trace. On the basis of formula (2.1) this is the case when

$$d(s) = 3 - d(a) \qquad (2.3)$$

It is important to remember this relationship because this opens the way to the efficient point counting methods of stereology. The formula here described for three-dimensional space is a special case of a general formula applying to n-dimensional space, as will be outlined in Chapter 2 of Volume 2.

2.1.2. GEOMETRIC PROBABILITY

Stereological methods are statistical in nature: we cut a structure at random and make our observations on a random sample of profiles obtained by this procedure, whereby we assume that each type of profile or trace obtained occurs with a certain well-defined probability. In order to understand these methods it is necessary that we discuss the special considerations on probability that apply to stereology. Since we deal with geometric objects, and since the section planes, test lines etc. are likewise geometric elements, the probabilities to be considered will also be geometric in nature.

The usual definition of probability is as follows: if one repeatedly performs an experiment of some kind there will be a number of outcomes of a type we expect, and a number of outcomes that are different. The probability of obtaining the expected outcome is the number of "positive" or "favorable" outcomes divided by the total number of trials, provided the number of trials is very large.

A probability of this kind can also be predicted by theoretical considerations. Take the two well-known examples of tossing a coin or throwing a dice. In tossing a coin you predict to have a "fifty-fifty chance" to have "head up". What you have done is the following: the coin has two faces, hence 2 outcomes are possible; with a "good coin" both outcomes are equally likely, so that you predict that both "head up" and "tail up" will have a chance of 50%. The probability that an event i occurs is usually given in values $\Pr\{i\}$ in such a way that

(1) $\Pr\{i\}$ is a number between 0 and 1

$$(0 \leqslant \Pr\{i\} \leqslant 1) \qquad (2.4)$$

(2) the sum of all the possible $\Pr\{i\}$ in one experiment is 1:

$$\sum_i \Pr\{i\} = 1 \qquad (2.5)$$

In the coin-tossing experiment there are two outcomes, both equally likely; hence the probabilities are

$$Pr\{head\} = Pr\{tail\} = \tfrac{1}{2}$$

which satisfies the two rules.

In throwing a dice you have six possible outcomes: 1, 2, 3, 4, 5, or 6 up. With a "good" dice all are equally likely, so that

$$Pr\{1\} = Pr\{2\} = Pr\{3\} = Pr\{4\} = Pr\{5\} = Pr\{6\} = \tfrac{1}{6}$$

which again satisfies the two rules for setting up probabilities.

In both examples we have proceeded the following way: we have grouped all possible outcomes into sets in such a way that each set is equally likely to occur, which was easy in these examples. We determined the number N of these sets of possible outcomes and obtained the probabilities by

$$Pr\{i\} = 1/N$$

If we were to ask for the probability that either a 1 or a 3 is up in the dice experiment, we would proceed the same way and find that, because $N = 6$ and $Pr\{1\} = Pr\{3\} = \tfrac{1}{6}$, the probability of having either 1 or 3 is the sum of the two

$$Pr\{1 \text{ or } 3\} = Pr\{1\} + Pr\{3\} = \tfrac{2}{6}$$

This little excursion into elementary probability theory is intended to show that one must find what is called an "element of probability measure" for which the probability is given by $p = 1/N$, N being the number of such elements making up the entire experiment. Other probability statements can then be obtained by grouping a certain number of such elements.

Now we can discuss geometric probability. In the above examples probability was given as the dimensionless ratio of two numbers; a geometric probability statement is also a dimensionless ratio, but this time the ratio of two geometric parameters, two lengths, two areas, two volumes etc. Let us look at an example.

Assume a structure in the form of a cube of side l which contains one object (Fig. 2.3); it is placed into an x,y,z coordinate system for convenience. Find the probability that a plane parallel to the x,z-plane cuts the object. To follow the subsequent development you may look directly at the front face of the structure and you will see the object projected onto the back plane of the cube; the section will appear as a vertical line. What is the element of probability measure? We may slice the cube into very thin slices of thickness dy; since all these slices are of equal width a random section parallel to the x,z-plane will hit each of them with equal likelihood. The number of possible slices is evidently $N = l/dy$ and the probability of hitting any one is

$$p = 1/N = dy/l \qquad (2.6)$$

Note that both dy and l are lengths measured along the y-axis, i.e. perpendicularly to the section plane. A section through the object will occur in all slices which contain a slice of the object. As you may easily verify from Fig. 2.3 these slices lie between two planes, parallel to the section, which touch the object tangentially, its two tangent planes. The number of slices within this range is $n = H/dy$ where H is the distance between the two tangent planes, also called the caliper diameter. In analogy to the last dice experiment we now find the probability of cutting the object a to be

$$\Pr\{a\} = n \cdot dy/l = H/l \qquad (2.7)$$

the ratio of caliper diameter of the object to the side of the containing cube; l defines the number of possible cuts, H those cuts which are "favorable" in that they hit the object.

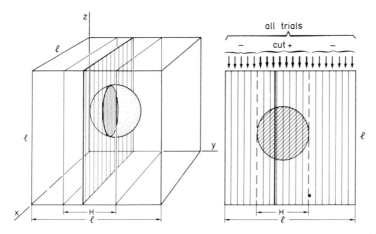

FIG. 2.3: Model structure, composed of sphere contained in cube, is used to derive probability of cutting an object. At right, projection of sphere and section onto front face of cube. Probability depends on ratio of tangent (or caliper) diameter H of object to the side length of the cube.

As a second example, let us now find the probability that the same object is cut by a random line which is parallel to the z-axis (Fig. 2.4). Look at the structure from the top so you see the object projected onto the base plane of the cube; the lines will appear as points. These lines can be displaced in two directions, in the x- and in the y-direction; the element of probability measure must therefore be represented by a little square of area $dxdy$: as you can now intuitively see each of these squares is equally likely to be hit by a test line. The number of such elements is given by the area of the cube face l^2, ($N =$

$l^2/dxdy$) and the number of those containing the structure by the projection area of the object a_p, ($n = a_p/dxdy$). Hence, the probability that a random test line intersects the object is

$$\Pr\{a\} = \frac{n}{N} = \frac{a_p}{l^2}$$ (2.8)

again a dimensionless ratio, this time of two areas.

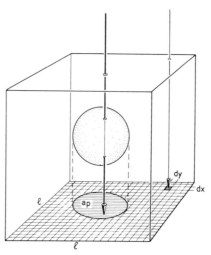

FIG. 2.4: Probability that a random vertical line intersects the object depends on the vertical projection area of the object.

You may now note that H has been measured perpendicularly to the section plane, and a_p perpendicularly to the test line. This is part of a general law, which is outlined in the mathematical section of Volume 2.

The last problem is now to find the probability that a random point lies in the object (Fig. 2.5). To find an element of probability measure we cut up the cube into small cubes of volume $dx \cdot dy \cdot dz$, and postulate that each one is equally likely to contain the random point (by this we actually define randomness of point location!). The numbers of cubic elements fitting into the object and into the cube is evidently proportional to the respective volumes $V(a)$ and $V(c)$, so that the probability we seek is

$$\Pr\{a\} = V(a)/V(c) = V_{Va}$$ (2.9)

which is equal to the volume density of the object in the cube. Note that by this statement we have given a justification of the point counting method for measuring volume density, to which we shall return: if you randomly deposit a number P_T of test points into the structure (the cube) a number

$$P_a = \Pr\{a\} \cdot P_T \tag{2.10}$$

of these points will be included in the object, and it follows that

$$\frac{P_a}{P_T} = P_{Pa} = V_{Va} \tag{2.11}$$

as you can easily verify.

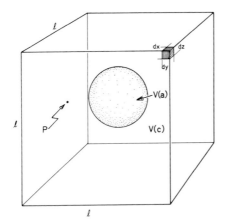

FIG. 2.5: Probability that a random point lies in the object depends on the ratio of the object to the cube volumes.

In summary, the rules for finding geometric probabilities are the following:

(1) Define exactly the type of experiment that is to be performed; this may not always be very easy, as is shown in the Volume 2, Section 2.2.

(2) Define an element of probability measure which is compatible with the experiment performed.

(3) Find the geometric parameters whose ratio defines the geometric probability; they are generally multiples of the element of probability measure and have the same dimensions.

A profound understanding of stereological principles requires a good background of geometric probability theory and integral geometry. Some of this background will be provided in the theoretical part of this book, presented in Volume 2. The reader is also referred to the excellent and concise treatise on geometric probability by Kendall and Moran (1963).

2.1.3.   AVERAGING WITH RESPECT TO ORIENTATION

In discussing geometric probability we have assumed a model structure with a spherical object, and have then cut this structure with planes or lines whose orientation was fixed. It is evident, however, from Fig. 2.6, that the caliper diameter of a non-spherical object varies depending on the direction of sectioning; consequently, the probability of sectioning the object depends on its orientation with respect to the section plane.

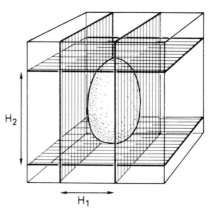

FIG. 2.6: The caliper diameter of a non-spherical object depends on the orientation of the object with respect to the tangent planes.

One of the fundamental postulates of stereology is, however, random sectioning, i.e. random orientation of the sectioning plane (or line) with respect to the objects. The measure of geometric probability relevant for stereology, therefore, is the average probability of hitting an object, obtained by averaging over all possible orientations of the sectioning plane. This is not a problem that can easily be demonstrated non-mathematically; we must therefore leave much of it to intuition. We shall merely describe the general approach to be taken for the determination of a mean caliper diameter of the objects.

What do we mean by random orientation of the sectioning planes? It means that all directions of sectioning must be equally likely.

The orientation in space of a plane is determined by the direction of its normal, i.e. of a line perpendicular to the plane (Fig. 2.7). If the plane is related to an x,y,z-coordinate system one considers the normal to the plane which passes through the origin of the coordinates; its direction is unambiguously determined by two angles: the angle θ between the normal and the vertical z-axis, and the angle ϕ between the x-axis and the vertical projection of the normal onto the x,y-plane. With this definition of plane orientation we can now specify the meaning of random section orientation: all orientations of the normal with respect to the fixed coordinate system must be equally likely.

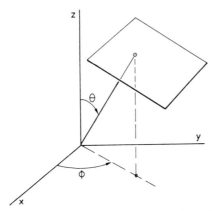

FIG. 2.7: Three-dimensional polar coordinates defining orientation of the normal to a section plane.

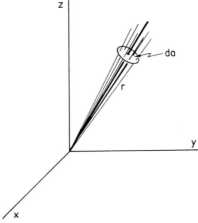

FIG. 2.8: Rays deviating by a small angle from the orientation of a main line form a small cone; the deviation is described by its cross-sectional area da at a distance r from the origin.

In analogy to the general considerations on geometric probability discussed in the previous paragraph we must now find a measure for orientation probability. It is evident from Fig. 2.8 that those lines which deviate by a very small angle from the direction of the normal form a slender cone of rays; they cut across a small area *da* whose size, at a given distance *r*, is a measure of the angular deviation. We could therefore define the probability to have an orientation in a certain range of angles by the ratio of two areas: the area *da* divided by an area which represents all possible orientations. What is this reference surface?

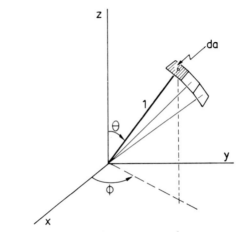

FIG. 2.9: Element of probability measure for orientation.

We can obtain it by the following procedure (Fig. 2.9). Take an *x,y,z*-coordinate system and let a line of length $l = 1$ arise from the origin of the system. Fix a small plane of area *da* perpendicularly at the tip of this line. Now add one line after the other, each fixed at the origin and each fitted with

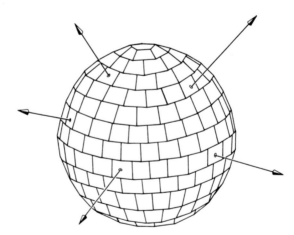

FIG. 2.10: All possible elements of probability measure for orientation fitted together form a sphere surface.

a little plane of area *da*, and arrange them in such a way that the little planes are in contact at their edges. If this is repeated until no further line can be fitted, the little planes will collectively form the surface of a sphere (Fig. 2.10).

This intuitive elementary approach is perhaps a bit too simplistic, for in reality it is not possible to completely "pack" a sphere surface with surface elements of constant size and shape. For practical purposes, however, a sphere can be approximated by a geodesic polyhedron with triangular faces, such as the geodesic domes of R. Buckminster Fuller (Pugh, 1976). In such domes the triangles form mostly hexagons, but a number of pentagons are required to allow packing of the triangles (Fig. 2.11).

FIG. 2.11: Geodesic dome of R. Buckminster Fuller (American pavillon at Montreal World Exhibition) is constructed of triangular elements. Arrows point to pentagon.

It is now evident that the sphere surface provides the area of reference we looked for, because by adding all the little areas *da*, each representing a particular orientation, we have exhaustively collected on the sphere surface all the possible orientations of the lines. Since in this example the radius of the sphere is $r = 1$, its surface is 4π. We can furthermore say that the elementary areas *da* are all equally likely to be hit by a ray emanating at any random angle from the origin of the coordinate system (Fig. 2.10). This area *da* is therefore a suitable element of probability measure for orientation. By definition, the probability is defined by the ratio of the element of probability measure to the measure of all possible events; hence, the probability of having a certain orientation of the normals to the section plane is given by the ratio of da to the sphere surface:

$$p = da/4\pi \tag{2.12}$$

Having defined the elementary probability for orientation of the section planes we can now proceed to outline the procedure for obtaining the average caliper diameter of a non-spherical object. Let us fix its orientation in space with respect to the x,y,z-coordinate system (Fig. 2.12). Now we pick one

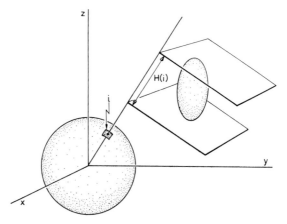

FIG. 2.12: Estimating tangent diameter $H(i)$ with respect to orientation on sphere model.

orientation of the normals after another and determine for each the distance between the two normal planes tangent to the object: this is the caliper diameter $H(i)$ at this orientation. If we define "one given orientation" by the elementary area da (Figs 2.8 to 2.10), then the total possible number of such measurements of $H(i)$ is

$$N = 4\pi/da \tag{2.13}$$

the number of elementary areas fitting onto the sphere surface. The average caliper diameter $\overline{H}$ is obtained by summing all values of $H(i)$ and dividing by the number N of measurements, so that we write

$$\overline{H} = \frac{\sum_{i=1}^{N} H(i)}{4\pi/da} = \frac{1}{4\pi} \sum_{i=1}^{N} H(i) \cdot da \tag{2.14}$$

Unfortunately, with this approach orientation is not precisely defined and the procedure cannot be performed in practice. To unambiguously define orientation the direction of the line arising from the origin must be stated with two angles, ϕ and θ, as shown in Fig. 2.13. The probability measure must then be defined in terms of ϕ and θ, and this is not very simple. Furthermore, instead of summing we must integrate—which is nothing else but summing over infinitesimally small steps—but we must integrate twice, once for θ and once for ϕ. The horrid formula applying is

$$\bar{H} = \frac{1}{4\pi} \int_0^\pi \int_0^{2\pi} H(\theta, \phi) \cdot [\sin \theta \cdot d\theta \cdot d\phi] \tag{2.15}$$

But if you compare this formula with equation (2.14) above the correspondence is evident: the term in square brackets corresponds to the probability measure da, and summation is replaced by integration over both angles.

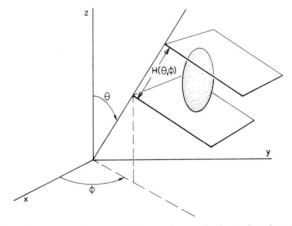

FIG. 2.13: Estimating tangent diameter in dependence of orientation described by polar coordinates.

We have engaged in this discussion on averaging for orientation because we have noted that the probability of sectioning a non-spherical object depends on its caliper diameter which, in turn, depends on the relative orientation of section plane and object. In stereology we will, in general, be interested in estimating the average probability of cutting the object with a random section, as defined above. This average probability is related to the mean caliper diameter $\bar{H}$. If we consider the object to be enclosed in a cube of side l, in analogy to the case treated above (Fig. 2.3), then the average probability that a section through this cube will cut a randomly oriented non-spherical object is

$$\overline{\Pr}\{a\} = \bar{H}(a)/l \tag{2.16}$$

We shall not pursue this any further. It has become evident that averaging for orientation requires mathematical methods (see Volume 2). There is only one solid for which such considerations are not necessary, and this is the sphere, because a sphere has no orientation. In subsequent parts of this chapter we may therefore take recourse to the sphere if we wish to deal with problems involving averaging for orientation. In fact, chosen of proper dimensions a sphere may often stand as a model for a large collection of randomly oriented non-spherical objects; the diameter of the "equivalent sphere" for problems involving probability of sectioning is evidently $\bar{H}$.

2.1.4. RANDOM SECTIONING

The term "random sectioning" occurs consistently in all discussions of stereological methods; it thus appears in place to give as good a definition as possible before deriving stereological principles.

The postulate of random sectioning in stereology stems from the fact that stereological methods are statistical in nature. A single profile is not representative of the object from which it was derived by sectioning; a large number of profiles stemming from a large number of objects must be studied to arrive at an estimate of the parameters characterizing the structure.

If we single out one object (Fig. 2.14) we note that it can be cut in an almost unlimited number of ways: at different levels, in different directions. In

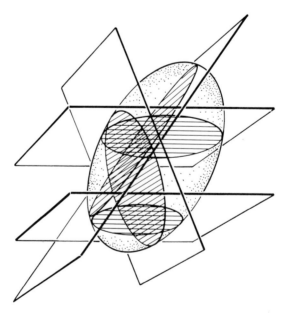

FIG. 2.14: A random section cuts the object in any direction and at any level.

deriving stereological principles we will assume that all these directions and levels of cutting will occur, and that they occur with equal likelihood or probability. As we will never be able to analyse all these possible profiles we must postulate that the sample which is included in our study is representative of all the possible cuts.

The conditions for random sectioning follow immediately: the section must produce a random sample of profiles. To obtain this the section may not be "aimed" at the object; we must cut blindly into the dark—just as in elementary statistics a blind-folded person picks a red or black ball out of a dark bag.

If a structure contains a very large number of the objects under consideration then a single section cut at random into the structure may produce a representative sample of profiles provided the objects have no preferred orientation within the structure (Fig. 2.15).

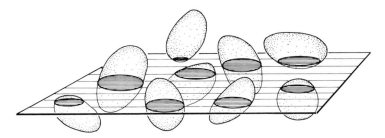

FIG. 2.15: Sectioning randomly oriented objects with a single plane.

The section must, however, also be random with respect to the structure containing the objects. This means that superficial parts of an organ must have the same likelihood of being cut as deep parts, and that the section should traverse the structure in any possible direction of space.

In Volume 2 (Chapter 2) we will show that giving the term "random sectioning" a proper meaning is a tricky matter, and that it is different for relatively small structures which we want to section totally, as compared to large specimens where we section small samples. The method for developing stereological formulae used in this elementary chapter are strictly valid only for the second case. For small "restricted" specimens the conditions on "random sectioning" are more strict if the results of a stereological analysis are to be unbiased.

Definition of random section: A section may be called "random" (in the sense of stereology) if it can cut (*a*) the structure at any level and in any direction of space with equal likelihood, and (*b*) the objects contained in the structure at any level and in any direction of space with equal likelihood.

2.2.   Volume density measurement

2.2.1.   THE PRINCIPLE OF DELESSE

In 1847 the French geologist Delesse proved that the volume density of the various components making up a rock can be estimated on random sections by measuring the relative areas of their profiles, also called the areal density of the profiles on section. This is one of the fundamental principles of stereology; it showed for the first time that a random section can be quantitatively representative of the composition of the material from which it is derived. We shall here present a plausible demonstration of this principle.

Take as a model structure a cube of tissue containing only one component in form of irregular objects embedded in a matrix (Fig. 2.16). These objects may have any shape and any size; they may be connected or separated, this is all irrelevant. But the cube should be cut out of a larger structure so that the objects may fill the cube in all parts, also in corners and on edges.

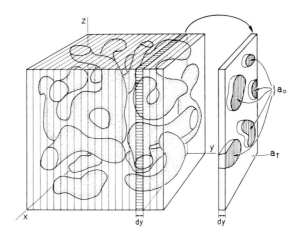

FIG. 2.16: Model for deriving Delesse principle.

The approach we shall take is very similar to that used in deriving the probability of cutting an object (Section 2.1.3.): We place the cube into an x, y, z-coordinate system and slice it, parallel to the x, z-plane, into very thin slices of thickness dy. On each slide surface, which for all slices has area a_T, one can measure an area a_0 to be covered by profiles of the object. The slice evidently contains a certain volume v_0 of the objects which must be equal to the profile area multiplied with the slice thickness:

$$v_0 = a_0 \cdot dy \tag{2.17}$$

whereas the volume of the slice is

$$v_T = a_T \cdot dy \tag{2.18}$$

Now take all the slices, add the slice volumes as well as the object volumes contained in them, and divide the summed object volumes by the summed slice volumes:

$$\frac{\Sigma v_0}{\Sigma v_T} = \frac{V_0}{V_T} = V_V \tag{2.19}$$

this defines the volume density. But now replace v_0 and v_T by the products of area times thickness (equations 2.17 and 2.18) and do the following:

$$\frac{\Sigma(a_0 \cdot dy)}{\Sigma(a_T \cdot dy)} = \frac{dy\Sigma a_0}{dy\Sigma a_T} = \frac{A_0}{A_T} = A_A \tag{2.20}$$

Since the slice thickness is constant it can be taken out of the sums, so that it cancels. The ratio of the sum of profile areas to the sum of section area is evidently the areal density of profiles on the section, A_A. It is evident that both equations (2.19) and (2.20) are equal to each other, so that we have proved

$$V_V = A_A \tag{2.21}$$

In practice, one will measure the areal density of profiles on only a small sample of these thin slices. It is easy to see that the ratio of the sum of profile area to the sum of section area will still be an unbiased estimate of V_V. If the section is large enough even a single section will estimate V_V, but evidently the reliability of the estimate will increase with the size of the sample, as in all measurements based on a sampling procedure.

2.2.2. LINEAR INTEGRATION

The measurements of areal density is a difficult procedure. Rosiwal (1898) has therefore proposed to estimate A_A by "linear integration". The procedure is to lay out some test lines onto the section and to measure the fractional length L_L of these lines which is included in profiles of the objects.

We can justify this procedure in the following way. To measure the area of a plane figure (Fig. 2.17) one can cut it into narrow strips of equal width d. The area of the figure is then equal to the total length of these strips, Σl_0, multiplied with d:

$$a_0 = \Sigma l_0 \cdot d \tag{2.22}$$

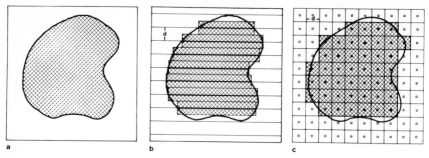

a   b   c

FIG. 2.17: Estimation of profile area by (a) planimetry, (b) linear integration, and (c) point counting.

This is no different if the square section area contains several such figures in the form of the profiles of some objects in the tissue (Fig. 2.18). If we cut the section into strips the area of the section a_T is equal to the sum of their length l_T times d, whereas the area of the profiles follows from the sum of the length of strips of profile. Hence

$$A_A = \frac{a_0}{a_T} = \frac{\Sigma l_0 \cdot d}{\Sigma l_T \cdot d} = \frac{\Sigma l_0}{\Sigma l_T} = L_L \tag{2.23}$$

in strict analogy to the derivation of the Delesse principle.

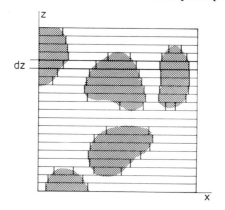

FIG. 2.18: Strip model to derive linear integration method.

2.2.3. POINT COUNTING METHOD

Glagolev (1933) and Thomson (1930) were the first to show that one can estimate volume density through a random point counting procedure. In biology this procedure was independently introduced by Chalkley in 1943.

In discussing in Section 2.1.3 the geometric probability that a random point lies in an object of volume $V(a)$ contained in a structure of volume $V(c)$, we presented a straight-forward proof that the volume density V_V can be estimated from the fraction of test points P_P lying in the structure:

$$V_V = P_P \qquad (2.24)$$

This demonstration need not be repeated. However, in practice one is usually applying the test points to a section and we must therefore briefly demonstrate that P_P also estimates A_A. Two ways to show this are possible.

In analogy to the aforementioned demonstrations we could define the following method for planimetry (Fig. 2.17c): One can cut up the square area containing the figure into small squares. It is immediately evident that the number of squares contained in the figure or profile related to the total number of squares in the section area is estimating A_A. There is one problem: that is to decide whether a square which intersects the boundary of the figure is "in" or "out". One usually says: if at least half the square is inside we count it for the figure. A simple trick to facilitate this decision is to mark the center point of each square: if it is in the figure we attribute this square to the figure. But then one may just as well count the center points contained in the figure and forget the squares; one has then plausibly adopted

$$A_A = P_P \qquad (2.25)$$

We could present a second demonstration of this principle by applying the rules of geometric probability, as outlined in Section 2.1.3, to the case of a plane section: we must define the experiment, find an element of probability measure, and derive the probability as the ratio of two geometric properties. The experiment is to randomly deposit a number $P(c)$ of test points on the section of area $A(c)$ and to count the number $P(a)$ of these points which fall

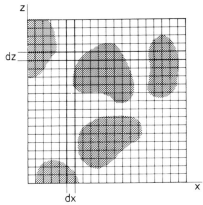

FIG. 2.19: Square grid model to derive point counting method.

on profiles whose total area on the section is $A(a)$. The element of probability measure is a small square of area $dxdz$ (Fig. 2.19); remember that randomness of point distribution is defined by the condition that each of the unit squares fitting into the section has an equal chance of containing a test point. This definition of randomness immediately leads to the statement that the probability that a point lies on a profile, $\Pr\{a\}$, is proportional to the number of unit squares which fit into the profiles, or proportional to the section area covered by profiles, $A(a)$. We leave it to the reader to show for himself that

$$\Pr\{a\} = P(a)/P(c) \qquad (2.26)$$

$$\Pr\{a\} = A(a)/A(c) \qquad (2.27)$$

from which it follows that

$$A_A = P_P.$$

2.3. Surface density measurement

The surface density expresses the amount of surface area $S(a)$ of a class of objects a that is contained in the unit containing tissue volume $V(c)$, such that

$$S_{Va} = S(a)/V(c) \qquad (2.28)$$

If such a tissue is sectioned with a plane the surface of the objects appears on the section as the boundaries of their profiles, whose length $B(a)$ can be related to the section area through the tissue, $A(c)$, to express the boundary density

$$B_{Aa} = B(a)/A(c) \qquad (2.29)$$

Furthermore if test lines are placed either on the section or directly into the tissue they will intersect the surface; the number of intersection points $I(a)$ can in turn be related to the test line length in the tissue $L(c)$ to form the intersection density

$$I_{La} = I(a)/L(c) \qquad (2.30)$$

It is intuitively plausible that these parameters are proportional to each other. Indeed it is found that

$$S_V = \frac{4}{\pi} B_A = 2I_L \qquad (2.31)$$

2.3.1.   INTERSECTIONS WITH TEST LINES IN SPACE

Let us first present a simple derivation of the relationship

$$S_V = 2I_L \qquad (2.32)$$

which has been independently derived by Saltykov (1946) and by Tomkeieff (1945), and was later rediscovered at least five more times (Smith and Guttman, 1953; Duffin *et al.*, 1953; Horikawa, 1954; Corrsin, 1954; Hennig, 1956b).

Let us choose a simple model structure (Fig. 2.20) composed of one single object contained in a cube of volume

$$V(c) = l^3 \qquad (2.33)$$

The surface of the object should be isotropic; i.e. the normals (perpendicular) to its surface should point with equal likelihood into all directions of space.

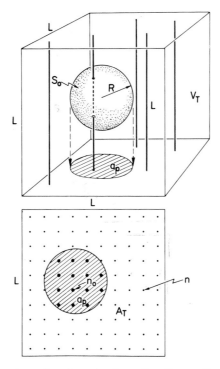

FIG. 2.20: Direct derivation of surface density estimation method by intersection counting.

The only single object to fulfill this requirement is the sphere; let its radius be R, so that its surface is

$$S(a) = 4\pi R^2 \qquad (2.34)$$

The surface density of the sphere in the cube is

$$S_{Va} = \frac{S(a)}{V(c)} = \frac{4\pi R^2}{l^3} \qquad (2.35)$$

The cube also contains a bundle of n straight lines parallel to the vertical edges; they are so arranged that their n intersections with the base plane of the cube are homogeneously distributed over the basal cube face, for example in a quadratic lattice (Fig. 2.20). All lines which pass through the sphere will also intersect the vertical projection of the sphere whose area is

$$a_p = \pi R^2 \qquad (2.36)$$

as one can immediately understand from our considerations on geometric probability (Fig. 2.4). From the arguments presented on p. 29 with respect to estimating areal density by point counting we can conclude that the relative number of lines intersecting the sphere, n_0/n, is equal to the relative area of the sphere projection a_p on the cube face of area l^2:

$$n_0/n = a_p/l^2 = \pi R^2/l^2 \qquad (2.37)$$

or

$$n_0 = n \cdot \pi R^2/l^2 \qquad (2.38)$$

Considering that each line passing through the sphere forms two intersections with its surface, one above and one below the equator, the number of intersections is

$$I(a) = 2 \cdot n_0 = 2 \cdot n \cdot \pi R^2/l^2 \qquad (2.39)$$

The length of test lines being evidently

$$L(c) = n \cdot l \qquad (2.40)$$

we obtain by dividing equation (2.39) by equation (2.40)

$$\frac{I(a)}{L(c)} = \frac{2\pi R^2}{l^3}$$

The term on the right side is one half of the surface density (equation 2.35). By rearranging we hence obtain

$$S_{Va} = 2 \cdot I(a)/L(c) \qquad (2.41)$$

which was the relationship to be demonstrated.

This relationship also holds if the structure is made up of any number of spheres of any size, as long as overlapping projections are separately considered in the derivation: the projection areas and the intersection numbers linearly increase with the surface area of the set of spheres, since all three are related to R^2 or rather to the linear sum of all R^2.

It is furthermore not necessary that the objects be spheres, as long as the entire structure, i.e. the multitude of objects contained in the structure, is isotropic. One could cut up the sphere's surface into small units and shift them around without changing their orientation; the principle would still hold. In Volume 2, Section 3.2, a derivation valid for general isotropic structures is given.

2.3.2. INTERSECTIONS WITH PROFILE BOUNDARIES AND THE BUFFON PRINCIPLE

The length of the boundaries $B(a)$ which appear on the section area can also be estimated by the number of intersections they form with test lines. The formula is

$$B_A = \frac{\pi}{2} I_L \qquad (2.42)$$

A demonstration of this principle can go as follows. Take a curve on a plane which is ruled with a set of parallel lines at distance d (Fig. 2.21). The curve can be cut up into n_0 short line segments of length $l < d$, so that the total length is given by

$$B(a) = n_0 \cdot l \qquad (2.43)$$

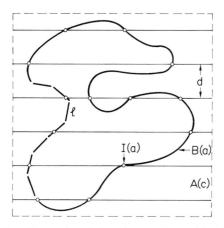

FIG. 2.21: Estimation of curve length from intersections with grid of parallel lines.

Now we can ask for the probability that one line segment forms an intersection with one of the test lines, which is the original problem of Buffon
(1777). This evidently depends on the orientation of the segment with respect
to the test lines: if it is parallel there are no intersections, if it is at right angles
the probability of having an intersection is l/d as is easily seen: shift the line
by short distances in the vertical direction (Fig. 2.22) and the number of times

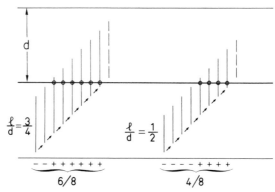

FIG. 2.22: Probability that needle at right angle intersects grid lines.

it crosses a test line is proportional to l. At any other angle the probability
will depend on the inclination of the segment to the test lines, measured by
the angle θ. In fact, it is given by the length of projection x onto a line perpendicular to the test lines (Fig. 2.23), and this is

$$x = l \cdot \sin \theta \tag{2.44}$$

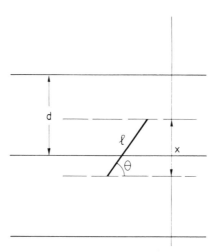

FIG. 2.23: Probability that needle inclined at angle θ intersects grid lines.

The probability of having an intersection hence is

$$\Pr\{I\} = \frac{x}{d} = \frac{l}{d} \cdot \sin \theta \tag{2.45}$$

To allow for all orientations one must now average $\sin \theta$ by integrating over all angles. This turns out to give

$$\Pr\{I\} = \frac{2}{\pi} \frac{l}{d} \tag{2.46}$$

the answer found by Buffon (1777), as shown in Volume 2, Section 3.2. But we can also derive this in a simple way. For the line segment l to represent all

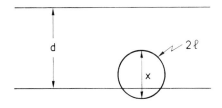

FIG. 2.24: Intersection of circle and grid lines.

orientations on the plane (for it to be isotropic) it can be taken to be a circle; because the circle forms two intersections whenever it lies on a test line, let us give it length $2l$. This circle has a diameter

$$x = \frac{2}{\pi} l \tag{2.47}$$

as is easily seen from Fig. 2.24. We therefore find

$$\Pr\{I\} = \frac{x}{d} = \frac{2}{\pi} \cdot \frac{l}{d} \tag{2.48}$$

as before. Returning to the original problem the number of intersections is found by

$$I(a) = n_0 \cdot \Pr\{I\} = n_0 \cdot \frac{2}{\pi} \cdot \frac{l}{d} \tag{2.49}$$

Since $n_0 \cdot l = B(a)$ (equation 2.43) we find

$$I(a) = \frac{2}{\pi} \frac{B(a)}{d} \tag{2.50}$$

The area which contains the line $A(c)$ is given by the total test line length $L(c)$ and the line distance d (Fig. 2.21):

$$A(c) = L(c) \cdot d \qquad (2.51)$$

We can now solve this for d and introduce it in equation (2.50) to obtain

$$I(a) = \frac{2}{\pi} \times \frac{B(a)}{A(c)} \cdot L(c) \qquad (2.52)$$

or, after rearrangement

$$\frac{B(a)}{A(c)} = \frac{\pi}{2} \frac{I(a)}{L(c)} \qquad (2.53)$$

which corresponds to equation (2.42).

2.3.3. SURFACE AND BOUNDARY LENGTH DENSITY

We have derived the two formulae:

$$S_V = 2I_L$$

$$B_A = \frac{\pi}{2} I_L$$

It is now easy to combine these two to derive indirectly the relationship between boundary length density and surface density:

$$S_V = \frac{4}{\pi} B_A \qquad (2.54)$$

This formula can however also be derived directly by using again the simple model structure of a single sphere of radius R contained in a cube of side l (Fig. 2.25). A plane section traversing the sphere will show a trace in form of a circle whose size will depend on the distance of the sphere center from the section plane. As will be shown later (equation 2.105) the average radius of circles produced by random sectioning is

$$\bar{r} = \frac{\pi}{4} R \qquad (2.55)$$

and the average length of circle perimeter

$$\bar{b} = 2\pi\bar{r} = \frac{\pi^2}{2} R$$

$$(2.56)$$

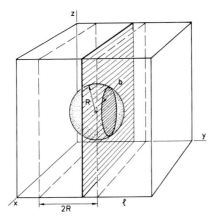

FIG. 2.25: Sphere intersected by plane generates circular trace.

Now we must remember that the relative number of random sections across the cube which will hit the sphere is given by the ratio of sphere diameter to length of the cube ($2R/l$; equation 2.7) so that the mean trace $\overline{B}$ of the sphere on all sections is

$$\overline{B} = \overline{b} \cdot \frac{2R}{l} = \frac{\pi^2 R^2}{l} \tag{2.57}$$

With section area $A(c) = l^2$ the mean boundary density is

$$\overline{B_A} = \pi \cdot \left[\frac{\pi R^2}{l^3} \right] \tag{2.58}$$

Considering that the surface density of the sphere in this structure is (equation 2.35)

$$S_V = \frac{4\pi R^2}{l^3} \tag{2.59}$$

it is evident that the term in brackets in equation (2.58) is $\frac{1}{4}$ of S_V, so that

$$\overline{B_A} = \pi \cdot \frac{S_V}{4} \tag{2.60}$$

or, by rearrangement

$$S_V = \frac{4}{\pi} \overline{B_A} \tag{2.61}$$

which we wanted to demonstrate.

2.4.   Length density measurement

If a curve is suspended in three-dimensional space we can express its length $J(a)$ with respect to the containing volume $V(c)$ as a length density

$$J_V = J(a)/V(c) \qquad (2.62)$$

A plane section across this structure (Fig. 2.26) will show the curve as transection points; their number being $Q(a)$ and the section area of the structure again $A(c)$ we define

$$Q_A = Q(a)/A(c) \qquad (2.63)$$

as the transection density on the section area. Saltykov (1946), Smith and Guttman (1953) and Hennig (1963a, b) have shown that

$$J_V = 2Q_A \qquad (2.64)$$

a formula which is identical to equation (2.32) establishing the relation between S_V and I_L. In fact, the two formulae are identical as can easily be seen by setting $S = A$, $L = J$, and $Q = I$, and slightly rearranging. It is also plausible that the two equations should be identical; in both instances a surface intersects a line and we require isotropy of the structure, i.e. the surface element of S_V must show all possible orientations, and the length elements of the curve in space must also assume all orientations with equal likelihood.

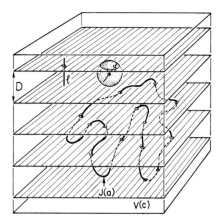

FIG. 2.26: Intersection of space curve by planes.

A demonstration of this principle would proceed similar to the derivation of the formula $B_A = (\pi/2)I_L$ on the basis of Buffon's principle, only that the little line segments, representing elements of the curve, would now have to intersect a set of test planes (Fig. 2.26); they must also be allowed to assume all orientations in space and it can be shown that the diameter of an equivalent sphere (instead of a circle), which allows us to calculate the probability of obtaining an intersection, is just half the length of the segment. With this information at hand the reader could well derive equation (2.64) for himself, following the pattern used in the two-dimensional example.

2.5. Mean curvature density measurement

One of the important physical properties of surfaces is their curvature. The curvature is measured at each point of a surface by the reciprocal of the two radii, r_1 and r_2, of the largest and smallest circle, respectively, that can be fitted to the surface. One distinguishes two kinds of curvature:

(1) the mean curvature

$$K = \tfrac{1}{2}\left(\frac{1}{r_1} + \frac{1}{r_2}\right) \tag{2.65}$$

(2) the total or Gaussian curvature

$$G = \frac{1}{r_1} \cdot \frac{1}{r_2} \tag{2.66}$$

It is quite clear that for a sphere of radius R, $K = 1/R$ and $G = 1/R^2$. For any other solid the mean curvature is obtained by integrating over the surface of the solid. It is left until the mathematical part presented in Volume 2 (Section 3.4) to exploit this further. In the present context it suffices to state that the density of these curvatures in the unit volume of structure are fundamental parameters of structure which can, in part, be estimated by stereological methods. Thus K_V is the integral of the mean curvatures of all points on the object surface $S(a)$ in the containing volume $V(c)$. According to De Hoff (1967) and Cahn (1967) it can be measured on random sections by sweeping a test line across the section area. As this test line meets the surface trace of a it forms tangents (Fig. 2.27); these are either "positive", $T_+(a)$, or "negative", $T_-(a)$, depending on whether the local curvature is positive or negative (note that spherical surfaces produce only positive tangents). The net tangent count

$$T(a) = T_+(a) - T_-(a) \tag{2.67}$$

is related to the test area $A(c)$ to give the tangent count density

$$T_A = T(a)/A(c) \tag{2.68}$$

It is now found (De Hoff, 1967; Cahn, 1967) that the integral mean curvature per unit volume is

$$K_V = \pi \cdot T_A \tag{2.69}$$

In the case of spherical particles (or convex particles in general) each profile will form two positive tangents so that we can write equation (2.69) as

$$K_V = 2\pi \cdot N_A \tag{2.70}$$

As will be discussed in more detail in Volume 2 Section 3.4, the Gaussian curvature per unit volume is closely related to the particle number. Its estimation on sections requires that shape and size of the particles must be known.

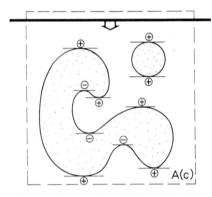

FIG. 2.27: Tangent count by sweeping test line yields positive $\oplus$ and negative $\ominus$ tangents.

2.6.   Numerical density of particles

2.6.1.   BASIC PRINCIPLE

The larger the number of particles contained in the unit volume, the larger should be the number of profiles observed per unit area of a section. On the other hand we have already discussed in Section 2.1.3 that a larger particle will have a greater chance to be hit by a random section plane. Hence, it is plausible *a priori* that the number of profiles per unit section area, N_A, must depend both on the number of particles per unit volume, N_V, and on some measure of particle size (Fig. 2.28).

We should now recall the discussion of the geometric probability that a

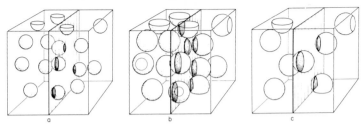

FIG. 2.28: Three cubes containing spheres. In (a) and (b) the number of spheres is identical but their size is different, in (b) and (c) the spheres are of equal size but differ in number. In each case the number of profiles found on a random section is different.

section will hit a particle (Section 2.1.3). We came to the conclusion in equation 2.16 that this probability was

$$\Pr\{a\} = \bar{H}(a)/l \tag{2.71}$$

where $\bar{H}(a)$ was the mean tangent diameter of the particle, averaged over all orientations, and l the extent or width of the sampling space perpendicular to the test plane. We derived this by studying a model containing one (spherical) particle in a cube of volume l^3, and keeping the section parallel to one cube face.

Let us now use the same type of model to derive the relationship between particle number and size, and profile number, but this time we shall enclose a large number, n_a, of spherical objects of diameter $H = D$ within the cube of volume l^3 (Fig. 2.29). When this cube is sectioned randomly—again with a plane parallel to the x, z-plane—a certain number n_p of sphere profiles can be expected to appear on such a section of area l^2.

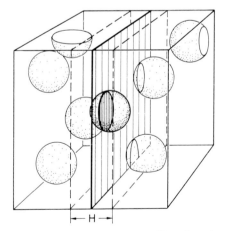

FIG. 2.29: Spheres of equal size in cube. Probability of cutting depends on H.

For each particle, the probability of being cut by the section plane is evidently

$$\Pr\{a\} = D/l = H/l \qquad (2.72)$$

A practical way of estimating the probability $\Pr\{a\}$ would be to make a large number of trials, say with n_a randomly distributed spheres, and to count the number of times the section will hit any one of these spheres. In such an experiment we may count n_p profiles or "successful cuts", and this yields as an estimate of this probability

$$\Pr\{a\} = n_p/n_a \qquad (2.73)$$

It hence follows that

$$n_p/n_a = D/l \qquad (2.74)$$

so that

$$n_p = n_a \cdot D/l \qquad (2.75)$$

is the expected or average number of profiles observed on a section of area l^2. If we now divide both sides of equation (2.75) by the section area l^2 we obtain

$$n_p/l^2 = (n_a/l^3) \cdot D \qquad (2.76)$$

or

$$N_A = N_V \cdot D \qquad (2.77)$$

since, by definition, the numerical densities of particles and profiles are, respectively:

$$N_V = n_a/l^3 \,;\, N_A = n_p/l^2 \qquad (2.78)$$

It can easily be seen that, if the spheres vary in size, D must be replaced by the average sphere diameter $\overline{D}$. It will be shown in Volume 2, Section 5.1 that a weighted average must be considered, i.e. the relative importance of each size class must be taken into consideration in calculating $\overline{D}$.

In practice one will usually determine N_A experimentally by counting profiles on sections; the particle number density then follows from

$$N_V = N_A/\overline{D} = N_A/\overline{H} \qquad (2.79)$$

a fundamental principle of stereology which is usually ascribed to De Hoff and Rhines (1961), but had already been given by Wicksell in 1925 and 1926.

It is to be noted that estimation of numerical particle density by this method requires two measurements to be obtained: the number of profiles per unit area, N_A, and the mean tangent diameter of the particles, $\overline{H}$. Whereas the determination of N_A poses no special problems the estimation of $\overline{H}$ may cause considerable difficulties, as will be discussed in subsequent chapters.

2.6.2. ALTERNATIVE METHODS BASED ON SIMPLIFYING ASSUMPTIONS

Because of the difficulties in reliably estimating $\overline{H}$, alternative solutions for deriving N_V have been sought. They all essentially make some assumptions as to particle shape and attempt to estimate N_V on the basis of N_A and one or two other stereological estimates such as volume or surface density which provide a measure of particle size. These methods will be justified in Volume 2, Chapter 5 but are briefly repeated here for the sake of completeness.

2.6.2.1. *Method of De Hoff* (1964) *for constant size particles*

De Hoff (1964) has proposed that N_V can be estimated from N_A together with a point count of intersections between test lines and the profile boundary, I_L, or of point hits of the profiles by a test point lattice, P_P, such that

$$N_V = \frac{N_A^2}{2I_L} \cdot \frac{\gamma_2}{\gamma_1^2} \qquad (2.80)$$

or

$$N_V = N_A \cdot \frac{2I_L}{P_P} \cdot \frac{\gamma_1 \gamma_2}{\gamma_3} \qquad (2.81)$$

These formulae require that the shape of the particles can be defined by some geometric form which determines the shape coefficients γ_1, γ_2, and γ_3; values of some of these coefficients are compiled in Table 2.2 for various solids. These coefficients actually mean that the particle parameters $\overline{H}$, v, and s can be related to a single characteristic dimension of the particle called x:

$$\overline{H} = \gamma_1 \cdot x$$

$$s = \gamma_2 \cdot x^2 \qquad (2.82)$$

$$v = \gamma_3 \cdot x^3$$

For the simple case of spheres we have $x = r$ and

$$\gamma_1 = 2$$

$$\gamma_2 = 4\pi$$

$$\gamma_3 = 4\pi/3$$

Table 2.2

Shape coefficients for particle counting methods by equations (2.80), (2.81), and (2.83).

Particle shape	Eqs. (2.80) and (2.81)			Eq. (2.83)
	γ_1	γ_2	γ_3	β
Sphere	2	12.57	4.19	1.38
Spheroid prolate (2:1:1)	2.76	21.4	8.38	1.58
oblate (2:2:1)	1.71	8.67	2.09	1.55
Cube	1.5	6	1	1.84
Octahedron	1.17	3.46	0.47	1.86
Icosahedron	1.74	8.66	2.18	1.55
Dodekahedron	2.64	18.05	7.66	1.55
Tetrakaidekahedron	3	26.8	11.3	1.55

2.6.2.2. *Method of Weibel and Gomez* (1962) *for constant shape particles*
A method which has been frequently used in biology relates N_V to N_A and V_V through a dimensionless shape coefficient β, such that

$$N_V = \frac{1}{\beta} \cdot \frac{N_A^{\frac{3}{2}}}{V_V^{\frac{1}{2}}}$$ (2.83)

where β is defined by the dimensionless relationship between the particle volume v and its mean cross-sectional area $\bar{a}$:

$$v = \beta \cdot \bar{a}^{\frac{3}{2}}$$ (2.84)

Values of β are also compiled in Table 2.2 for various solids. Figure 2.30 shows a graph of the change in β for ellipsoids and right circular cylinders of various axial ratios. The value for spheres is $\beta_s = \sqrt{(6/\pi)} \sim 1.382$. The main virtue of this coefficient is that it varies relatively little for short-axed ellipsoids so that the restrictions on applicability of this method due to shape variations are not so critical for particles which can be considered "nearly spherical" (see Volume 2, Chapter 5).

Variation in size of the particles will however introduce a certain systematic error. This can be accounted for by introducing a size distribution coefficient K which is proportional to the ratio of third to first moment of the size distribution. The complete formula is

$$N_V = \frac{K}{\beta} \sqrt{\left(\frac{N_A^3}{P_P}\right)}$$ (2.85)

where volume density has been replaced by its directly obtainable estimator P_P. Values of K can be read from the graph in Fig. 2.31. It is seen that for particle size distributions with standard deviations less than 20% of the

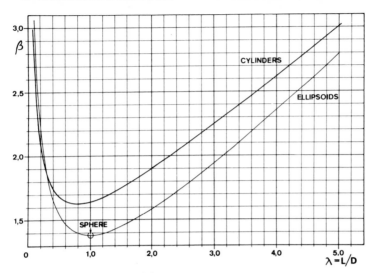

FIG. 2.30: Shape coefficients β for ellipsoids and cylinders of varying axial ratio. (Reproduced by permission from Weibel, 1969.)

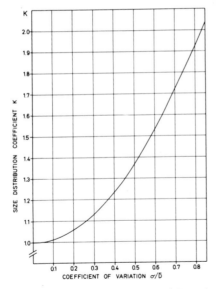

FIG. 2.31: Size distribution coefficients K (Equation 2.85).

mean the coefficient K varies between 1 and 1.05; neglect of this coefficient would hence introduce errors of no more than 5% in the estimation of N_V. For many biological applications K can therefore be neglected.

2.6.3.   COUNTING IN THICK SLICES

All the counting methods discussed in the preceding paragraphs assumed that the section is a plane of no thickness. This is an acceptable assumption only if the particles are very large compared with section thickness, a condition which often does not apply in practice. If the section thickness t is relatively large a greater number of particles will be observed on the section plane due to projection of the entire slice content (Fig. 2.32).

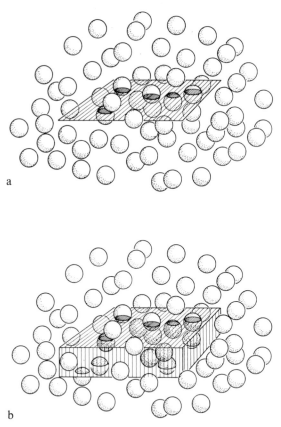

FIG. 2.32: Population of spheres intersected by plane (a) and slice of finite thickness (b).

If we look at the slice from the side, as shown in Fig. 2.33 we can derive the relationship between observed profile number per unit section area, N_A, and the numerical density N_V. Let us consider that the center of each sphere is marked by a point. Now we observe two kinds of profiles: (1) profiles

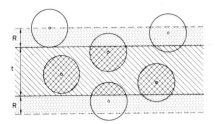

FIG. 2.33: Spheres with centers in superslice of thickness $(t + 2R)$ form profiles (total projection).

whose center point is within the slice; let us call their number $N(w)$, and we find that this is N_V multiplied with the slice volume $A \cdot t$:

$$N(w) = N_V \cdot A \cdot t \tag{2.86}$$

(2) profiles whose center is outside, either above or below the slice, $N(o)$; these centers are however contained in two thin slices of thickness R, the sphere radius (if the center is above or below the dotted line no profile is formed!). Thus we find, in analogy to (1)

$$N(o) = N_V \cdot A \cdot 2R \tag{2.87}$$

Consequently the total number of profiles observed is

$$N = N(w) + N(o)$$
$$= N_V \cdot A(t + 2R) \tag{2.88}$$

or, dividing by the area of the section, we find the numerical density of profiles to be

$$N_A = N_V(t + 2R) \tag{2.89}$$

from which we derive with $D = 2R$ the practical formula for calculating N_V as

$$N_V = N_A/(t + D) \tag{2.90}$$

This formula was first derived by Abercrombie (1946). It means that we are counting profiles in a "super-slice" of thickness $(t + D)$. It is easily seen that it reduces to equation (2.77) if t becomes very thin. As in the basic equation, D must be replaced by the mean tangent diameter of all the particles, $\bar{H}$, for the general case.

In trying to count profiles in thick slices we meet with one additional problem: if the particles are not perfectly opaque (black) and the matrix in which they are embedded is not perfectly translucent a certain number of small cap sections will not be visible, so that they are lost from the count. Floderus (1944) has therefore introduced a correction which considers that

C

part of the spheres which have their center outside the slice will not be visible (Fig. 2.34). If a profile has to reach a "height" of the cap h before it is visible, then particles contained in a "super-slice" of thickness $(t + D - 2h)$ are counted, and the appropriate formula for deriving N_V is

$$N_V = N_A/(t + D - 2h) \qquad (2.91)$$

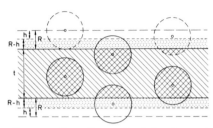

FIG. 2.34: Loss of small cap sections (truncation) due to contrast deficiency reduces superslice thickness by $2h$.

In using this method it is hence necessary to estimate the particle diameter, D (or $\bar{H}$ for the general case), the section thickness, t, and the height of the minimum visible cap, h. Methods for doing this will be discussed in Chapters 4 and 5. Cap height is related to the radius r of the smallest visible profile and can be calculated, for spherical particles, by

$$h = R - \sqrt{(R^2 - r^2)} \qquad (2.92)$$

This is evidently a rough approximation, mainly if the particles are not spherical or vary greatly in size. Aherne (1967) has proposed an alternative method for estimating N_V in thick slices; this is discussed in detail in Volume 2, Chapter 5.

An interesting method has been proposed by Ebbeson and Tang (1965). Taking two sections of different thickness, $t_1 > t_2$, and counting profile numbers on each, N_{A1} and N_{A2}, one may obtain N_V by

$$N_V = (N_{A1} - N_{A2})/(t_1 - t_2) \qquad (2.93)$$

It is evidently not necessary to know particle size, but the difference between t_1 and t_2 must be large enough. This formula is easily derived from equation (2.89); see also Volume 2, Chapter 5.

2.7.   Shape and mean tangent diameter of convex solids

The mean tangent or caliper diameter, $\bar{H}$, of a particle determines the probability that this particle is cut by a random plane; it is therefore also a most important piece of information when counting particles on sections, as outlined in the preceding chapter.

For spherical particles of radius R the mean tangent diameter is simply $\bar{H} = 2R$, because the diameter of a sphere is equal in every direction of space. For a particle of any other shape, however, the tangent diameter will depend on the orientation of the particle with respect to the sectioning plane, as discussed in Section 2.1. The mean tangent diameter is then obtained by integration over all orientations.

In practice this can very often not be performed. One will have to be satisfied with an approximation by finding a convenient solid which describes the shape of the particle as well as possible. Such convenient solids may be ellipsoids, cylinders, polyhedra, or even spheres. We will here briefly discuss how the mean tangent diameter is related to the geometric measures of the solid's dimensions. This is dealt with in more detail in Volume 2, Section 5.5.

Let us look at one simple example in some detail: a right circular cylinder. It is generated by rotating a rectangle with the dimensions $l \times 2r$ around its axis "l"; the cylinder's length will be l, its diameter $2r$ (Fig. 2.35). Let us now place this cylinder into an x, y, z-coordinate system and incline its axis by an angle θ with respect to the vertical z-axis (Fig. 2.35b). The caliper diameter $H(\theta)$ at this angle is the distance between the two horizontal planes which touch the uppermost and lowermost points of the cylinder. For convenience we have placed the bottom end of the cylinder's axis into the origin of the coordinate system. If we now rotate the cylinder around the x-axis, while keeping the inclination by angle θ constant, the uppermost point of the cylinder will describe a circle which lies always in the same horizontal plane. It is easily seen that this rotation does not change the caliper diameter of the cylinder with respect to horizontal planes; the caliper diameter, $H(\theta)$, is hence solely dependent on the angle of inclination θ.

For further analysis we can therefore confine ourselves to the y, z-plane and define the caliper diameter of the rectangle from which the cylinder had been generated by revolution (Fig. 2.35c). The caliper diameter of this rectangle at any angle θ can be split into two parts: an upper portion z_1, which is the horizontal projection of the cylinder length onto the z-axis, and a lower segment z_2, which is the projection of the diameter $2r$ onto the z-axis. By simple trigonometric analysis it is found that

$$z_1 = l \cdot \cos \theta$$

$$z_2 = 2r \cdot \sin \theta \tag{2.94}$$

and hence

$$H(\theta) = l \cdot \cos \theta + 2r \cdot \sin \theta \tag{2.95}$$

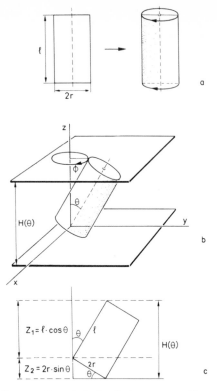

FIG. 2.35: Caliper diameter (b and c) of cylinder generated by rotation of rectangle (a).

It is evident, that if the cylinder stands upright the angle θ will be 0; since $\sin 0 = 0$ and $\cos 0 = 1$ the caliper diameter will then be l. Conversely, if it lies flat so that $\theta = 90°$ the caliper diameter is $2r$. To obtain the mean caliper diameter we must integrate over all orientations according to equation 2.15; but since ϕ has no effect on H a simpler formula can be used (see Volume 2, Sections 2.3 and 5.5):

$$\bar{H} = \int_0^{\pi/2} H(\theta) \cdot \sin \theta \, d\theta \qquad (2.96)$$

If this is performed (see Volume 2, Section 5.5.1) we obtain for a cylinder

$$\bar{H} = \tfrac{1}{2}l + \frac{\pi}{2}r \qquad (2.97)$$

The mean tangent diameter is hence directly related to length and diameter of the cylinder in a simple fashion.

Similar derivations are performed in Volume 2, Section 5.5 for ellipsoids.

Likewise a general formula for deriving the mean tangent diameter of convex solids is derived. Some specific formulas are given there, together with the relationships between mean tangent diameter and geometric measure for some typical polyhedra, as derived by Hilliard (1967a).

2.8.   Profile size and particle size

In the preceding sections we have counted profiles found on random sections, and have related them to the particle number in the structure; we have also estimated their total relative area and have related this to total relative volume of the objects, and so on. We have, however, never been concerned about the size of individual profiles and how this might be related to particle size. This aspect shall now be pursued, but we shall restrict the analysis to a simple solid, the sphere.

2.8.1.   PROFILES DERIVED FROM SPHERES OF EQUAL SIZE

Let us first look at spheres of equal size. As is evident from Fig. 2.36 the diameter of the profiles cut by a random plane will vary, depending on the the level at which the sphere is cut: it will be largest around the equator and become smaller as the sectioning plane is displaced towards the pole. We

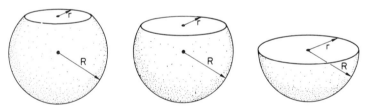

Fig. 2.36: Profile radius r depends on level at which sphere is sectioned.

must now try to formulate the dependence between profile size and location of the section. This is quite easy: place the sphere into an x, y, z-coordinate system with the sphere center at the origin (Fig. 2.37). The location of the horizontal sectioning plane is then determined by the distance z from the origin. The profile radius r forms a triangle with the sphere radius R and the distance z; the theorem of Pythagoras says that

$$R^2 = r^2 + z^2 \qquad (2.98)$$

so that the profile radius is evidently

$$r(z) = \sqrt{(R^2 - z^2)} \qquad (2.99)$$

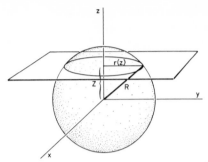

FIG. 2.37: Location of section plane with respect to sphere center.

In order to characterize the total population of profiles let us now cut up the sphere into horizontal slices of equal thickness dz (Fig. 2.38); if dz is small we may represent each slice by a cylindrical disc of radius $r(z)$ which will vary with z according to equation (2.99). We must now remember that the probability of hitting a body with a horizontal sectioning plane is determined by the vertical thickness of the body, as outlined in Section 2.2. Since the slices have the same vertical thickness each slice has evidently the

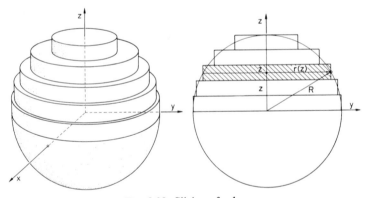

FIG. 2.38: Slicing of sphere.

same likelihood of being sectioned by a random horizontal sectioning plane. If we now number the slices from 1 (at the origin) to n at the top of the sphere we can immediately derive the mean profile radius as

$$\bar{r} = \frac{1}{n} \sum_{i=1}^{n} r_i = \frac{1}{n} \sum_{i=1}^{n} \sqrt{(R^2 - z_i^2)} \qquad (2.100)$$

From the way we had sliced the sphere it is evident that the slice thickness is

$$dz = R/n \qquad (2.101)$$

and it is also evident that the slice distance from the origin could be written as

$$z_i = i \cdot dz = i \cdot R/n \qquad (2.102)$$

If we now substitute this for z_i in equation (2.100) we find

$$\bar{r} = \frac{1}{n} \sum_{i=1}^{n} \sqrt{\left[R^2 - R^2 \cdot \left(\frac{i}{n}\right)^2 \right]} \qquad (2.103)$$

Since the sphere radius is a constant this leads to

$$\bar{r} = R \left\{ \frac{1}{n} \sum_{i=1}^{n} \sqrt{\left[1 - \left(\frac{i}{n}\right)^2 \right]} \right\} = k \cdot R \qquad (2.104)$$

This tells us that the mean profile radius $\bar{r}$ is linearly related to the sphere radius R; the term in brackets is a constant which could be worked out by using any sufficiently large value of n.

However, there is an easier graphical way of determining the value of this constant k. In Figs. 2.37 and 2.38 we restricted the graphical demonstration of the relationship between R, z and r to the main circle of the sphere in the y, z-plane of the coordinate system, in fact to only one quadrant of this circle. This quadrant is shown again in Fig. 2.39. It is immediately evident that the

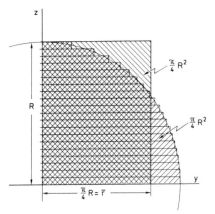

FIG. 2.39: Relationship between sphere radius R and mean profile radius r; graphical derivation.

area of this quadrant is equal to the sum of the area of the strips representing the slices. The average strip length $\bar{r}$ is then immediately obtained by dividing this area by R. The area of a quadrant evidently being $\pi R^2/4$ we obtain

$$\bar{r} = \frac{\pi R^2/4}{R} = \frac{\pi}{4} R \qquad (2.105)$$

This could be formalized; going back to equation (2.100) we multiply r_i

with dz (which is constant) to express the area of the strip and then we divide the sum by dz:

$$\bar{r} = \frac{1}{n dz} \sum_{i=1}^{n} r_i dz \qquad (2.106)$$

The sum of the strips is the area of the quadrant, $\pi R^2/4$, and $n \cdot dz = R$ which leads to equation (2.105).

The result we have obtained is very important because it establishes a linear relationship between sphere size and mean profile size. This can be used to calculate particle radius from a measurement of profile radii

$$R = \frac{4}{\pi} \bar{r} \qquad (2.107)$$

Evidently, the same factor applies directly to diameters:

$$D = \frac{4}{\pi} \bar{d} \qquad (2.108)$$

A more refined characterization of the population of profiles produced by random sectioning calls for a description of the frequency by which profiles of various sizes are encountered; we must hence characterize the frequency distribution of profiles derived from cutting spheres of equal size. To do this we must classify the profiles into size classes, whereby the class interval Δr is constant; for each class we must determine the frequency.

We can use a graphical approach to achieve this. Look again only at one quadrant of the main circle of the sphere in the y, z-plane; for convenience we shall now use the lower right hand quadrant (Fig. 2.40). We cut the sphere with a horizontal plane, and hence we measure the profile radius along the y-axis; the largest possible profile radius is evidently R, the sphere radius. Draw a scale for r at the bottom of the sphere and subdivide the range from 0 to R into, for example, 10 classes. Now draw vertical lines at the class limits up to the circle, and then horizontal lines from the intersections between the circle and the vertical lines to the z-axis. These horizontal lines delimit a sequence of 10 horizontal slices of the sphere and it is evident that each one represents one size class of profiles: if a horizontal section plane hits slice 8, the profile radius will fall into size class 8, if it hits slice 10, the profile radius will fall into size class 10, and so on.

This situation is in some way similar to what we had done before in order to derive the mean profile radius (Fig. 2.38), with one important difference: it is evident from Fig. 2.40 that the slices vary in thickness, becoming thinner and thinner as we go down to the bottom half of the sphere. But now remember that the probability of hitting a slice by a horizontal plane is directly proportional to its vertical thickness: the thickness of each slice hence

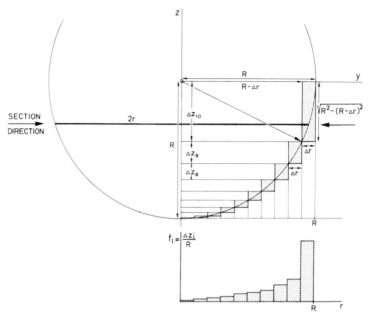

FIG. 2.40: Graphical construction of frequency distribution of profile size classes.

determines the frequency of the profiles falling into the corresponding size class! The frequency distribution of profile radii can hence be graphically constructed from Fig. 2.40 by simply plotting the slice thickness as the height of the corresponding column in the histogram. The relative frequency f_i of profile radii in each class is given by the ratio of slice thickness Δz_i to the sphere radius R:

$$f_i = \Delta z_i / R \tag{2.109}$$

The value of Δz_i can be easily formulated if we examine one typical slice (Fig. 2.40). Taking r_i as the upper limit of all profile radii falling into class i, the lower limit is evidently $(r_i - \Delta r)$. The value of Δz_i is the difference between the values of z at the two points where the lower and upper slice boundaries intersect the z-axis; this is directly given again by the theorem of Pythagoras:

$$\Delta z_i = \sqrt{[R^2 - (r_i - \Delta r)^2]} - \sqrt{(R^2 - r_i^2)} \tag{2.110}$$

and hence, considering all possible values of r, the frequency becomes

$$f(r) = \frac{\Delta z(r)}{R} = \frac{1}{R} \cdot \{ \sqrt{[R^2 - (r - \Delta r)^2]} - \sqrt{(R^2 - r^2)} \} \tag{2.111}$$

The first square root within the braces is the value of z corresponding to the lower class limit, the second root that corresponding to the upper class limit.

We may carry this one step further. Note that we obtained this histogram by subdividing the sphere radius R into ten classes of equal width Δr. In more general terms we may subdivide R into any number j of classes of width Δr; it is then evident that the sphere radius may be written as

$$R = j \cdot \Delta r \qquad (2.112)$$

In this case we may then number the profile size classes with i, so that the profile radius corresponding to the upper limit of the size class is

$$r = i \cdot \Delta r \qquad (2.113)$$

If we now substitute these values into equation (2.111) we obtain

$$f(i \cdot dr) = \frac{1}{j\Delta r} \{ \sqrt{[j^2 \cdot \Delta r^2 - (i \cdot \Delta r - \Delta r)^2]} - \sqrt{(j^2\Delta r^2 - i^2\Delta r^2)} \} \qquad (2.114)$$

which may be simplified by pulling Δr out of the root to give

$$f(i) = \frac{1}{j} \{ \sqrt{[j^2 - (i-1)^2]} - \sqrt{(j^2 - i^2)} \} \qquad (2.115)$$

This now defines the frequency of profiles in each size class in a way which is independent of the absolute sphere size. It is hence easy to construct histograms of this frequency distribution for any number of profile size classes. In Fig. 2.41 this is done for up to 10 classes. Note that the sphere diameter is always given by the upper limit of the largest size class; this is plausible because no profile may be larger than the sphere diameter! Note further, that the largest class always has the highest frequency. An interesting result is that in which only two classes of profiles are considered: the larger class, into which all profiles with a diameter larger than $D/2$ fall, comprises 86.6% of all profiles, whereas only 13.4% are smaller than $D/2$. We shall return to this in the application section.

It should be remarked here that some authors use a different approach to classifying profile sizes by using the class midpoint or even the lower class limit as the characteristic profile radius. This slightly modifies equations (2.110) to (2.115).

While this is generally a proper procedure, in this particular case the use of the upper class limit makes more sense, because the profile radius r can be expressed as an integer fraction of sphere radius R, and because the largest observable profile radius cannot be larger than R.

Finally, it should be noted that in some formulations the frequency of

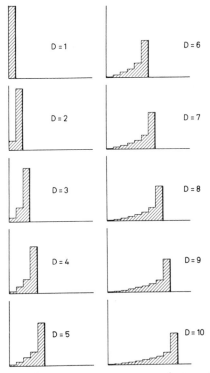

FIG. 2.41: Profile size distribution for spheres of varying size D.

profile radii is derived by differentiation of the equation of the circle. In this case the frequency $f(r)$ is expressed as

$$f(r)dr = \frac{r}{R\sqrt{(R^2 - r^2)}}dr \qquad (2.116)$$

Which formulation is used depends on the procedure of analysis.

2.8.2. PROFILE SIZE DISTRIBUTION IF A MIXTURE OF SPHERES OF VARYING SIZE IS SECTIONED

Let us now consider the more realistic situation where the spheres making up our population of particles vary in size. Look at Fig. 2.41 and imagine that spheres of three size classes, e.g. 4, 5 and 6, are present in the population. If these are cut, profiles are generated which fall into the size classes 1 to 6. We note that classes 1 to 4 will contain profiles derived from all three sphere sizes, whereas spheres of size 4 are missing in classes 5 and 6, and spheres of size 5 are also missing in class 6. In a general way we can say that each profile class will receive profiles from spheres whose diameter is larger or equal to the size of these profiles.

Let us now see how many profiles will be provided by each sphere size class. This evidently must depend on the probability of obtaining a profile from the spheres in each class. Consider spheres of radius $R_j = j \cdot \Delta r$. We have found in equations (2.77) that the numerical density of profiles derived from such spheres is proportional to the numerical density of the spheres multiplied with their diameter; therefore

$$N_{Aj} = N_{Vj} \cdot 2R_j \qquad (2.117)$$

It is now quite plausible that these profiles will be distributed among the size classes according to equation (2.115) defining the frequency of each size class. We can then immediately derive the numerical density of profiles of a certain size class i derived from spheres of size j to be

$$N_A(i, j) = N_{Aj} \cdot f(i) \qquad (2.118)$$

and by appropriate substitution we obtain

$$N_A(i, j) = N_{Vj} \cdot 2R_j \cdot \frac{1}{j} \cdot \left[\sqrt{(j^2 - (i - 1)^2)} - \sqrt{(j^2 - i^2)} \right] \qquad (2.119)$$

But since $R_j = j \cdot \Delta r$ this becomes

$$N_A(i, j) = N_{Vj} \cdot 2\Delta r \left[\sqrt{(j^2 - (i - 1)^2)} - \sqrt{(j^2 - i^2)} \right] \qquad (2.120)$$

To obtain the numerical density per unit section area of profiles of size class i contributed by spheres of class j we must hence simply multiply the term in brackets (a simple coefficient!) with the numerical density of particles of size j, and with the class width. In Fig. 2.42 this has been performed for three size classes, assuming certain numerical densities for the sphere classes. The pattern of the histograms is similar, but they differ in relative height of the columns which is proportional to N_{Vj}.

Now that we have expressed the profile frequency as numerical densities per unit section area it is a simple matter to construct the corresponding overall profile size distribution: we simply add the values of N_A in each column i to give the profile number per unit area in size class i; this is shown in the last histogram of Fig. 2.42.

2.8.3   DERIVATION OF SPHERE SIZE DISTRIBUTION FROM A MEASURED PROFILE SIZE DISTRIBUTION

The practical situation is usually reversed: We are given a profile size distribution which we have obtained by measurement on sections, and would like to know the sphere size distribution from which it was derived. This

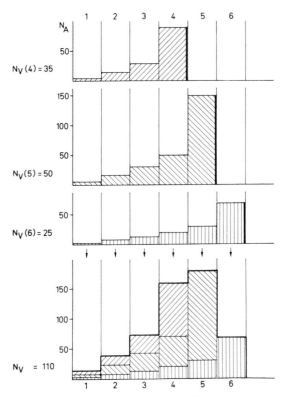

FIG. 2.42: Mixture of spheres of three size classes results in compound size distribution of profiles.

problem is not easy to solve; it is dealt with in more detail in Volume 2, Chapter 6.

The general approach to this problem is the following (Fig. 2.43): we know that the largest profile class (12) has received profiles from the largest sphere class only; and we know that these spheres have contributed to all profile classes in proportion to equation (2.120). From the numerical density of profiles in the largest class, call it m for maximal, we can derive N_{Vm} by setting $j = i = m$ in equation (2.120); this yields

$$N_{Vm} = N_{Am}/[2\Delta r \sqrt{(2m - 1)}] \qquad (2.121)$$

Then we can calculate through equation (2.120) the profile numbers contributed to the smaller size classes and subtract them from the histogram. We are then left with a residual histogram with the largest profiles now occurring in class 11, or $(m - 1)$ in general terms. We can proceed in exactly the same way to now estimate the numerical density of spheres in class 11, subtract their contribution from the histogram, and so on, until the histogram is exhausted (Fig. 2.43).

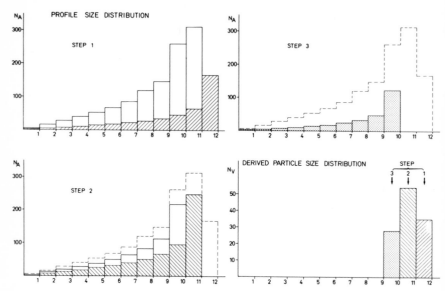

FIG. 2.43: Principle for deriving particle size distribution from profile size distribution.

In practice this is a cumbersome procedure and it requires a precise knowledge of the large size classes; because they are subtracted first, any error in their estimation will introduce large errors in the calculations. Wicksell (1925) and later Saltykov (1958) have proposed solutions which apply a set of coefficients to the measured profile frequencies in order to obtain, by simple but tedious computation, the frequency distribution of spheres. These principles will be developed in detail in Volume 2, Chapter 6. Practical procedures for their application will be discussed in Section 5.2.9 of this volume. The reader is also referred to the excellent presentation of these methods by Underwood (1970).

2.9.   Particle size and chord or intercept length

If a test line traverses a spherical particle the length of the chord, l_i, of the line intercepted by the particle depends (1) on the size of the particles measured by its diameter $D = 2R$, and (2) on the distance x_i of the chord from the sphere's center. From Fig. 2.44 it is easily seen that

$$l_i = 2\sqrt{(R^2 - x_i^2)} \qquad (2.122)$$

It will be left to the mathematical development to be given in Volume 2, Section 7.1 to show that the probability of obtaining an intercept of length l is

$$\Pr\{l\} = l/2R^2 \qquad (2.123)$$

which means that the frequency distribution of l is linearly proportional to l;

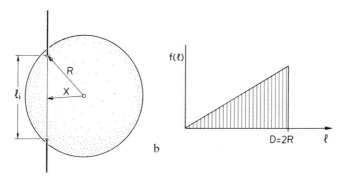

FIG. 2.44: (a) Relation between intercept length l and sphere radius R; (b) frequency distribution of intercept lengths.

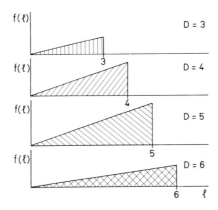

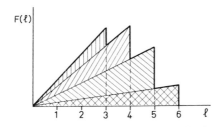

FIG. 2.45: Summation of intercept length distributions for four sphere size classes.

since no intercept can have length $l > D$ the frequency distribution $f(l)$ is represented by a straight line (Fig. 2.44b). This is evidently a very simple relationship which should offer great analytical advantages, particularly if the particles have different sizes.

Indeed, if a mixture of spherical particles of varying size D is probed with random test lines each size class of particles will give rise to a chord length distribution in the form of a triangle, as shown in the top part of Fig. 2.45. It can be shown that the area of the triangle is proportional to the number of spheres in this size class (see Volume 2, Chapter 7). The overall frequency distribution of intercept lengths $F(l)$ obtained from all spheres in this population is then found by simply adding up all the triangles, maintaining the area relationship, as shown in the bottom graph of Fig. 2.45. This is evidently identical to the procedure used in Fig. 2.42 to construct the overall frequency distribution of profile diameters obtained from cutting a population of spheres by a plane.

This simple relationship will be exploited in Section 5.3 when we introduce the graphical procedure of Lord and Willis (1951) for deriving particle size distribution from a measured distribution of intercept lengths.

CHAPTER 3

Sampling of tissue

In the general practice of microscopic research little attention is usually paid to the process of obtaining micrographs. One is seeking a certain observation and is satisfied when it is obtained in a reproducible fashion: this is the nature of a qualitative investigation. In morphometric work, however, quantitative relationships are sought; it is hence evident that meaningful results can only be obtained if each step of sampling—from the selection of experimental animals, through the removal of an organ from the body, the collection of tissue blocks, the recording of micrographs, to the counting of points—is rigorously controlled. Before discussing the practical application of stereological principles we shall therefore start off by considering sampling principles as they apply to a stereological analysis.

3.1. Hierarchy of reference spaces

The notion of measuring with respect to a reference system is the basis of all stereological principles: each parameter is obtained as the ratio of two measurements, one estimating the size of the objects under investigation, the other the size of the space in which they are contained, which we may call the reference space. Thus, in speaking of "relative volume" or "volume density", or of "surface density", we imply that the number given is valid only if it is related to the appropriate reference space.

Let us consider an example: we have found the volume density of mitochondria in hepatocyte cytoplasm to be 0.23; if this same parameter is now related to liver parenchyma it is reduced to 0.18, because hepatocyte cytoplasm amounts to only 78% of liver parenchyma. If this is carried one step further, namely to the total organ, the volume density of mitochondria in

63

the liver becomes about 0.16 because some 10% of the liver is made up of components other than parenchymal tissue.

It emerges from this example that a variety of reference spaces may be selected to suit the purpose of the investigation, and it is also clear that the specific nature of the reference system is not self-evident but must be stated very clearly.

A stereological study must therefore begin with a careful qualitative analysis of the various compartments of an organ and their hierarchical order. The result of such an analysis is shown in Fig. 3.1: the liver has been sequentially divided into smaller and smaller units, which are ordered in the form of a pedigree chart. At the first level, parenchyma—i.e. the main functional tissue—is divided from non-parenchyma, which comprises larger vessels, connective tissue, bile ducts, etc. Then cells and extracellular spaces are identified; this leads to some fifteen to twenty or more different compartments, depending on how far the differentiation of cell types is pushed. The further subdivision of cells follows a rather stereotyped pattern: nuclei are separated from cytoplasm, the latter being decomposed into the various spaces with their bounding membranes, with a further indication of where the notion of discrete particles could be applied. In Fig. 3.1 this has been

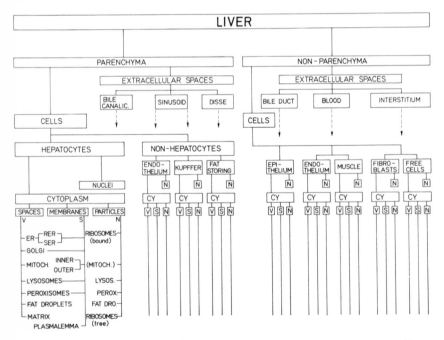

FIG. 3.1: Hierarchical model of possible reference spaces for liver as an example.

detailed for hepatocytes, but it is quite evident that the subdivision of other cell types would follow the same general pattern.

In practice it is often not necessary to push the analysis as far as that: one can concentrate on one main axis leading from the organ to the target cell of the specific study. In Chapter 8 the reader will find various examples of such reduced hierarchical models specified towards the study of some particular aspects of the different organs. Figure 8.1.1. is a simplified version of the diagram of Fig. 3.1.

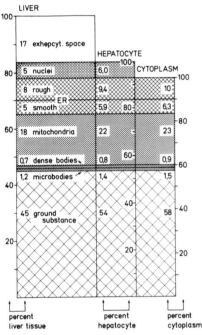

FIG. 3.2: Volume densities of hepatocyte organelles expressed with respect to the three reference spaces liver, hepatocyte and cytoplasm. (Reproduced by permission from Weibel *et al.*, 1969.)

It is quite evident from looking at Fig. 3.1. that a variety of reference spaces can now be selected to suit the purposes of a specific study. But there are basically two types of dependencies between them: (1) vertical or hierarchical, and (2) horizontal or parallel dependencies. The result of hierarchical dependency of reference spaces is shown in Fig. 3.2, where, step by step, measurements of cytoplasmic composition are related to hepatocyte cytoplasm, to total hepatocyte, to liver parenchyma and, finally, to the entire liver (Weibel *et al.*, 1969). This type of dependency is closely related to the procedure of multiple stage sampling which will be discussed in Section 3.2. Figure 3.3 alternatively shows how vertical and horizontal dependencies are combined. In a study of the four cell types making up liver parenchyma (Blouin, *et al.*, 1977) the relative volume of each cell type was first determined (horizontal dependency) and then the composition of each cell in terms of organelles was estimated (vertical dependency). It is evident that Fig. 3.3 is nothing else than a quantification of the lower left hand quadrant of the chart shown in Fig. 3.1, whereby the relative area of each rectangle within the large square indicates the relative volume of that component within liver parenchyma.

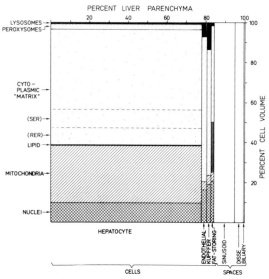

FIG. 3.3: Combination of vertical and horizontal dependencies: composition of individual cell types (vertical) related to composition of liver parenchyma (horizontal). (Reproduced by permission from Blouin *et al.*, 1977.)

From such data it is, however, also possible to derive purely "horizontal" comparisons. One may, for example, ask how all the mitochondria of liver parenchyma are distributed among the various cell types. The reference space is taken to be the volume of "all parenchymal mitochondria", or "all parenchymal lysosomes" etc.; it hence represents a horizontal slice of the chart in Fig. 3.1, running right across the various cell types. Fig. 3.4 shows that this can also be done for membranes, the reference system now being the "surface of all endoplasmic reticulum membranes" etc.

Whilst the possibility of choosing from a variety of possible reference systems affords the investigator with a great degree of flexibility in the design of his study—and gives his ingenuity free play—the question may justly be asked which principles should govern the selection of an optimal reference system. In very general terms one could say that the optimal reference system should have a high degree of stability from experiment to experiment. This is to say that we should have independent estimates which show that the reference system does not change between controls and experimental animals, for example. This is a most important point because it can be easily conceived that a change in volume density of a certain organelle may be due as well to a change in the organelles as to a change in the reference space: if we find the volume density of mitochondria in hepatocyte cytoplasm to be

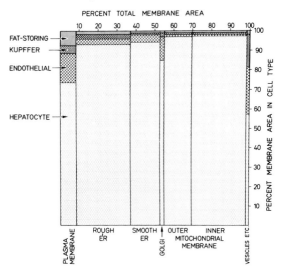

FIG. 3.4: Combination of vertical and horizontal dependencies: various membrane types are allocated to the four cell types of liver parenchyma. (Reproduced by permission from Blouin *et al.*, 1977.)

increased this cannot be interpreted as a true increase in mitochondrial volume, unless we know that the volume of hepatocyte cytoplasm has not been reduced by some other process. If the surface area of endoplasmic reticulum is induced by drug treatment (Stäubli *et al.*, 1969) the surface density in hepatocytes changes very little; the important change is an increase in the total volume of cytoplasm containing these membranes expressed in an enlargment of liver volume. In this case the true changes could be assessed by calculating all data with respect to "specific liver volume", i.e. to the volume of liver per 100 gm body weight of the experimental animals. The "stable" reference system in this case was therefore the body weight of the animals. It can be predicted that this situation may occur more often than not, and it is hence mandatory that the size of all reference systems is recorded, body weight and organ volume, in particular.

This brings up another point of great importance: the potential non-linearity in the proportion of various reference systems, particularly with respect to body weight. The allometric relation states that any parameter Y characterizing some features of an organism can be expressed as a power function of body mass W:

$$Y = a \cdot W^b \tag{3.1}$$

There exists an extensive literature on allometry (Kleiber, 1961; Stahl, 1967; Brody, 1968; a.o.) and many important data are found in the Biological Handbooks (e.g. Altman and Dittmer, 1971). It should be pointed out that most of these data apply to an *interspecies* comparison and that *intraspecies* relations may be different. It is fortunate, however, that many organ volumes —the pre-eminent basic reference system—vary linearly with body weight, at least within the same species and under normal conditions.

Concluding this discussion we then find that any quantitative study on biological structure must be preceded by a systematic evaluation of the hierarchical order of the organ under investigation in order to find the most informative reference system with respect to which the stereological data are expressed. The fundamental condition for a valid reference system is that it should be stable under the conditions of the study. To satisfy this condition one may have to go beyond the "organ" studied and consider body size as a reference system.

By considering, at the beginning of this chapter, the high complexity of the organization of a biological organism we pursued the purpose of pointing out the need for very careful planning of a stereological study in biology. Stereological principles are based on the concept of random sampling; can they be applied to biological "materials", where very little is left to chance, where order is one of the fundamental principles of structure (D'Arcy Thompson, 1942; Needham, 1968; Weibel, 1967b, 1978)? The answer to this question is yes and no. *Yes* in the case where a reference space can be found which contains the units (cells etc.) in very large number, where all these units are about alike, irrespective of their position in the hierarchy of structure; this is usually the case with the terminal units of any gland-like structure such as liver parenchyma, lobules of salivary glands, or the alveoli of the lung. *No* in all those instances where "similar" but non-identical units occur at various levels of the hierarchy of structure and where this hierarchy is one of the fundamental features of the system; the system of bile ducts or of hepatic blood vessels, branching systematically from the hilum towards the parenchyma, are examples. In studies on the lung (Fig. 3.5), to quote an extensively investigated example, the morphometric properties of the gas exchange apparatus can be obtained by applying stereological principles (Weibel, 1963, 1973a), whereas the conducting airways and blood vessels

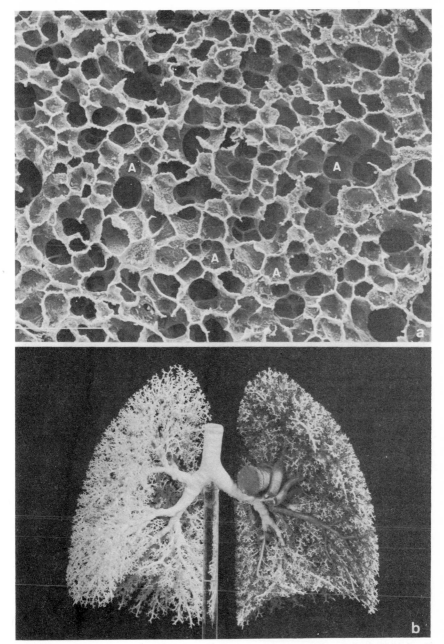

FIG. 3.5: (a) Scanning electron micrograph of human lung parenchyma shows "random" arrangement of alveoli (A). (b) Bronchial tree is however a highly ordered branching system.

must be measured by methods that admit an ordering of the elements with respect to the branching pattern (Weibel, 1963; Horsfield and Cumming, 1968). It is hence evident that a careful analysis of the hierarchy of structure must form the basis for the planning of any morphometric study, allowing the selection of proper methods by carefully defining the reference spaces with respect to which measurements are to be obtained.

3.2. Multiple stage sampling

In the preceding section we have seen that a systematic hierarchy of reference spaces exists in biological organisms because of the hierarchical nature of the organization of living matter. Here we shall discuss an operational hierarchy of sampling which is of a purely practical nature: stereological methods are applied at the microscopic scale whereas the organs we intend to study are macroscopic, thus necessitating a link between detailed point counts on micrographs and the organ or organism by means of a rigorously hierarchical sampling procedure.

Figure 3.6 shows the sequence of steps which must be followed to lead from a population of biological organisms to the final point counts on micrographs. The sampling operation can be subdivided into seven steps, but it is interesting to note that only four of these are sampling operations in the ordinary sense of the term, where the investigator has some freedom to influence the size of the sample by determining the number of sampling units used:

(1) In sampling from the population he may choose the number of organisms which are to be retained as a population sample.

(2) In sampling tissue blocks he will have to pick a certain number of them.

(3) In recording micrographs on the sections he will have to choose the number of fields to be analysed.

(4) In preparation for point counting he may decide on how many points, how much test line length etc. are to be used.

The other operations can be considered mere reductions of the preceding sample size, imposed by the very nature of the system:

(1a) Each organism will yield only a limited number (or volume) of a certain organ.

(2a) From each tissue block only one section is in general obtained for further analysis.

(3a) The entire micrograph is covered by the test system, thus reducing the set of test points or lines etc. to those required for point counting.

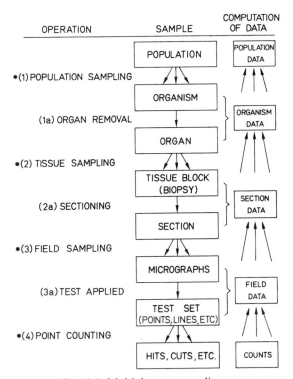

FIG. 3.6: Multiple stage sampling.

These ancillary steps are hence of little consequence for the subsequent considerations on sample size and on the computational and statistical procedures used to generate the final data where only the four main steps need to be considered.

There is however another type of operational sampling hierarchy which is more closely related to the systematic hierarchy described above. It is related to the fact that the units which make up an organ are of very different orders of magnitude, decreasing in size as we go down the pedigree of Fig. 3.1: parenchymal lobules measure about 1 mm, hepatocytes 10 μm, mitochondria 0.5 μm, and mitochondrial cristae 20 nm, for example. Reliable measurements of these objects will require very different magnifications; it is hence not possible to obtain them in one step.

The multiple stage sampling procedure we propose is shown diagrammatically in Fig. 3.7. It consists of a cascade of several sampling stages which are all linked in such a way that the measurements obtained in the last stage can be referred back to the reference space of the first stage, in this case the organ volume.

(1) In a first set of low power micrographs the volume density of parenchyma is estimated:

$$V_{Vp,L} = V_p/V_L \qquad (3.2)$$

all other measurements that may or may not be performed at this stage need not concern us in the current context.

(2) In a second set of higher power micrographs the volume density of hepatocytes in parenchyma

$$V_{Vh,p} = V_h/V_p = V_{Vp}* \qquad (3.3)$$

is determined; we may label this parameter with an asterisk to denote that it is obtained with respect to a restricted reference space, viz. parenchyma.

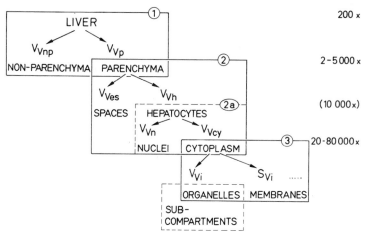

FIG. 3.7: Cascade of sampling stages for liver with suitable microscopic magnifications indicated.

In general it will be possible at this level of magnification to differentiate adequately between cytoplasm and nuclei, so as to obtain directly the volume density of hepatocyte cytoplasm in parenchyma:

$$V_{Vcy,p} = V_{cy}/V_p = V_{Vcy}* \qquad (3.4)$$

If this does not yield a sufficiently accurate result one may have to insert an additional step (2a) in which hepatocytes are subdivided into nuclei and cytoplasm, with the resulting volume density of cytoplasm in hepatocytes being

$$V_{Vcy,h} = V_{cy}/V_h \qquad (3.5)$$

(3) In the third step we restrict the sample to hepatocyte cytoplasm, exclud-

ing all other components from the reference system; the technical procedures for doing this are outlined in Section 4.2. On these high power micrographs which afford adequate resolution we now obtain the measurements on the organelle population, such as volume density of the various organelles, surface density of membranes, or length density of filaments, etc. The parameters obtained are "densities in cytoplasm":

$$V_{Vi,cy} = V_i/V_{cy}, \quad S_{Vi,cy} = S_i/V_{cy} \tag{3.6}$$

It is now a simple procedure to refer these parameters back to any of the other reference spaces. To refer them to hepatocytes they must be multiplied with $V_{Vcy,h}$, the volume density of cytoplasm in hepatocytes; further multiplication with $V_{Vh,p}$ relates them to parenchyma, and, finally, multiplication with $V_{Vp,L}$ yields densities in liver. This is easily understood if we simply write out the product of ratios:

$$V_{Vi} = \frac{V_i}{V_L} = \frac{V_i}{V_{cy}} \cdot \frac{V_{cy}}{V_h} \cdot \frac{V_h}{V_p} \cdot \frac{V_p}{V_L} \tag{3.7}$$

or, for surface densities,

$$S_{Vi} = \frac{S_i}{V_L} = \frac{S_i}{V_{cy}} \cdot \frac{V_{cy}}{V_h} \cdot \frac{V_h}{V_p} \cdot \frac{V_p}{V_L} \tag{3.8}$$

In stereological notation this would read

$$V_{Vi} = V_{Vi}{}^* \cdot V_{Vcy}{}^* \cdot V_{Vh}{}^* \cdot V_{Vp} \tag{3.9}$$

$$S_{Vi} = S_{Vi}{}^* \cdot V_{Vcy}{}^* \cdot V_{Vh}{}^* \cdot V_{Vp} \tag{3.10}$$

whereby one should actually specify the meaning of the asterisk, namely the reference space,

$$V_{Vi,L} = V_{Vi,cy} \cdot V_{Vcy,h} \cdot V_{Vh,p} \cdot V_{Vp,L} \tag{3.11}$$

$$S_{Vi,L} = S_{Vi,cy} \cdot V_{Vcy,h} \cdot V_{Vh,p} \cdot V_{Vp,L} \tag{3.12}$$

This notation (see Chapter 1) evidently becomes somewhat clumsy and it may be advisable to just use the plainly written ratios of volumes, as used in equations (3.7) and (3.8).

We have used the liver studied by electron microscopy as an example. The sampling schemes for other tissues will be discussed in Chapter 8. If we want to generalize this procedure we could simply replace the phase indicators i, cy etc. by numbers indicating the sequence of sampling stages. Thus in the first stage we would measure volume density of phase 1 in the starting material (0); in the second stage volume density of phase 2 in 1, and so on. With four phases estimated in sequence we have

$$V_{V4} = \frac{V_1}{V_0} \cdot \frac{V_2}{V_1} \cdot \frac{V_3}{V_2} \cdot \frac{V_4}{V_3} = \frac{V_4}{V_0} \qquad (3.13)$$

It should be noticed that the same phase is used as object space in one ratio, but as reference space in the next one; this is the whole basis of multiple stage sampling.

The ultimate generalization says that we may perform sampling through n subsequent stages, and that at the last stage we may estimate any parameter Y with respect to the last phase n. We then find the density of Y in the starting material as

$$Y_{Vn} = \frac{Y_n}{V_0} = \frac{V_1}{V_0} \cdot \frac{V_2}{V_1} \cdot \frac{V_3}{V_2} \cdots \frac{V_{n-1}}{V_{n-2}} \cdot \left(\frac{Y_n}{V_{n-1}} \right) \qquad (3.14)$$

$$= \frac{Y_n}{V_{n-1}} \prod_{i=1}^{n-1} \frac{V_i}{V_{i-1}}$$

The procedure of multiple stage sampling here described may look rather complicated and difficult to perform. In practice, however, it is very efficient and time saving. Its main virtue is that it allows us to exploit to our advantage the hierarchical order inherent in biological structure. Figure 3.8 shows a simple model of an organ; it could be a gland, and we might be interested in estimating a certain parameter of the small elements "a" marked by black dots. Evidently we will require a fairly high magnification giving micrographs of small field, marked by a square. In Fig. 3.8a we see the phase elements a in relation to the whole organ; they appear clustered in small groups which are themselves again clustered in certain regions. If we refer to Fig. 3.8b we find the reason: the phase a occurs only in some of the "cells", and these are not randomly dispersed across the organ but are contained within groups of cells, which we might call lobules. The inhomogeneity apparent on the over-view picture of Fig. 3.8 is hence due to an essential property of structural order, the grouping of cells with similar function, and we must anticipate this to be of frequent occurrence in biology.

If we now were to apply the small test field necessary to achieve adequate resolution to the organ as a whole, it is evident from Fig. 3.8a that the majority of these fields will fall onto "empty" space; sampling becomes very inefficient. Resorting to the multiple stage sampling here proposed we can take advantage of the pattern of organization and estimate the parameter Y_a with respect to the volume V_c of the cells containing a; with a small number of micrographs at fairly high power the parameter $Y_{Va,c} = Y_a/V_c$ will be very reliably measured because it is evident that, within space (c), the phase elements a

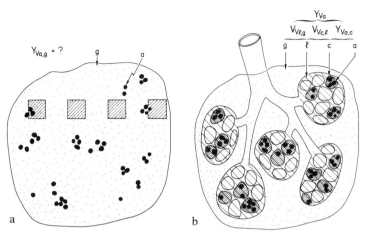

FIG. 3.8: Clustering of phase a in gland g is due to grouping of cells c in lobules l, imposed by branching ducts of the gland.

are rather homogeneously distributed. It will then only be required to estimate at two lower magnifications the volume density of (c) in (l), and of (l) in (g) in order to derive the parameter density Y_{V_a} in the entire organ.

In conclusion, simple sampling procedures are appropriate only if the objects of interest are distributed with sufficient homogeneity throughout the material. If this is not so—as will often be the case in biology—sampling in sequential stages may remedy the difficulties if each stage can be so selected as to yield samples representative of the composition of the reference space. The basic rationale of sampling for stereology can be summarized as follows:

(1) Measurements should be performed at optimal magnification to obtain adequate resolution.

(2) If the reference space is not representatively sampled at this magnification, the measurements should be referred to a "sub-reference space".

(3) The relative volume of the sub-reference space in the over-all reference space is then estimated at a lower magnification.

The rule of thumb would be as follows: if, at the magnification chosen, you find too many "empty" fields, i.e. fields not containing the object of interest, revert to multiple stage sampling.

3.3.   The practice of random sampling for stereology

We have repeatedly stressed that stereological procedures are based on the concept of "random sampling" of structures by the sections, the micrographs, and the test lattices applied to them. In Section 2.1.4. we specified what is meant by randomness. Since we are dealing with spatial specimens which we are cutting primarily with a plane section the geometric requirements on sampling are the following: (1) the section must be an *isotropic random* sample, i.e. all section orientations in space must be equally likely; and (2) it must be a *uniform random* sample, i.e. once the orientation is determined all positions of the sample section in the specimen (the location between the two tangent planes at this orientation) must be equally likely.

There are two ways of generating such a random section: (1) one can apply the section purely *from outside* the specimen, which leads to what has been called an *IUR-section* (Isotropic Uniform Random); (2) one can attempt to locate the section within the specimen, which yields an *A-weighted random section* (Miles and Davy, 1976; Miles, 1978a, b). As will be discussed in Volume 2, Chapter 2, only A-weighted sections lead to unbiased estimates of stereological parameters; but for the sake of generality we shall explain both random sectioning procedures.

How to obtain an IUR-section is shown in Fig. 3.9 (Miles, 1978a). The specimen is kept in a fixed position in space.

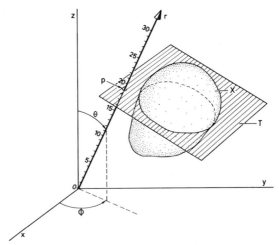

FIG. 3.9: Construction of IUR (isotropic uniform random) section of specimen X.

(1) It is first required to determine the direction of sectioning given by the two angles ϕ and θ with respect to an x, y, z-coordinate system which is arbitrarily fixed in space too. Remember that the orientation in space is proportional to $\sin \theta$ and to ϕ (Section 2.1.3); read a pair of (say two-digit) random numbers from a random number table: the first determines ϕ, the second $\cos \theta$ from which θ can be derived. The orientation of the section plane is determined.

(2) The random position of the random plane is determined by picking a random number p which determines the distance of the plane from the origin of the coordinate system (Fig. 3.9) measured in some convenient units. If the plane hits the specimen, i.e. if it falls between the two tangent planes in direction (θ, ϕ), it is accepted as an IUR section; if it does not hit the specimen a new random number p is picked, until the specimen is hit.

In performing this sampling procedure difficulties may arise particularly with respect to orientation of the section, mainly if the section is to be obtained on a microtome. It is evidently also possible to keep the section plane in a fixed horizontal position (in the x, y plane) and to orient the specimen: In order to obtain the same section plane we can imagine an orientation axis of the specimen which we first place vertically (in the z-axis): if the specimen is tilted by (θ, ϕ) (but in the opposite directions, although this does not matter!), any horizontal cut will then be an isotropic random section. All we need to do is to determine the level at which the section needs to be placed to make an IUR sample. This procedure can "fairly easily" be practiced if the microtome chuck is fitted to a goniometer; the sectioning position is obtained by trimming down to the required level, p then being the number of "wasted" sections.

How to obtain an A-weighted section is shown in Fig. 3.10 (Miles, 1978a). The procedure is to begin, as for an IUR section, with the specimen in fixed position with respect to a coordinate system:

(1) Determine the sectioning direction (ϕ, θ) from random numbers for ϕ and $\cos \theta$.

(2) The position of the section is however not fixed uniformly with respect to a linear sequence of points on the ray at (θ, ϕ), but rather uniformly with respect to the *specimen space*. For that purpose we must identify and number a "cloud" of uniformly distributed points within the specimen; with a cubic lattice each point could be designated, for example, by a three-digit number (from 000 to 999) corresponding to x, y, z. Pick a three-digit number from a random number table; if the point lies within the specimen the position of the section is determined. All that is now required is to position the (ϕ, θ) section plane such that it passes through the identified specimen point.

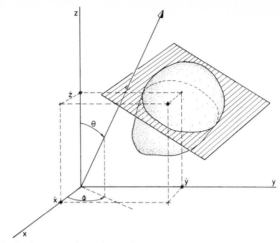

FIG. 3.10: Construction of A-weighted random section of specimen X.

Here again it seems possible to realize the technology required to perform A-weighted sampling; one needs again a goniometer stage for the specimen and then some kind of 3-D coordinate system for locating the random specimen point.

What is the difference between IUR and A-weighted random sampling? In IUR sampling a section that grazes the specimen near one of the two poles is equally as likely as one which passes near the center; the second section will produce a much larger specimen section than the former one. If the stereological estimates on such sections are averaged the small marginal section will have the same statistical weight as the much larger central section, and this introduces a bias. In contrast, A-weighted sampling gives the larger central sections a greater chance than the peripheral cuts; it is shown by Miles and Davy (1976) that this alone leads to an unbiased estimate. It should be mentioned that this bias has been recognized by Mayhew and Cruz-Orive (1974) who proposed essentially that the calculation of stereological parameters must consider the statistical weight of the section area (see Section 3.4). However, this computational procedure will only eliminate the bias if the sample has already been weighted for section area; a biased sample cannot lead to an unbiased estimate. For this reason, Mayhew and Cruz (1974) specified that ratio estimates, in general, are not necessarily unbiased, but that the ratio of mean profile areas yields a consistent estimate of the volume ratio.

The procedure of A-weighted sampling is particularly critical and difficult to implement with small specimens that one wants to cut totally, the so-called "restricted case"; for example, the thyroid gland of a rat sectioned for light microscopy. In the—much more frequent—"extended case" the specimen to be cut is a (small) sample of a much larger body. In this case it is fairly easy to implement A-weighted random sampling.

How to obtain an A-weighted random sample from a large specimen is shown in Fig. 3.11. The procedure is the following.

(1) Locate a uniform random point in the specimen space. This can be done by slicing the entire specimen into slices of equal thickness. The slices could be numbered (corresponding to z) and a grid of numbered points (x, y) deposited on each of them; find from a random number table a point (x, y, z) which lies in the specimen. This marks the location of the A-weighted random sample. If several sample blocks are required repeat the procedure with the grid unchanged as many times as necessary. This procedure can be simplified in practice if all slices are laid out flat and covered with one large grid of numbered points; this is actually the way we are practicing this sampling method (Fig. 3.11).

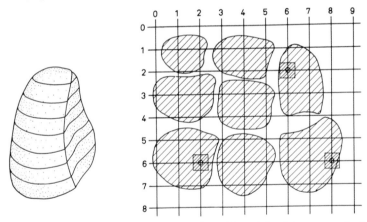

FIG. 3.11: Slicing procedure for practical location of A-weighted tissue samples in large specimen.

(2) Obtain a small block of tissue around the location identified by the random point in such a way that an isotropic section can be cut. An optimal procedure would be to determine (ϕ, θ) and to cut out a block of perhaps pyramidal or prismatic shape with the long axis coinciding with (ϕ, θ); a cross-section of such a block is then the isotropic section required. This is certainly feasible for light microscopy, at least with a certain degree of approximation; it will be more difficult to handle the minute chips of tissue processed for electron microscopy in this way. Fortunately, however, these small blocks usually settle in any orientation within the embedding resin

D

and one can safely assume that an arbitrary section cut from such embedded blocks is an isotropic random sample, at least if no effort is made to orient one of the block faces with respect to the section plane. Indeed, such a section is now an A-weighted random section and yields unbiased stereological ratio estimates.

3.4.   The choice of test quadrats on sections

Once an A-weighted section is obtained, the next step is to measure the feature characteristics on this section. In general, the section is not measured completely but one is rather forced to select small areas of the section for measurement, e.g. micrographs, and apply the test field or "quadrat" to them. Miles and Davy (1977) have extensively discussed the rules governing the positioning of this quadrat with respect to the section so as to obtain unbiased ratio estimates. We shall present their conclusions without going through the mathematical reasoning (see Volume 2, Chapter 2).

From what has been said above it is plausible that the quadrat sample must again be weighted for an unbiased ratio estimate to result. Figure 3.12 shows how a weighted FUR ("fixed orientation uniform random") sample is obtained. (1) Determine by random numbers two points, one in the specimen section X (W_1) and one in the quadrat area T (W_2); the random numbers could identify coordinates within X and T. (2) Shift the quadrat T into X so that W_2 is superimposed on W_1. The resulting intersection ($X \cap T$) between the section and the quadrat is an A-weighted sample. Note that if T is small

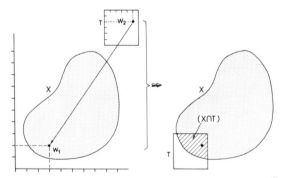

Fig. 3.12: Procedure for placing "quadrat" T (microscope field) in specimen X in accordance with A-weighted sampling.

as compared to X then this method will ensure that most of the time T will be entirely within X, so that the reference area $A(X \cap T)$ will vary little. If the "quadrat" is a test line of finite length then W_2 is taken uniformly along this line; super-position of W_2 onto W_1 yields an L-weighted FUR sample. Finally, if the "quadrat" is a grid of P_T test points a P-weighted FUR sample is obtained by choosing one of the grid points as W_2 and superimposing it on W_1.

Such FUR samples lead to unbiased results for all stereological ratio estimators that are not sensitive to anisotropy, such as A_A, L_L, P_P, B_A and also the mean curvature density C_A of the boundary trace. It should be noticed, however, that $B_A = (\pi/2) \cdot I_L$ does not belong to these estimators because I_L is sensitive to anisotropy; an unbiased estimator requires that the test line be selected as a weighted IUR quadrat, that is to say that the orientation θ of the test line must be chosen uniformly between 0 and 2π. Alternatively, an isotropic test line system such as a circle or the curvilinear test system of Merz (1967) discussed below can be used and applied as a weighted FUR quadrat. If the structure itself is isotropic then even a weighted FUR straight line "quadrat" is acceptable.

In practice, it is feasible to obtain weighted FUR quadrats, but it may be a quite time-consuming procedure. If the microscope stage is fitted with a scale giving coordinates, then one can choose from random numbers W_1 as a pair of stage coordinates; the center point of the field then marks W_1. A second coordinate system is required in the quadrat, i.e. in the micrograph field; a second random number marks W_2. A weighted FUR sample is obtained if the section is shifted until W_1 coincides with W_2. With light microscopy it should also be possible to introduce isotropy by turning the concentric specimen stage by θ around the point W; in the electron microscope this is hardly possible.

To approximate this type of sampling we have marked a grid of 9 points on the screen of our electron microscope, numbered 1–9; these serve the determination of W_2 by random numbers. We have fitted the microscope stage with a coordinate indicator on the x and y stage drives (some microscopes have these commercially built in) which allows us to locate W_1 in the section. This may be a rather painful procedure for certain specimens, such as lung tissue, so that we are also using as an alternative procedure a systematic rather than random point grid for W_1: we place the microscope screen into the corners of the support grid (see Fig. 3.15); if the center point is within the specimen we accept this point as W_1 and shift it to coincide with W_2 picked by random numbers from the 9 screen marker points. This is not FUR but weighted "FUS" (FU systematic) sampling.

The main advantage of this procedure is that it reduces edge effects which introduce a bias into the ratio estimates. This is evidently of importance with small specimens, whereas large specimens are not as critically affected. We have found that "spongy" structures such as the lung may preferably be sampled by the weighted FUR (or FUS) procedure described. In analysing the internal composition of the alveolar septa we require fairly high magnifications (see Section 8.10); the micrograph will only be partly filled by the septum (see Fig. 8.10.7). Since we take the area of the septum (A_s) as a reference for estimating capillary volume and surface densities etc., unbiased estimates require that the sample fields be weighted for A_s. Thus we proceed to a weighted "FUS" sampling as described above and record only fields for which W_1 is in the septum. This sampling procedure is somewhat more laborious than simple systematic sampling, but the efficiency is increased in the sense that one obtains a larger specimen area on a smaller number of micrographs, and the standard error is reduced.

3.5.   Random and systematic sampling

Only rarely does the sample considered for estimating a stereological parameter consist of only one section or quadrat. If multiple samples are taken from one section (micrographs) or from one specimen (sections) then there are various ways of obtaining them.

(1) *Simple random sampling*: The statistical theories used to predict or calculate sample variance postulate that the samples should be *independent*; such a sample is obtained by locating each sample unit, e.g. a tissue block in the specimen or a microscope field on the section, independently by means of random numbers, as described in the preceding sections. If the units are independent, then it is evidently possible—and the more likely the greater the sample—that two sample units may be very close together or that they can even (partially) overlap. (Fig. 3.13a); clustering of fields in certain regions is likely, whereas other regions are undersampled. This disadvantage of simple random sampling can be overcome by spacing the samples in a more or less regular fashion, by either systematic or stratified random sampling in order to get a more representative sample of the entire specimen.

(2) *Systematic sampling*: By spacing the sample units according to a predetermined regular lattice which is then randomly applied to the material, the sample becomes quite evenly dispersed throughout the material; an overlap of fields with consequent double sampling is excluded. It has been repeatedly shown that systematic sampling with a random start is superior to simple random sampling because it yields smaller errors. This can be shown on theoretical grounds (Cochran, 1953; Hilliard and Cahn, 1961; Hennig, 1967); but it has also been demonstrated experimentally. Ebbeson

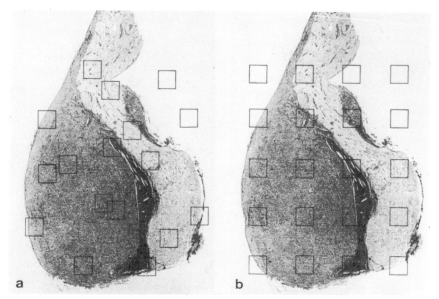

FIG. 3.13: Random (a) versus systematic (b) sampling of fields from a pituitary gland.

and Tang (1967) have shown that the standard error of the estimate of the number of nucleoli is nearly ten times as large if the microscope fields are obtained by simple rather than systematic sampling (Fig. 3.14). The main danger with systematic sampling is the following. If the structure contains a certain periodicity which coincides with that of the sampling lattice then the variance may be dramatically increased. One must therefore carefully evaluate the specimen for such periodicities before proceeding to systmematic sampling.

(3) *Stratified random sampling*: The material is first divided into regions or strata of more or less equal size; from these regions one or several blocks are picked by a systematic sampling procedure or, preferably, by a weighted FUR sampling technique. This method is intermediate between simple random and systematic sampling; as seen from Fig. 3.14 it is found that stratified sampling yields nearly the same standard error as systematic sampling. Miles and Davy (1977) have discussed the advantages of "partitioning" and stratified random sampling, particularly for cases where the specimen structure may vary importantly from one region to another.

There are hence theoretical and empirical reasons for choosing either systematic or stratified sampling procedures, rather than simple random methods. There are further practical reasons: it is indeed a difficult undertaking to attempt a rigorous simple random sampling process at any stage

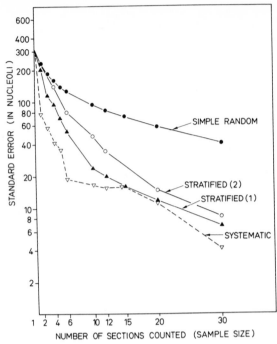

F<small>IG</small>. 3.14: Reduction of standard error in counting nucleoli in sympathetic ganglion achieved by random, stratified, and systematic sampling. (After Ebbeson and Tang, 1967.)

of sampling for stereological analysis whereas the two other forms are easily applied.

In our laboratory, the standard procedure for sampling micrograph fields is to follow a strictly systematic process: in light microscopy by programming an automatic stage drive (Weibel, 1970b), in electron microscopy by recording the micrographs with respect to one specified corner of the supporting grid (Fig. 3.15). We have mentioned above that for certain difficult cases we may combine the systematic with a weighted random sampling procedure. The sampling of tissue blocks from the organ usually follows a stratified sampling scheme (Fig. 3.16): a few sample slices are cut from different and clearly specified regions of the organ; these slices are diced and processed in individual vials, and, from these stratified pools of blocks, a few are randomly picked before embedding or sectioning. This way it is easy to compensate for regional differences, such as are known to exist in the human lung, for example (Glazier et al., 1967). Recently we have begun to improve on this procedure by using an A-weighted sampling procedure within each stratum, as described in Section 3.3.

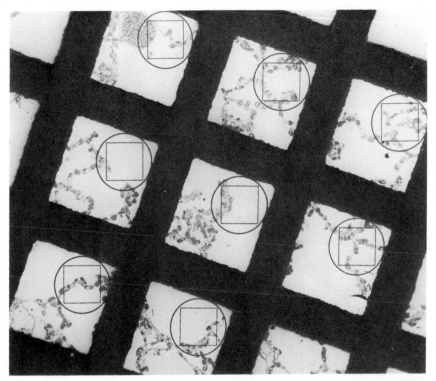

FIG. 3.15: Systematic sampling of electron micrographs by locating screen in pre-determined corners of supporting copper grid. (Reproduced by permission from Weibel *et al.*, 1966.)

3.6.   Sampling in ordered systems

The choice of an appropriate sampling method becomes particularly im-portant if the objects in the structure are not randomly and uniformly spread throughout the entire structure, i.e. if the structure shows some order—as is indeed the case in many biological materials!

There are basically five patterns of structural order which must concern us:

(1) *Layered structures*, usually occurring when some tissue is arranged between two more or less flat boundary surfaces, as is the case for all surface epithelia or for cartilage in the epiphyseal growth plate of long bones (Fig. 3.17). Furthermore, brain tissue is an exquisite example of layered structures (see Section 8.9).

(2) *Fasciculated structures*, i.e. the more or less parallel bundling of elongated elements, as for example in muscle or nerve tissue (Fig. 3.18).

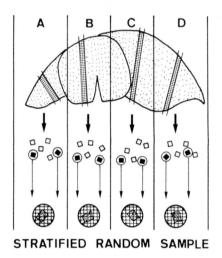

STRATIFIED RANDOM SAMPLE

FIG. 3.16: Stratified random sampling of tissue blocks. (Reproduced by permission from Weibel, 1969.)

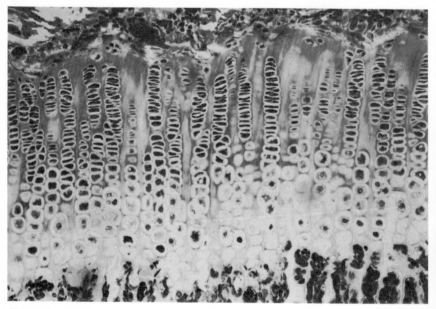

FIG. 3.17: Cartilage of epiphyseal plate of long bone as example for layered structures. (Courtesy R. K. Schenk.)

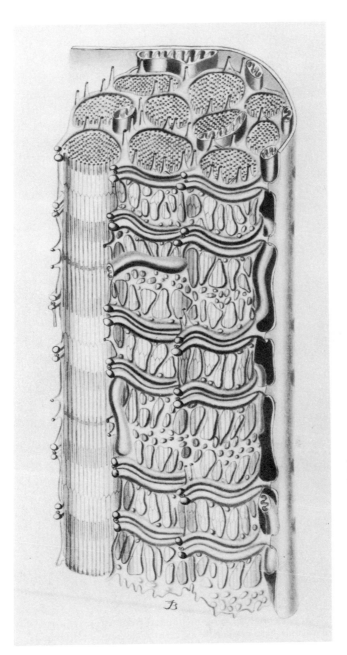

FIG. 3.18: Skeletal muscle as example for fasciculated and periodic structures.

(3) *Branching structures*, as they occur in the "vascular tree" as arteries branch out to finally form capillaries which are then converging back to fewer and fewer vessels on the venous side. Other examples are the duct system of nearly all exocrine glands, or the airways of the lung (Fig. 3.5).

(4) *Polarized structures*, occurring mainly where two sides (or ends) of a structure are differentiated towards different functions. This is generally the case for layered structures such as surface epithelia. This polarity usually results in the formation of a *gradient structure* within this layer (Fig. 3.17) because the cells at one surface have a composition different from those at the other. Branching systems always show this type of polarized or gradient structure.

(5) *Periodic structures* resulting from the sequential arrangement of identical units in a pattern of translational symmetry. An extreme example from biology is striated muscle (Figs. 3.18 and 19): the myofibrils are built of

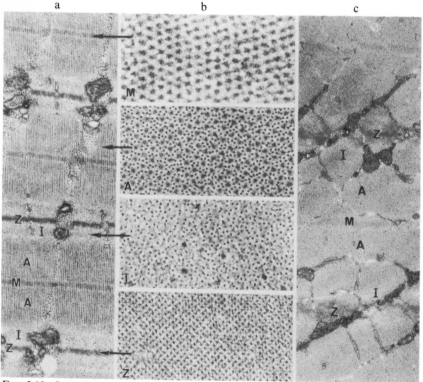

Fig. 3.19: Sampling muscle. (a) Longitudinal section reveals periodicity. (b) Cross-sections of various bands will have different appearance. (c) Oblique section evenly samples all levels of the periodic structure and is representative for composition across the cell as well. (Courtesy H. Claassen.)

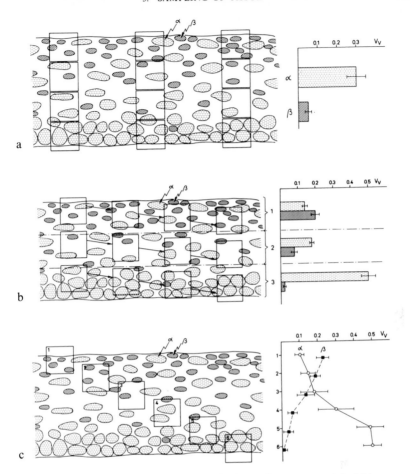

FIG. 3.20: Three different sampling schemes for layered structures: (a) yields average density; (b) and (c) yield information on composition gradient in terms of the two components α and β.

strictly identical elements, the sarcomeres, which show bilateral symmetry by the interlacing of actin and myosin and are then lined up in the following pattern:

$$*A \cdot M - M \cdot A * A \cdot M - M \cdot A * A \cdot M - M \cdot A *$$

There is, in addition, a hexagonal symmetry in the plane normal to the longi-tudinal direction due to the packing of actin and myosin filaments (Fig. 3.19). The length of these units is exactly identical: during contraction they change their length in synchrony, and they impose this order on all other cytoplasmic elements such as sarcoplasmic reticulum and mitochondria (Fig. 3.18). This is further discussed in Chapter 8.4. Periodicities also occur in layered structures, such as in the stacking of lamellae in lamellar bone.

It is evident that such ordered patterns will require sampling methods which account for the particularities of the system. Consider, as an example, a polarized layered structure with an internal gradient which has been cut by a section perpendicular to the surface (Fig. 3.20). We can conceive three basically different sampling schemes which serve different purposes:

(a) We can arrange the sample units (fields etc.) in form of a continuous strip extending from top to bottom; by this procedure we can determine the *average* composition of the structure, the gradient being averaged out by the sampling scheme. Miles and Davy (1977) show that this type of "partitioning" of the specimen is best for reducing the sample variance.

(b) In order to obtain information on the gradient itself—often an important property of the system—we can partition the structure into sublayers from each of which we secure an adequate sample by distributing the sample units homogeneously within this sublayer, i.e. more or less parallel to the surface, perhaps slightly staggering them to sample the entire depth of the sublayer, or even allowing for some overlap between the sublayer samples. This amounts to stratified sampling. The result is a histogram of the density of each component in each sublayer (Fig. 3.20b); over-all results will then have to account for the relative weight of each stratum or sublayer.

(c) An alternative sampling scheme, essentially identical to (b), would consist in recording, for example in an oblique row, a large number of fields that extend systematically throughout the gradient. For each such strip sample the progression of the parameters as a function of distance down the gradient is computed, and the profiles produced by several such samples are compared, as shown in Fig. 3.20c. This is evidently a very demanding procedure.

Sampling periodic structures poses some severe problems, particularly if there is also anisotropy, as in muscle. The picture, and hence the sample, will be quite different depending on the orientation of the section to the direction of periodicity and/or anisotropy (Fig. 3.19). A longitudinal section of muscle will reveal the periodicity; all parts of the periodic structure are sampled equally. A transverse section, however, will cut across only one region of the element of symmetry; but the composition of the muscle fiber in that plane is more faithfully represented than in a longitudinal section. An oblique section is a compromise: it will at once sample all levels of the periodic structure and reveal the composition in the transverse direction; the angle of sectioning will however influence, in part, the measurements obtained on such a section, as outlined below (Sections 4.3, 8.4 and 8.5).

By far the greatest sampling problems are encountered in dealing with branched structures showing a gradient down the tree. Stereological methods are usually not applicable to this situation in a simple way. The gradient in the dimension of the (usually tubular) elements must be determined by some way of hierarchical sampling which accounts fully for the order of the elements (Weibel, 1963; Horsfield and Cumming, 1968). If one can then find some reference parameter, which is measurable by stereological methods and which can be linked to the hierarchical data, then it might be possible to use stereological methods for estimating, for example, wall thickness or wall composition of the tubes at various levels of the hierarchy.

We have discussed the problems encountered in sampling from ordered structures in a very general way. It is difficult to propose broadly valid procedures, as these will always be determined by the particular conditions of the material and by the type of information sought. Some additional suggestions are found in Chapter 8 among the specific examples.

3.7. Stereological parameters and representative sample size: general considerations on statistics and sample planning in stereology

In the preceding sections we have loosely used the term "representative sample"; this now needs some comment. Primary stereological data are collected in terms of point counts, or feature counts; the stereological parameters we seek are ratios of such counts. Looking at a typical table of primary data (Table 3.1) we note a considerable variation in the counts per micrograph: the reference point number, P_c, varies by a factor of two, and the object point counts, P_a, are zero on two occasions. If we took the ratio of P_a/P_c for each micrograph we would obtain values from 0 to 0.5, the average being 0.305. An alternative procedure would be to sum all the points and to take the ratio of these sums, which is equivalent to taking the ratio of the average point counts: the result is 0.313. Which is the correct value?

It is common high school knowledge that the ratio of two means is not equal to the mean of the ratios:

$$\frac{1}{n} \cdot \sum_{i=1}^{n} (X_i/Y_i) \neq \left(\frac{1}{n} \sum_{i=1}^{n} X_i\right) \Big/ \left(\frac{1}{n} \sum_{i=1}^{n} Y_i\right) \qquad (3.15)$$

One of the fundamental problems arising is therefore the question of at which point of the complex sampling procedure the primary counts should be used to calculate the stereological parameters.

The problem becomes immediately evident if we look at one—perhaps slightly atypical—example (Mayhew and Cruz-Orive, 1974). Take a cell type in which we would like to estimate the volume density of nuclei within

Table 3.1

Example of primary data on volume density estimate for component a in cell c; four micrographs from five sections.

Section	1				2				3				4				5				Σ	$\bar{P}$
Micrograph	1	2	3	4	1	2	3	4	1	2	3	4	1	2	3	4	1	2	3	4		
P_a	2	8	8	3	12	6	8	0	0	9	11	7	6	7	6	9	7	3	8	5	125	6.25
P_c	21	22	17	24	23	16	21	13	12	25	22	19	20	18	24	21	16	21	21	23	399	20
ΣP_a	21				26				27				28				23					
ΣP_c	84				73				78				83				81					

the cells by measuring their profiles on sections. It is evident that some profiles will contain nucleus as well as cytoplasm, whereas others will contain only cytoplasm (Fig. 3.21). Any one of these profiles alone can therefore impossibly be considered representative of the cell; we need counts on several sections in order to calculate nuclear volume density; we must hence sum the hits on nuclear profiles and those on the cell and then take the ratio of these sums. This notion was, by the way, implied in the derivation of the basic stereological principles, since we found that a stereological parameter, e.g. V_V or S_V, is the ratio of the measure of the object to the measure of the containing space. Both measures are obtained by integration over the sampling range; the consistent estimate of a stereological parameter is therefore the ratio of the sums of primary data such as point hits, intersections etc. Hence, for volume density:

$$\hat{V}_V = \frac{\sum^n P_a(i)}{\sum^n P_c(i)} \tag{3.16}$$

and for surface density

$$\hat{S}_V = 2 \cdot \frac{\sum^n I_a(i)}{\sum^n L_c(i)} \tag{3.17}$$

where $P_c(i)$ and $L_c(i)$ are, respectively, test point number and test line length contained in the reference space c, and where i numbers the individual observations, viz. micrographs. Note that this is only of consequence if P_c and L_c vary from micrograph to micrograph.

The use of the ratio of sums as estimator for a parameter may appear as a disadvantage because the statistical reliability of such an estimate cannot be

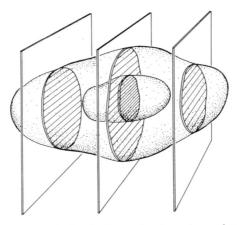

FIG. 3.21: Sectioning an isolated cell to estimate nuclear volume density.

calculated by conventional statistics. This is no limitation, however, since the variance of a ratio estimate can easily be obtained by other procedures (Cochran, 1953, p. 124), as shall be shown below.

In setting up a computation procedure for stereological parameters we can now make use of the operational sampling hierarchy outlined in Fig. 3.6. The ultimate aim of such a study is, in general, the estimation of some parameter with respect to either one of the organisms or to the population from which these organisms were sampled. A basic procedure would hence be to simply sum up all the primary data points, take the ratio of the sums and estimate the variance of this ratio estimate by the appropriate formula. An alternative procedure, which has more appeal to the biologist, is to set up the sampling procedure in such a way that a representative sample of field data is obtained with respect to that sampling step for which a first estimate of the parameters is desired. In most of our own work we usually attempt to record as many micrographs from one section, i.e. from one tissue sample, as are required to obtain an estimate of the parameters representative with respect to each of the tissue samples. The primary data from these micrographs are summed—we like to call this "pooling" of the data—and the parameter is computed as the ratio of sums. The estimate of the parameters with respect to the organism is then obtained by averaging the tissue sample estimates, and the variance is obtained by conventional statistics, as described below.

This procedure, however, requires that the notion of representative sample size is carefully considered in setting up the sampling protocol. What is the precise meaning of "representative sample"? A more or less philosophic definition would be that a representative sample should "faithfully" reflect the composition of the material, but evidently this kind of statement must be qualified further by indicating statistical bounds or confidence intervals that are acceptable. We could, for example, say that a sample is representative if it has a coefficient of variation (square root of variance divided by mean) which is no larger than, say, 10 % or 20 %. Or, that the 95 % confidence interval should be no larger than 20 % etc. Quite evidently, the minimum requirement for a representative sample is that none of the parameters that exist in the material may be zero, at least in 99 % or 95 % of these samples.

It is hence clear that the qualitative design of a sampling scheme, as outlined in Section 3.2, must be followed by quantitative considerations on acceptable error and an attempt to define how many micrographs are required to fulfill these requirements. Such considerations are, however, of a basic statistical nature; they are identical to all other investigative procedures that involve sampling and are not specifically related to stereological methods, except for the requirement that parameters should only be computed from the primary data with respect to a representative sample of micrographs.

In practice, it will generally be necessary to proceed to a pilot study in which the sampling errors inherent in the stereological procedures chosen are empirically evaluated. One good way of doing this is to record a relatively large number of micrographs on which point counts are made; one can then calculate and plot cumulative averages and cumulative variances, as shown in Fig. 3.22, from which it is then easy to find the number of micrographs required to satisfy the confidence bounds adopted (Chalkley, 1943).

This procedure may often be too laborious. One can alternatively attempt to predict on theoretical grounds the number of observations required to achieve a certain precision, and then devise a sampling scheme that will yield this number of observations on each of the representative sample units.

The easiest case to deal with is that for volume density estimation by point counting; we shall adopt Cochran's (1953) development. Here we are sampling for proportions between two spaces, the object space a and the containing space c, by applying a certain number P_c of test points to the containing space, and determining for each point whether it is in phase a or not. In general statistics we obtain the mean $\bar{y}$ by averaging all measurements y and dividing

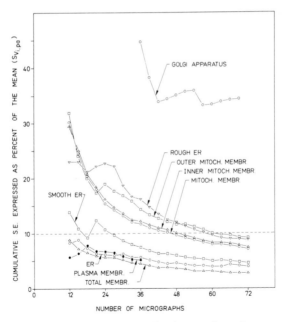

FIG. 3.22: Plot of cumulative standard error of estimate of membrane surface density (S_V) for rat liver cell organelles as the number of micrographs is increased. Note that the number of micrographs required to reduce S.E. to $<10\%$ of the mean varies considerably; it remains high for Golgi apparatus which is rare and inhomogeneously distributed. (Reproduced by permission from Bolender, 1978.)

by the total number of observations n. In the present particular case the total number of observations n is the number of test points in the containing space, P_c; the "measurements" y are either 1 if the point is in phase a, or 0 if it is not in a. The mean of the volume density is then found very simply by

$$\bar{y} = \frac{1}{n}\sum_1^n y_i = \frac{P_a}{P_c} = P_{Pa} \tag{3.18}$$

because, evidently, the sum of the variable y is just the number of points falling on a. The variance is defined as

$$\sigma^2(y) = \frac{1}{n-1}\sum_1^n (y_i - \bar{y})^2 = \frac{1}{n-1}\left(\sum_1^n y_i^2 - n \cdot \bar{y}^2\right) \tag{3.19}$$

We must now note that y^2 is also either 1 or 0 depending on whether the point hits a or not. Hence

$$\sum_1^n y_i^2 = P_a \tag{3.20}$$

By appropriately substituting we find for the variance of a point counting estimate of volume proportions

$$\sigma^2(y) = \frac{1}{P_c - 1} \cdot (P_a - P_c \cdot P_{Pa}^2) \tag{3.21}$$

The error variance of a sample estimate being

$$\sigma^2(\bar{y}) = \frac{\sigma^2(y)}{n} \tag{3.22}$$

we find that the error variance of a volume density estimate is

$$\sigma^2(\overline{P_{Pa}}) = \frac{1}{(P_c - 1)P_c} \cdot (P_a - P_c \cdot P_{Pa}^2)$$

$$= \frac{1}{(P_c - 1)} \cdot P_{Pa} \cdot (1 - P_{Pa}) \tag{3.23}$$

The variance of a point count estimate for volume proportion hence depends on the product of the volume fraction of $\{a\}$ with the volume fraction of its complement $\{c - a\}$; this product has highest value for $P_{Pa} = 0.5$, i.e. when both phase a and its complement occupy half the space. The variance also depends on the number of points applied to the containing space P_c, minus 1 to give an unbiased estimate of the variance. However, since P_c is usually a very large number it is permissible to drop the 1 and to write

$$\sigma^2(\overline{P_{Pa}}) = \frac{1}{P_c} \cdot P_{Pa}(1 - P_{Pa}) = \text{S.E.}^2(\overline{P_{Pa}}) \tag{3.24}$$

Note that the square root of this value is usually referred to as the standard error of the estimate S.E. $(\bar{y})$. With respect to sampling one is more interested in knowing the relative variance of the estimate, i.e. the coefficient of variation, which turns out to be

$$\frac{\text{S.E.}^2\,(\overline{P_{Pa}})}{\overline{P_{Pa}}^2} = \frac{1 - P_{Pa}}{P_c \cdot P_{Pa}} = \frac{1 - V_{Va}}{P_c \cdot V_{Va}} \tag{3.25}$$

The denominator is evidently the number of points falling on phase a, so that

$$\frac{\text{S.E.}\,(\overline{P_{Pa}})}{\overline{P_{Pa}}} = \sqrt{\left(\frac{1 - P_{Pa}}{P_a}\right)} = \sqrt{\left(\frac{1 - V_{Va}}{P_a}\right)} \tag{3.26}$$

The coefficient of variation hence increases with the fraction of containing space occupied by components other than phase a, and it decreases in proportion to the square root of the number of points falling onto phase a.

From this it would now be easy to predict the number of test points that are required to achieve a certain precision measured by the standard error of the mean. However, remember that we are sampling and that repeated estimates of the same parameter will always differ to some extent. We must therefore specify a required precision by defining a confidence interval, d, and the acceptable error probability, α, that an estimate may fall outside this interval. For example, we could say that d should be $\pm 10\%$ of the true mean, and that we accept a 5% chance that an estimate will deviate by more than 10% from the true mean. General theory of statistics says that such a confidence interval is related to the standard error of the mean as

$$d = t_\alpha \cdot \text{S.E.}\,(\bar{y}) \tag{3.27}$$

where t_α is the abscissa of the normal curve which cuts off an area fraction α at the tails.

For a volume proportion estimate we hence find the confidence interval d, expressed as fraction of the mean, to be

$$d = t_\alpha \cdot \sqrt{\left(\frac{1 - V_{Va}}{P_a}\right)} = t_\alpha \cdot \sqrt{\left(\frac{1 - V_{Va}}{V_{Va} \cdot P_c}\right)} \tag{3.28}$$

This can now easily be solved for P_c, the number of points on containing space required to achieve a certain precision with an error probability α:

$$P_c = \frac{t_\alpha^2}{d^2} \cdot \frac{1 - V_{Va}}{V_{Va}} \tag{3.29}$$

The values of t_α can be obtained from tables that are found in any book on statistics; for $\alpha = 0.05$ (95% probability of being within confidence interval) $t = 1.96\ (\sim 2)$, for $\alpha = 0.01\ t = 2.57$; usually a 95% probability is a very good requirement. Taking the above example of $\alpha = 0.05$ and $d = \pm 0.1$ of the mean, we would obtain

$$P_c = \frac{4}{0.01} \cdot \frac{1 - V_{Va}}{V_{Va}} = 400 \cdot \frac{1 - V_{Va}}{V_{Va}} \tag{3.30}$$

For a volume density of 20% a number of 1600 test points are required to satisfy this condition, if V_{Va} is 10% we need 3600 points. Note that the number of points required falls by a factor of 4 if we double the 95% confidence interval to $\pm 20\%$ of the mean. Such a relaxation of the confidence interval is very often acceptable, particularly if P_c thus estimated is the number of test points applied to one representative sample unit, as described above, and if the total sample comprises several such units from which the over-all average and standard error are calculated. This amounts to a two-stage sampling procedure for which the following statistical considerations apply (Cochran, 1953, 1977).

Assume that the total sample investigated comprises m representative sample units, and that on each of these units we make n observations; the measurements *within* the units will have a variance S_w^2, whereas the variance between the units (i.e. between the means calculated for each unit) is S_u^2. The over-all variance then is

$$V(\bar{\bar{y}}) = \frac{S_u^2}{m} + \frac{S_w^2}{n \cdot m} \tag{3.31}$$

An increase in observations within one representative sample unit hence reduces only the variance within this unit, whereas an increase in the number m of units reduces both variances. The conclusion therefore is that one gains more over-all precision by using a greater number of representative sample units than by increasing the number of observations within each unit. This point has recently been emphasized by Nicholson (1978). Since the variance defined by equations (3.22) to (3.26) refers to test points applied to each unit, i.e. S_w^2, it is hence admissible that in the over-all error estimate this will be reduced by about m, the number of such units evaluated.

If we now take this into consideration we can introduce the number of representative sample units m into equation (3.29) to estimate the contribution of error within one unit to the over-all error and then use $\sigma^2(P_{Pa})/m$ for calculating the test points to be applied to one representative sample unit:

$$P_c^* = \frac{t_\alpha^2}{m \cdot d^2} \cdot \frac{1 - V_{Va}}{V_{Va}} \tag{3.32}$$

It is immediately seen that P_c^* is reduced by m; if we use $m = 5$ sample units, then the number of test points required to achieve a 95% confidence interval of $\pm 10\%$ of the mean is 320 for $V_{Va} = 0.2$ and 720 for $V_{Va} = 0.1$. This amounts essentially to distributing the required test point number over the total sample.

It has been shown (Cochran, 1953), that an unbiased estimate of the overall variance in such a two stage sampling procedure is obtained simply by

$$\sigma^2(\bar{\bar{y}}) = \frac{\sum_{i=1}^{m}(\bar{y}_i - \bar{\bar{y}})^2}{m(m-1)} \tag{3.33}$$

where $\bar{y}_i$ is the mean of the observation in each representative sample unit. Although the variance within these primary units does not explicitly appear in this formula it has been accounted for in its derivation (Cochran, 1953, p. 218).

We have in this presentation strictly followed the concept that we wish to determine sample size by (1) specifying the precision required through a statement on the range of error we accept in estimating the true mean, i.e. by indicating the confidence interval d, and (2) specifying the chance, α, we are willing to take that the true mean will be outside this range. This amounts to the following probability statement:

$$\Pr\{(\bar{x} - d) < \mu < (\bar{x} + d)\} = (1 - \alpha) \tag{3.34}$$

where μ is the true mean and $\bar{x}$ its unbiased estimate. This is necessary in order to know exactly what we are after. Note however that the required number of test points is actually solely dependent on the standard error; we find by equation (3.27) that

$$\frac{t_\alpha^2}{d^2} = \frac{1}{\text{S.E.}^2(\bar{y})} \tag{3.35}$$

so that equation (3.29) simplifies to

$$P_c = \left(\frac{1 - V_{Va}}{V_{Va}}\right) \cdot \frac{1}{\text{S.E.}^2\,(\overline{V_{Va}})} \tag{3.36}$$

We should finally mention that the notion of *probable error* is sometimes used to predict the number of test points required. This was first proposed by Hennig (1956a, 1967), and we have adopted this approach in previous work (Weibel, 1963; Weibel and Bolender, 1973). The probable error P.E. $(\bar{y})$ is directly related to the standard error S.E. $(\bar{y})$ by

$$\text{P.E. } (\bar{y}) = 0.6745 \cdot \text{S.E. } (\bar{y}) \tag{3.37}$$

where

$$t_{0.5} = 0.6745$$

is the abscissa which cuts each half of the normal error curve into halves, i.e. the range $\bar{y} \pm t_{0.5} \cdot \text{S.E. } (\bar{y})$ contains one quarter of all the estimates on each side of the mean (quartile). This simply means that there is a 50% chance that the true mean is within ±1 P.E. Note that the 95% confidence interval is about 3 P.E. On this basis we could estimate the test point number by (Weibel, 1963; Weibel and Bolender, 1973):

$$P_c^* = \left(\frac{1 - V_{Va}}{V_{Va}} \right) \frac{0.455}{\text{P.E.}^2 \, (V_{Va})}. \tag{3.38}$$

The question evidently arises which kind of error statement should be used to calculate the test point number required. It really does not matter, as long as we are conscious of the different meanings, and as long as the actual error is calculated *a posteriori* on the basis of the data obtained. We must simply remember that the confidence interval may be roughly three times the probable or average error; this shall be discussed further below (Section 4.2.2.1) when we deal with the actual choice of point lattices. We should however mention that the probability considerations at the basis of these formulations pertain to the case of simple random point sampling (essentially using a single point) in a system of randomly distributed objects. We shall however show that systematic point counting is to be preferred to simple random points; furthermore the distribution of objects is in reality often much more homogeneous than one would predict on the basis of a pure random process. Consequently, the errors predicted by equations (3.29) to (3.38) are often pessimistic estimates, because the actual error, as calculated *a posteriori* from the data, is very often smaller. This has been demonstrated both theoretically and empirically by Hennig (1967) who had also shown earlier (Hennig, 1956a) that the standard errors obtained for point counting estimates are generally well within the probable error, and are hence smaller than the theoretical prediction.

In summary, we have outlined a general sampling strategy which complies to a two stage (or multiple stage) sampling procedure. The basic sampling unit is a set of micrograph fields that we considered to be a representative sample for the tissue. The total sample consists of several such sample units. The number of observations required within each such representative sample unit can be estimated by specifying the precision requirements, and by taking into account the number of such sample units used to make up the total sample. The case has been specified in some detail with respect to volume density estimation by point counting, as an example; the conditions prevailing with other types of measurements will be specified in Section 4.2.

PHILLIP H. SMITH, Ph.D.

CHAPTER 4

Point counting methods

4.1. Review of basic principles of stereology

4.1.1. A COHERENT SYSTEM OF FORMULAS

The first set of fundamental principles of stereology derived in Sections 2.2 to 2.4 and in Volume 2, Chapters 3 and 4 aim at measurement of *structure parameters* such as volume density, surface density, length density, and density of mean curvature. The general property of these principles is that they require no assumptions as to the shape or topological property of the object phase; indeed the only requirement is that the various phases be clearly separated and distinguishable on section. These principles are therefore applicable to a very broad class of structures, which we could describe by two models (Fig. 4.1): in *model A* the elements of the object phase may have any shape, they can be connected, and can show as many non-convexities as is conceivable; in *model B* the elements of the object phase are discrete particles, and one in general has to assume convex shape, although this is not in every respect a binding condition. It is evident that the *structure parameters* V_V, S_V, J_V and K_V can be defined for both of these models, and that, accordingly, the fundamental principles of stereology can be applied to their estimation in both cases.

The major difference between these models is that counting of individual elements, i.e. the estimation of the structure parameter N_V, is only possible in a structure complying to model B. Likewise, the elements must be represented by discrete objects of known shape in order to justify the determination of mean diameter, volume, surface, or length of the objects, i.e. the assessment of *particle parameters*.

There are, however, a few *object parameters* which do not depend on the

101

presence of discrete particles. We should mention the mean surface curvature $\bar{K}$, and the mean intercept length $\bar{L}$ which is related to the volume-to-surface ratio or to the mean free path within the object; both these parameters are determined by the ratio of two fundamental stereological measurements (see Chapter 6).

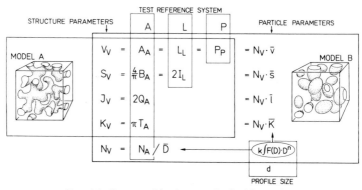

FIG. 4.1: System of basic stereological formulas.

Stereological measurements are performed with respect to *test reference systems*, which are either the section area A, a test line length, L, or a test point set P. We had shown that there is a lower limit for the dimensionality, $d(s)$, of this test reference system which depends on the dimension of the object feature studied, $d(a)$, such that (Section 2.1.1)

$$d(s) + d(a) \geqslant 3$$

otherwise no trace is formed. Thus, the object volume can be tested with section area, line length and point set, whereas its surface is only measurable with respect to area and line length, the length of space curves only with respect to area (Fig. 4.1).

On the other hand, numerical density N_V can only be estimated from counts on a section plane if the mean diameter of the particles—which determines their probability of being cut—is known, thus requiring an estimation of particle size.

4.1.2. RELATIVE MERITS AND POWER OF BASIC PRINCIPLES

Stereology hence offers two fundamentally different approaches to the assessment of quantitative properties of the structure: one can proceed to a direct measurement of structure parameters, or one can choose the path of determining particle parameters. In choosing an appropriate procedure we must therefore evaluate the relative merits of these two approaches.

The direct estimation of a structure parameter such as V_V, S_V, or J_V by one of the fundamental methods (frame A in Fig. 4.1) often conceals a lot of information on the structure, because it is only able to produce *average* estimates: the "field of operation" of these methods is not really the structural element of the object phase under consideration, but rather the aggregate of all elements, that is the total structure; this approach hence tests the content of the structure in terms of the phase as a whole.

In contrast, the estimation of particle parameters operates, at least theoretically, on the population of individual elements. It therefore reveals not only average size, but may also give some indication on the variability in the dimensions of the structural elements. Through the numerical density of the particles this dimensional analysis of the individual objects may again be linked to the total structure, in that volume density, for example, follows from the product of N_V and mean particle volume $\bar{v}$, and so on (Fig. 4.1).

It is hence evident that the adoption of model B, though restricted in terms of compatibility with the real structure, may be considered a more powerful approach than the application of basic stereological principles on the basis of model A because it produces a wider range of parameters for description of the structure. However, this gain in information is often forfeited by a considerable loss in efficiency, due to the higher cost of the analysis, and to a greater risk of erroneous assumptions.

In view of this it may be worth considering some rules for choosing between these two fundamental approaches. The following questions will have to be answered with respect to this choice:

(1) Which are the most informative parameters?

(2) Which are the geometric properties of the structure?

(3) How much may the analysis cost?

4.1.2.1. *Definition of most informative parameters.*

In planning a stereological analysis one will first have to decide which parameter provides the most relevant information. Intuitively, or perhaps naively, one will primarily consider number, shape and size of the component objects to be the best descriptors of the system. Very often, however, this "primary" information bears no relevance to the actual problem, but is rather used to estimate indirectly some other dimensional properties, such as the bulk volume of the phase.

Let us consider some simple examples from biology. If a muscle is trained to sustain higher performances its muscle cells will develop a greater capacity for producing ATP (adenosine triphosphate), the cell's fuel, by burning more oxygen. ATP is produced by an enzyme system contained in the mito-chondria, which usually appear as "elliptic" profiles on a section. In testing whether the augmented functional capacity for ATP production is related to a proportional increase in mitochondria one can evidently ask whether the number and the size of these organelles is increased and one finds that this is the case (Kiessling *et al.*, 1971). However, this is of secondary relevance to the problem. One may consider the mitochondria to be little bags which are stuffed with the ATP producing enzyme system, some of the enzymes being bound to the folded inner membrane. What really determines the bulk level of ATP production is, therefore, the over-all volume of mitochondria and the over-all surface of inner mitochondrial membrane available to the muscle cell. This suggests that the most directly relevant parameters are the volume density of mitochondria and the surface density of their inner membranes in muscle cells. These two parameters could evidently be calculated from the size and numerical density of mitochondria, provided the shape of the organelle and the disposition of inner membranes are known with sufficient reliability; they can, however, also be directly and more efficiently estimated by simple point counting procedures (Hoppeler *et al.*, 1973) thus avoiding all assumptions on the shape of the organelle. In this example the adoption of model A adequately produces the desired information.

On the other hand, in their study on structure–function correlation in subcellular fractions obtained by differential centrifugation Baudhuin and Berthet (1967) asked how the mitochondria were distributed between the fractions. To compare the biochemical analysis of enzyme concentrations with the structural composition of the fractions, evidently, the bulk volume of the organelles would again have been an adequately informative para- meter, and this could have been easily obtained from a direct estimation of volume density. However, in this case the size of the individual organelles, or rather their size distribution, must also be considered because the sedi- mentation rate is largely governed by organelle size, and this would hence determine the distribution of the organelles into the different fractions. In this study an analysis of size distribution and numerical density of mito- chondria succeeded in providing the relevant information; this was facilitated because the shape of these organelles in fractions is essentially spherical.

The lung is made of many millions of small air chambers, the alveoli, all of roughly the same size and shape. Although number and size of these structural units could be used as basic descriptors of lung structure (Weibel, 1963), these parameters are not particularly relevant in attempts to relate lung structure to lung function. Rather, the gas exchange properties of this organ are defined by the over-all size of the gas exchanging surfaces, the volume of capillary blood, and the harmonic mean thickness of the tissue barrier, as follows from a model analysis of pulmonary structure–function correlation (Weibel, 1970a). The lung's mechanical properties are on the other hand greatly governed by surface tension phenomena in alveoli, and these are directly related to mean surface curvature (Gil and Weibel, 1972). All these parameters can be estimated directly by stereological methods based on model A without the need to determine size distribution and number of alveoli.

Very similar conditions can be found in materials sciences. Fischmeister (1972) and Amstutz and Giger (1972) have reviewed the importance of selecting appropriate stereological parameters for describing the properties of metals, and of rocks and ores, respectively. Both metals and rocks are basically made up of "grains" of different phases, and it would hence appear logical to characterize them structurally through an analysis of number and size distribution of these granular phases. Functional considerations show however that many of the physical properties of these materials are more appropriately related to bulk parameters such as interfacial surface density, contiguity (relative contact surface), mean free path, volume composition, etc., most of which can again be obtained by simple methods based on model A.

An analysis of grain size and number becomes, however, highly relevant when morphogenetic problems prevail: the growth of mineral aggregates needs to be described by changes in grain size and number. Similarly, biological growth processes cannot be satisfactorily described by bulk parameters alone; it is rather important to ask to what extent changes in the size of the component elements and/or changes in their number contribute to over-all growth of the system. Furthermore, the quantitative characterization of brain structure calls for an estimation of the number of nerve cells as one of the descriptors of the complexity of the neuronal network (Haug, 1972).

4.1.2.2. *Limitations due to geometrical properties.*

As became apparent from the discussion on the two basic groups of stereological methods, those based on model B crucially depend on the condition that certain fundamental geometrical properties of the component elements (objects), most importantly those of convexity and geometric similarity, prevail in the real structure. These conditions are usually very difficult to test if only sections are available for study, for the following reasons.

(a) Particles of similar shape can produce quite heterogeneous profile shapes when randomly sectioned. For triaxial ellipsoids this will include ellipses of widely varying axial ratios and of a wide size range. In practice, therefore, it is extremely difficult if not impossible to test for shape similarity from an analysis of the section profiles. Indeed, if the triaxial ellipsoids vary in shape, the problem is indeterminate (Cruz-Orive, 1976c).

(b) If a slender torus-shaped structure or a convoluted cylinder are sectioned the vast majority of the profiles will be ellipses of varying axial ratio, and one would easily conclude that the objects are convex. However, the rarer ring or crescent-shaped profiles constitute the crucial finding which leads to rejection of the hypothesis that the condition of convexity is fulfilled. Considering the example of muscle mitochondria cited above, a careful analysis of electron micrographs reveals that they are by no means convex bodies but that their shape is rather complicated. Similar examples are easily found in materials sciences.

If one or the other of these conditions—convexity and geometric similarity—is not fulfilled those stereological methods which are based on model B should be avoided, or the results obtained must be interpreted with greatest caution. In case of doubt it is certainly safer to approach the problem by using one of the methods based on model A. These methods should also be used as a check on the validity of the assumptions if methods based on model B are nevertheless employed.

4.1.2.3. Cost-effectiveness of the methods.

It is plausible that the more powerful methods of model B will mean higher costs of the analysis. If reduced to the point counting mode the methods of model A are very efficient even without full automation because the "measuring process" is reduced to mere differential counting. Estimations of size distribution, however, require actual measurements to be performed on a very large number of profiles, and this is evidently costly in terms of time required.

4.1.3. SELECTION OF BEST METHOD FOR ESTIMATING STRUC- TURE PARAMETERS

In trying to choose the best method for estimating structure parameters by one of the fundamental principles summarized in Fig. 4.1, one has to decide which test reference system gives the most reliable results. For estimating V_V one has three choices (Section 2.2): (1) estimate directly the relative area occupied by profiles, (2) apply a test line and measure the intercept lengths (linear integration), and (3) perform a point count. Hilliard and Cahn (1961) have evaluated these methods and arrived at the conclusion, both by theoretical considerations and empirical verification, that a systematic point count gives the best results.

To check on this conclusion we have performed the following experiment: We have prepared five model sections on which an areal fraction $A_A = 20\%$ was covered by profiles of a gray component (Fig. 4.2). Seven collaborators have then independently estimated A_A by the following six methods:

A_{A1} : measurement of profile areas using a polar planimeter;

A_{A2} : measurement of profile areas by fitting a circle of equal area to the profile and counting profiles;

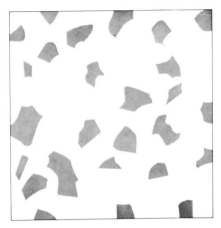

FIG. 4.2: Model Section ($A_A = 0.2$).

A_{A3}: tracing the profiles, cutting them out and weighing them;

L_L: summing intercept lengths using a very rapid graphical procedure using a long paper strip;

P_{P1}: point count with a random point grid;

P_{P2}: point count with a systematic lattice.

These methods were so designed as to yield approximately the same number of observations. The time was also recorded.

Table 4.1 shows the results. The lowest error was obtained by planimetric measurements, but this also required the longest time. Linear integration is worst, and random point counts are worse than systematic counts, which took the shortest time and gave an error close to that for the planimetric methods.

Table 4.1

Comparison of coefficient of variation and (average) time requirement in measuring areal density ($A_A = 0.2$) with different methods.

Method	Coefficient of variation	Average time
Planimetry by:		
polar planimeter	3.84%	323 sec.
circle fitting	3.32%	148 sec.
cut out and weigh	2.64%	461 sec.
Line integration	7.54%	105 sec.
Point count by:		
random points	6.20%	79 sec.
systematic lattice	4.74%	73 sec.

To assess the cost effectiveness of the methods we assumed that the precision (measured by the reciprocal of the standard error) increased with the
square root of the number of observations, and that the latter was proportional to time required. On this basis the methods were compared in Fig.
4.3: it is evident that systematic point counts and the method of circle fitting
yield by far the most reliable results on a time–cost basis. The method of
circle fitting has worked remarkably well in this model situation where the
elements were very prominent; it is doubtful, however, if it would be as good
with the more complex structures found in actual micrographs. We conclude
therefore that systematic point counting is the best method for estimating
volume density by stereological methods; this shall guide us in designing
appropriate test systems.

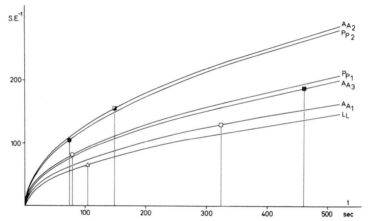

FIG. 4.3: Relative precision achieved with six different methods for estimating A_A.

By analogy, it is safe to conclude that intersection counts per unit test line
length should also be superior to the actual measurement of boundary length
of profiles, e.g. by a map measuring wheel, although this has not been verified
experimentally.

The result that a coarse point count should yield a better result than a
total planimetric measurement may appear astonishing at first. However,
we must reiterate what has been said above about sampling in the various
stages: it does not make sense to measure an individual profile with excessive
precision if the variation of profile size, or of their areal density on a section,
is large from section to section. It is better to work up, in a certain amount of
available time, a larger number of fields with rough precision than a small
number of fields with high precision, because the number of fields has a
greater effect on reducing over-all error than increasing the density of observation on one field.

4.2. Point counting using coherent test systems

The last section ended with the conclusion that point counting methods afford greatest efficiency in the estimation of structure parameters by stereological methods. The methods of choice therefore are:

Volume density:

Test reference system

$$V_V = P_P = P_a/P_T \qquad\qquad P_T \qquad\qquad (4.1)$$

Surface density:

$$S_V = 2I_L = 2 \cdot I_a/L_T \qquad\qquad L_T \qquad\qquad (4.2)$$

Length density:

$$J_V = 2Q_A = 2 \cdot Q_a/A_T \qquad\qquad A_T \qquad\qquad (4.3)$$

Curvature density:

$$K_V = \pi T_A = \pi T_{net}/A_T \qquad\qquad A_T \qquad\qquad (4.4)$$

Numerical density:

$$N_V = \left(\frac{1}{\overline{D}}\right)N_A = \left(\frac{1}{\overline{D}}\right) \cdot N_a/A_T \qquad\qquad A_T \qquad\qquad (4.5)$$

Just three basic reference systems therefore suffice to estimate these relations, as well as the combination parameters derived from them: A_T, L_T, and P_T. It is evident that optimal test systems should comprise all three probes in such a way that A_T, L_T and P_T are quantitatively related to each other. Such test systems shall be called "*coherent test systems*". Their main virtue is that they permit an easy determination of the length of test lines or the area of the test field from a simple count of test points when the measurements are to be obtained with respect to a restricted containing space, e.g. with respect to a certain cell type only.

4.2.1. BASIC DESIGN OF COHERENT TEST SYSTEMS

Since the test point lattice is optimally made of points systematically spaced at a distance d we would strive to construct coherent test systems that fulfill the following relationship:

$$L_T = P_T \cdot k_1 \cdot d \qquad\qquad (4.6)$$

$$A_T = P_T \cdot k_2 \cdot d^2 \qquad\qquad (4.7)$$

The three probes A_T, L_T, and P_T are realized simultaneously in a simply related fashion with a square lattice of lines of distance d as shown in Fig. 4.4a. If the intersections of the lines are considered as "points" for point counting, it is easily seen that there are as many squares of area d^2, and twice as many line segments of length d, as there are points. This test system is thus defined as follows:

test points: P_T

test line length: $L_T = P_T \cdot 2d$; $k_1 = 2$

test area: $A_T = P_T \cdot d^2$; $k_2 = 1$

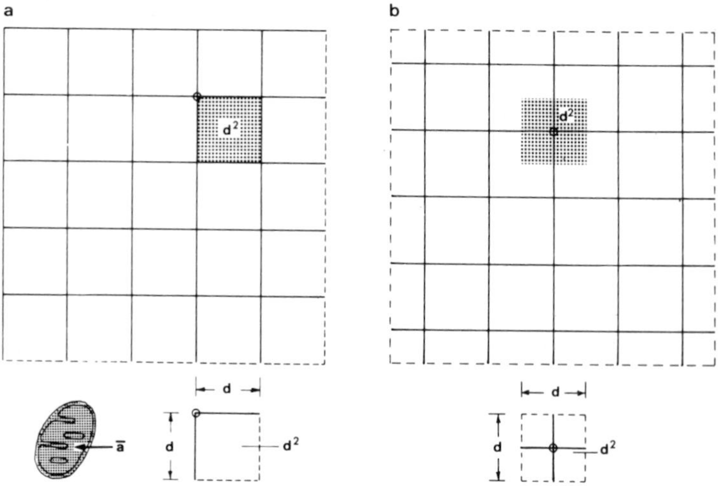

FIG. 4.4: Coherent quadratic test grids. (Reproduced by permission from Weibel and Bolender, 1973.)

It should be noted that the broken line of the frame square in Fig. 4.4a is not part of the test lines and produces no test points. To avoid this difficulty, the same test system can be drawn differently, as shown in Fig. 4.4b, where the test lines cross in the center of the small unit squares to form one central test point per square. The two test systems are quantitatively identical.

4.2.2. TEST POINT SYSTEMS FOR POINT COUNTING

4.2.2.1. *Point density*
In determining the optimal density of test points in a test system we must consider each point as an individual probe. Each point constitutes a sample, and one asks the question whether the point lies in the component or not. Such a sampling procedure is statistically described by a binomial distribution, and, therefore, it is necessary that each sample, i.e. each point, is independent of the others. Consequently, the points should be spaced in such a

manner that no more than one point can fall on the same profile. This is achieved if d is chosen such that

$$d^2 > a_m \tag{4.8}$$

the area of the largest profile encountered. The inaccuracy with which individual profiles are "measured" with such coarse point lattices is compensated by the fact that the point set covers a very large sample of profiles, thereby securing a representative sample of the structure as a whole.

This point may require some further comment. In using point counting methods of stereology we are not "*measuring*" profile areas; we are rather looking at each point asking whether it is "in" or "out", the quantitative measurement being either 1 or 0 respectively. By doing this we exploit the fact that the probability of a random point falling into a certain component is given by its volume density, irrespective of the size of the individual profiles. This type of testing for probabilities looses power if the points are too close together: if $d \sim \frac{1}{2}\sqrt{a_m}$ two neighbouring points are very dependent on each other: if the first point is "in" the second point has a chance of about 50% of being also in, although the areal density of the profiles may only be a few percent. The second point is therefore not very useful as a random probe. On the other hand for two well separated points the second point will have a higher chance of also being "in" if the areal density is high.

One may have to deviate from this general rule in studying rare components where one needs to make sure that no profiles remain undetected by the test system; in this case one may choose d^2 similar to or even smaller than the mean profile area,

$$d^2 \sim \bar{a} \text{ or } d^2 < \bar{a} \tag{4.9}$$

4.2.2.2.   *Coherent double lattice test systems*

It is evident from what has been said in the preceding section that very different point densities may be required if the volume density of more than one component is to be estimated on the same micrograph. It would hence be desirable to have test systems with at least two (or more) lattices of different point density superimposed. Such double lattices should preferably be coherent; that means that the two lattices should be simply related.

The principle for constructing coherent double lattice test systems is quite simple; an example is shown in Fig. 4.5. The basic test system is a square lattice of (heavy) lines whose intersections mark P_T test points. The test point distance d is subdivided q times and finer lines are drawn whose spacing is evidently d/q; in the example of Fig. 4.5 $q = 3$. Clearly then, the total number of test points—marked by the intersections of *all* lines—is

$$P'_T = q^2 \cdot P_T \tag{4.10}$$

E

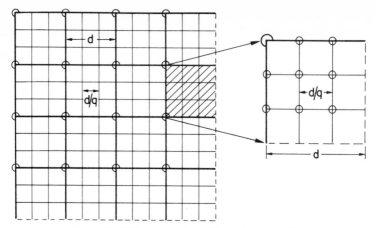

FIG. 4.5: Unit square of double lattice test grid.

It is evident that this test system is coherent: the lattice unit of the coarse square grid contains a constant number q^2 of evenly spaced additional points.

A number of double lattice test systems which have proven useful in practice are shown in Appendix 3. They are coded as follows:

Code	Lattice ratio q^2	P_T	P'_T
A 100	1	100	—
B 25	4	25	100
B 36	4	36	144
B 100	4	100	400
C 16	9	16	144
C 64	9	64	576
D 16	16	16	256
D 64	16	64	1024
E 36	25	36	900

The range of combinations of test point numbers offered by these lattices is evidently large, but it is very easy to construct additional variants if a specific task requires it.

4.2.2.3. *Technical design of double lattices*

If a test lattice is superimposed on a micrograph it is sometimes not easy to recognize the coarse points. In the test systems shown in Appendix 3 these coarse points are enhanced by a 3/4 circular arc (Fig. 4.6). The precise test point is then taken to be the intersection of the line edges in the open quadrant of the circle. This rule can be applied to all test points; it avoids uncertainties about point localization due to the thickness of the test lines (which in fact are broad bands!).

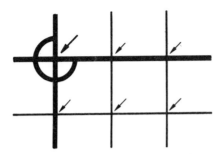

FIG. 4.6: Definition of true test point.

4.2.2.4. *Practical procedure to select a test system*

The practical procedure to determine the test point number required for a particular task and the optimal test point system is as follows:

(1) Determine the test point number P_T^* needed to satisfy the specified precision requirement. This is obtained either through equations (3.29), (3.30) or (3.36), or by means of the nomogram of Fig. 4.7 which is based on these same equations. Note that for this the order of magnitude of V_V of the component of interest must be known; it is obtained by a rough point count over a few low power fields.

(2) Determine the optimal magnification M^* at which the measurements are to be performed. This defines the micrograph area A_{T1} available for analysis.

(3) With M^* chosen measure approximately the area a_m of the largest profiles of interest (a_m is given in cm^2 as profile size at M^*).

(4) Choose a first value for the spacing of points, d_1, such that the condition $d_1^2 > a_m$ is conveniently satisfied, e.g. $d_1 \sim 1.2 \sqrt{a_m}$.

(5) The test point number P_{T1} applied to one micrograph follows from

$$P_{T1} = \frac{A_{T1}}{d_1^2} \qquad (4.11)$$

(6) Finally the ratio

$$P_T^*/P_{T1} = n \qquad (4.12)$$

determines the number of micrographs required to form one representative sample (see Section 3.7).

If both n and P_{T1} are considered acceptable then the next step is to find among the available test systems the one which most closely approximates P_{T1}. If P_{T1} is very high (> 100) it may be advantageous to choose a double lattice test system. If either n or P_{T1} are considered too large, resulting in excessive work, one may either relax the precision requirements, or lower the magnification M^*. Note that reducing magnification by 30% doubles the area of the section included in the sample!

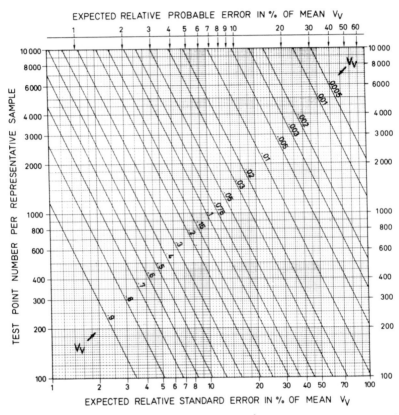

FIG. 4.7: Nomogram relating test point number to $\dot{V}_V$ and expected relative standard error (in $\%$ of mean).

If the volume density of more than one component is to be measured then the above listed procedure should be followed for each of them. A suitable compromise is then sought; this is often facilitated by making use of the collection of double lattice test systems as presented in Appendix 3. Sometimes it will however be unavoidable to work at different magnifications for different components.

A further word should be said about the meaning of P_T^*, A_{T1}, and P_{T1} if the reference volume, e.g. cytoplasm, does not cover the entire micrograph area. In this case A_{T1} should be given as the average area of the restricted containing space (cf. Section 4.2.5) on each micrograph. Accordingly P_T^* and P_{T1} are the test point numbers applied to the restricted reference space, so that the test system chosen must have a correspondingly greater number of test points.

Let us look at a specific example: the estimation of V_V for mitochondria (mi) and dense bodies (db) in liver cells. From previous work (Weibel et al., 1969) we know the orders of magnitude of these volume densities expressed per unit volume of liver parenchyma (pa):

$$V_{Vmi,pa} \approx 0.20$$

$$V_{Vdb,pa} \approx 0.01$$

To achieve a standard error of the order of 10% of the mean we find from equation (3.29) or from Fig. 4.7 that we require about 400 test points on cytoplasm for mitochondria, whereas 10,000 are needed for dense bodies. Clearly, a very different density of test points will be required for dense bodies than for mitochondria. There are various solutions to this problem. (1) One can decide on a compromise, e.g. 2,000 points; the error in V_{Vmi} will then be smaller and that for dense bodies larger than the one required. (2) One can use the larger one of the predicted test point numbers (10,000) and invest more work than needed on the mitochondria. (3) One can estimate V_{Vmi} and V_{Vdb} separately using two different test systems of appropriate point density, thus assuring equal precision with least work. (4) A final alternative is to use a coherent double lattice test system in which two different point lattices are combined.

Table 4.2 shows how the practical procedure outlined above leads to the selection of an appropriate test system and to the determination of the number of micrographs required to form one representative sample. A magnification of 25,000 × on the screen of the projector unit used in our laboratory (see Chapter 7) was found to be convenient. Since dense bodies are rare components we chose $d_1^2 \sim a_m$, thus establishing P_{T1} as 920 test points per micrograph, whereas < 165 would be required for mitochondria. The estimation of V_V for dense bodies is evidently more critical; the test point number requirement is best satisfied by choosing test lattice D 64 (Appendix 3) in which there are a total of 1024 test points, 64 being marked as "heavy points". In Fig. 4.8 this test lattice is shown applied to a liver cell micrograph. Using 10 micrographs the total test point number required for estimating V_{Vdb} is fulfilled, but we would have to count 640 points for estimating V_{Vmi}. A compromise is to use 8 micrographs; this increases the expected S.E. for dense bodies to 11% which is still acceptable. Even with only 7 micrographs the S.E. for dense bodies should be no more than 12%.

Table 4.2
Example for test system selection for V_V estimation on two components (liver cells).

	Mitochondria	Dense bodies
Rough estimate of V_V	~0.20	~0.01
Required test points P_T^* (S.E. 10%)	400	10,000
Magnification, optimal M_1	25,000 ×	25,000 ×
Micrograph area A_{T1} (27 × 27 cm)	729 cm^2	729 cm^2
Largest profile area at M_1	3.1 cm^2	0.8 cm^2
Point spacing d_1	$\geqslant 1.2 \cdot \sqrt{a_m} = 2.1$ cm	$\sim \sqrt{a_m} = 0.9$ cm
Points required per micrograph	< 165	~920
Micrograph number $n = P_T^*/P_{T1}$	> 2.4	~10.9

Test system options:	Coarse lattice	Dense lattice
C 64, $n = 10$		
Test points applied	640	5,760
D 64, $n = 10$		
Test points applied	640	10,240
S.E. (predicted)	7%	10%
D 64, $n = 8$		
Test points applied	512	8,192
S.E. (predicted)	8%	11%

Using five representative samples the practical procedure would then be to take five sections and to record on each 8 micrographs at 25,000 × final magnification, to do the point counts on each with test system D 64 using the dense lattice for dense bodies and the coarse lattice for mitochondria. On each group of 8 micrographs the volume densities are calculated, the mean and S.E. being calculated between the five representative samples.

4.2.3.   TEST LINE SYSTEMS FOR INTERSECTION COUNTING

4.2.3.1.   *Square lattice test systems*

The lines of square lattices may be used to estimate the surface density S_V of components by counting the intersections I, with profile boundaries. Since we have found that, in a coherent square lattice, the test line length is simply connected to the test point number.

$$L_T = P_T \cdot 2d \qquad (4.13)$$

we obtain the surface density by

$$S_V = 2 \cdot I_L = 2 \cdot I/L_T = I/(P_T \cdot d) \qquad (4.14)$$

If one wishes to determine the surface density of a component i within a re-

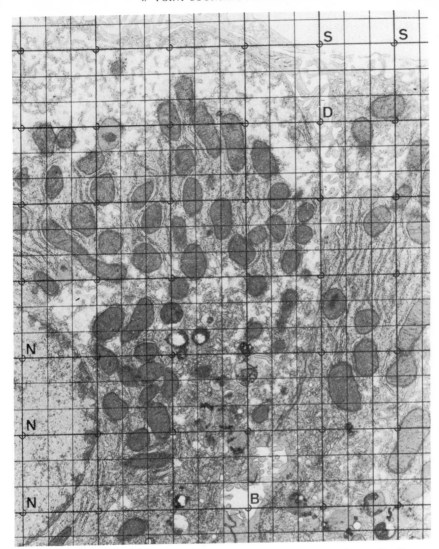

FIG. 4.8: Double lattice test grid (C 64) superimposed on section of liver cells. Coarse points on sinusoid (S), nucleus (N), Disse space (D), and biliary capillary (B) are excluded if reference space is restricted to cytoplasm. Coarse grid spacing 1.7 μm.

stricted space, e.g. cytoplasm, the test line length contained within cytoplasm follows directly from the number of test points falling on cytoplasm, P_c:

$$L_c = P_c \cdot 2d \qquad (4.15)$$

so that

$$S_{Vi,c} = I_i/(P_c \cdot d) \qquad (4.16)$$

It is evident that intersection counting on double lattice test systems follows the same rules. In using the fine line grid of spacing d/q for counting intersections, I_i', and the coarse points for estimating cytoplasmic area, the surface density in cytoplasm is obtained from

$$S_{Vi,c} = I_i'/(P_c \cdot q \cdot d) \qquad (4.17)$$

One of the technical difficulties encountered is that the test "lines" drawn in these test systems are in reality broad bands; this often introduces uncertainties as to whether the "line" intersects a membrane trace or not. To eliminate this uncertainty one can take *the upper and the right hand edge* of the lines to represent the *actual test line*.

In analogy to point counting volumetry the total length of test lines required for estimating S_V depends on the surface density of the component and on the acceptable error. The relationship between these parameters is not fully established yet. Hilliard (1976) has examined the situation and has come up with the following error prediction formula for a line intersection count:

$$\frac{\text{S.E. } (S_V)}{S_V} = \sqrt{\left(\frac{2}{I_T}\right)} \qquad (4.18)$$

where I_T is the total number of intersections counted. The coefficient 2 in this formula applies to cases of non-contiguous convex particles of low volume and surface density; in practical cases it must therefore be used with caution, but it seems that the order of magnitude is correct. Considering that

$$I_T = \frac{1}{2} S_V \cdot L_T \qquad (4.19)$$

the total test line length required to achieve a prescribed relative standard error in S_V

$$\text{R.S.E. } (S_V) = [\text{S.E. } (S_V)]/S_V \qquad (4.20)$$

can be estimated by

$$L_T = \frac{4}{S_V \cdot \text{R.S.E.}^2 (S_V)} \qquad (4.21)$$

Figure 4.9 shows a nomogram relating required test line length to error and surface density. It must be noted that L_T is the test line length which must be applied to the reference space; if, for example, the reference space is cytoplasm then this test line length corresponds to L_c as defined above. It furthermore indicates the test line length that is to be applied to a representative sample—in general several micrographs—for which the error requirement has been specified.

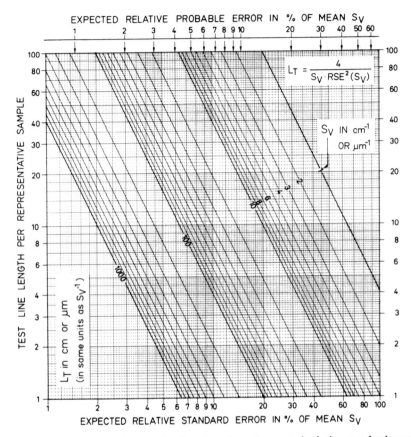

FIG. 4.9: Nomogram relating test line length to S_V and expected relative standard error (in % of mean).

Let us illustrate this, as an example, by the estimation of surface density of outer mitochondrial membranes in hepatocyte cytoplasm, $S_{Vom,cy}$, which is found to be of the order of $2\mu^2/\mu^3$ (Weibel *et al.*, 1969). If we would like to achieve a relative standard error of 10% we require a test line length of 200μ, as found by applying equation (4.21). We have decided to work at a magnification of $25,000 \times$ which means that we must use 500 cm of test line on the micrographs. In the practical example given in Table 4.2 we have decided to use test lattice D 64 on 8 micrographs; thus requiring about 62.5 cm of test line on each micrograph. In test lattice D 64 (Appendix 3, D 64) the spacing of the coarse points measures $d = 3.38$ cm for a 27×27 cm screen; the total test line length per field represented by the heavy lines is hence $L_T = 64 \cdot 3.38 = 216$ cm. Since only about 80% of the field is covered by cytoplasm the lines in cytoplasm are about $L_c = 172$ cm. Using all these lines to count intersections with mitochondrial outer membrane on the eight micrographs would estimate S_V with 1380 cm of test line which would reduce the expected S.E. to 6%. Alternatively one could use only the horizontal heavy lines, i.e. 690 cm, which would give an expected S.E. of 9%.

4.2.3.2. *Multipurpose test system*

Square lattices often contain an excessive line density for intersection counting. An alternative coherent test system, shown in Fig. 4.10, which was originally called the "multipurpose" test system (Weibel *et al.*, 1966), overcomes this difficulty. It is composed of discrete short test lines of length d whose end points are arranged in a regular triangular lattice. The lattice unit is not a square, but rather an equilateral rhombus, with angles of 60°, and with area

$$a_T = \frac{\sqrt{3}}{2} \qquad (4.22)$$

It is evident from Fig. 4.10 that there is only one test line segment of length d per every second rhombus, and hence the test line unit per test point is

$$l_T = \frac{1}{2} \qquad (4.23)$$

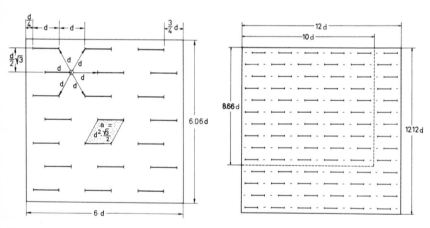

FIG. 4.10: Short line "multipurpose" test system. (Reproduced by permission from Weibel *et al.*, 1966.)

This basic unit allows a number of coherent test systems to be designed. The smallest test area that is (nearly) square can be obtained by using an array of seven rows of three test lines, as shown in Fig. 4.10a; it contains 42 test points and 21 lines. A larger square test system is obtained by joining four such units to contain 168 points and 84 lines (Fig. 4.10b). It may be convenient to have just 100 points in the test system, as shown in the rectangular area in Fig. 4.10b. This test system has the following characteristics:

test points: P_T

test line length: $L_T = P_T \cdot \frac{1}{2} \cdot d$

test area: $A_T = P_T \cdot \frac{\sqrt{3}}{2} \cdot d^2$

and it is coherent in the sense defined above. Two practical variants of this test system are given in Appendix 3 (M 42 and M 168).

For point counting volumetry, the end points of the test lines are used as markers. The same rules as given above for square lattices apply; namely, the spacing of points is optimally chosen so that, in general,

$$a_T > a_m$$

the maximal profile area of the structure of interest. Also the same requirements concerning the total number of test points to be used on a representative sample apply. It is evident that this test system is less flexible with respect to point spacing than square lattices, because double lattice test systems are not easily designed, although it is not impossible.

This test system has its main advantage in surface area estimations. If we count the intersections of a surface trace with the test line segments, I_i, the following formula is obtained:

$$S_V = 2 \cdot I_L = 4 \cdot I_i/(P_T \cdot d) \tag{4.24}$$

Since the test system is coherent, it is possible to estimate S_V in the cytoplasm, for example, by introducing cytoplasmic points, P_c, instead of P_T:

$$S_{Vi,c} = 4 \cdot I_i/(P_c \cdot d) \tag{4.25}$$

The total test line length needed to satisfy specified precision requirements is evidently derived from equation (4.21) or from the nomogram of Fig. 4.9.

This test system was originally designed (Weibel and Knight, 1964) to apply the principle of Chalkley et al. (1949), which uses short "needles" to estimate the volume-to-surface ratio from a count of intersections (I_i) and point hits (P_i) on the structures:

$$\frac{v}{s} = \frac{d \cdot P_i}{4 \cdot I_i} \tag{4.26}$$

This parameter has been shown to be directly related to the mean intercept length and thereby to mean particle size. It is especially useful in estimating the arithmetic mean thickness, $\bar{\tau}$, of sheets of irregular thickness (Weibel and Knight, 1964) where

$$\bar{\tau} = \frac{d \cdot P_i}{2 \cdot I_i} \tag{4.27}$$

The coefficient 4 has been replaced by 2 in this formula, because the tissue volume is related to both surfaces of the sheet and, consequently, intersections with both surfaces must be counted (see also Section 6.1).

4.2.3.3. Curvilinear test system of Merz

The straight lines used in all the preceding test systems have a high degree of anisotropy. This introduces a great hazard when they have to be used in estimating the surface of components which themselves also show a certain degree of anisotropy (preferred orientation of membranes or boundaries). To overcome this difficulty Merz (1967) proposed the use of a test line system composed of semicircles disposed in a square lattice. The element of this test system is shown in Fig. 4.11 to be a semicircle whose diameter d corresponds

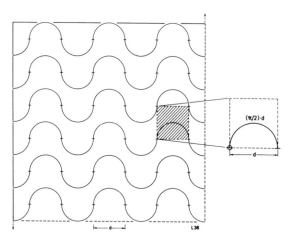

Fig. 4.11: Isotropic curvilinear test system of Merz with unit square containing a hemicircle. (After Merz, 1967.)

to the spacing of test points so that this test system is also coherent in the above sense with the following characteristics:

$$\text{test points:}\quad P_T$$

$$\text{test line length:}\quad L_T = P_T \cdot \frac{\pi}{2} \cdot d$$

$$\text{test area:}\quad A_T = P_T \cdot d^2$$

Two variants of this test system are shown in Appendix 3 (L 36 and L 100). Here again, the test point number and test line lengths required to satisfy precision requirements are derived from equations (3.29) and (4.10) or from the nomograms in Figs. 4.7 and 4.9 respectively.

4.2.4.   TEST AREA FOR TRANSECTION AND PARTICLE PROFILE COUNTING

In order to estimate the length density, J_V, of lines or curves in space, one must count the number of times these objects are transected by the unit area of section plane. In a square test system the test area A_T is well defined by its frame which is related to the test point number, so that

$$J_{Vi} = 2 \cdot Q_i/A_T = 2 \cdot Q_i/(P_T \cdot d^2) \tag{4.28}$$

or, if expressed in relation to a restricted containing space, e.g. cytoplasmic volume,

$$J_{Vi,c} = 2 \cdot Q_i/(P_c \cdot d^2) \tag{4.29}$$

Similarly, the numerical density of profiles per unit section area can be determined by counting the profile number N_i within the frame of the test system:

$$N_A = N_i/A_T = N_i/(P_T \cdot d^2) \tag{4.30}$$

In this connection, one must consider carefully an error introduced by the fact that some profiles are intersected by the frame of the test area (Fig. 4.12). Basically, one counts all profiles entirely within the frame and those intercepting two of the edges.

To facilitate this task the frame delimiting the test area in the various test systems (see Appendix 3) has been marked with long dashes on two sides and short dashes on the other two sides:

A first convention proposed is the following:

—*Count* all profiles totally within the frame and all those intersected by the boundary marked by long dashes (shaded in Fig. 4.12).

—*Disregard* all profiles intersected by the boundary marked by short dashes (unshaded in Fig. 4.12).

A further convention is needed for those profiles hit by one of the two ambiguous *corner points*. Gundersen (1977) has called my attention to the fact that the rules we have previously given in this respect yield overestimates, because some of the profiles that lie on the corners of the frame may be counted twice. He proposes a "forbidden line", shown as a short-dashed line in Fig. 4.12. Any profile that intercepts this line—which extends infinitely

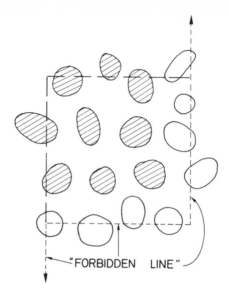

FIG. 4.12: Test frame for profile counting: short-dashed line represents "forbidden line". (After Gundersen 1977.)

beyond the counting frame—must be rejected; this holds even for profiles that also intercept the "accept line" shown in long dashes. The revised rule hence is: *Accept and count* all profiles totally or partially within the frame that *do not* intersect the "forbidden line" at any point.

The test area required to obtain Q_A and N_A again depends on the magnitude of the parameters to be estimated, J_V and N_V respectively, and on the relative error accepted. Hilliard (1976) has derived the following relation for J_V:

$$\text{R.S.E.}(J_V) = \frac{\text{S.E.}(J_V)}{J_V} = \frac{1}{\sqrt{Q_T}} \tag{4.31}$$

where Q_T is the total number of transections counted. Since

$$J_V = 2Q/A_T$$

this becomes

$$\text{R.S.E.}(J_V) = \sqrt{\frac{2}{J_V \cdot A_T}} \tag{4.32}$$

from which we derive the required test area to be

$$A_T = \frac{2}{J_V \cdot \text{R.S.E.}^2(J_V)} \tag{4.33}$$

The relative error obtained in estimating N_V from a count of profiles must depend on N_V, the mean tangent diameter $\bar{D}$, and the test area A_T; it has the following form:

$$\text{R.S.E.}(N_V) = \sqrt{\left(\frac{k}{N_V \cdot \bar{D} \cdot A_T}\right)} \tag{4.34}$$

so that

$$A_T = \frac{k}{N_V \cdot \bar{D} \cdot \text{R.S.E.}^2(N_V)} \tag{4.35}$$

The value of k is not yet established; it appears however that it should be 1.

4.2.5. COHERENT TEST SYSTEM AND RESTRICTED CONTAINING SPACES

One of the principal advantages of coherent test systems is the possibility of performing a stereological analysis with respect to containing spaces which do not cover the entire test field. By counting the test points contained in the containing space of interest—e.g. within the cytoplasm, if the volume or surface density of organelles is to be expressed with respect to cytoplasm—it is possible to determine the size of the test system enclosed within this restricted space. If P_c is the test point number within this restricted space it follows that the test line length and the test area enclosed are

$$L_c = P_c \cdot k_1 \cdot d \tag{4.36}$$

$$A_c = P_c \cdot k_2 \cdot d^2 \tag{4.37}$$

In double lattice test systems the number of points of the dense lattice contained in the restricted space is approximately

$$P'_c = P_c \cdot q^2 \tag{4.38}$$

Alternatively it is evidently also possible to count the points P_0 falling *outside* the containing reference space under consideration and to calculate P_c simply from

$$P_c = P_T - P_0 \tag{4.39}$$

4.2.6. CALIBRATION OF TEST SYSTEMS

The set of test points P_T is dimensionless; it serves for the estimation of the dimensionless parameter V_V so that no calibration is required.

Test line length L_T however serves for the estimation of S_V which has dimension cm^{-1}. Likewise the test area A_T is used to derive parameters such as length density J_V or numerical density of profiles N_A which have dimension cm^{-2}. In these instances the test system needs to be calibrated.

In coherent test systems the unit for calibration is the spacing of main test points, d (Figs 4.4, 5.4, 4.10, 4.11); it has two values, namely

d_a: the *apparent* length in the actual test system superimposed on the micrograph;

d_t: the *true* length related to the real material or tissue.

We are evidently interested in knowing d_t so that we may express S_V, for example, in real dimensions. These two values of d are connected by the *linear magnification* M between the material and the final micrograph used for analysis, such that

$$M = d_a/d_t \qquad (4.40)$$

If magnification is accurately known and if it is reproducible—as is mostly the case in light microscopy—then d_t can b derived by dividing d_a by M:

$$d_t = d_a/M \qquad (4.41)$$

In electron microscopy, or in light microscopy combined with photography, the magnification is not always consistently reproducible. It is then necessary to record a calibration specimen along with the micrographs, which then allows M or directly d_t to be estimated.

In our laboratory electron micrographs intended for stereological analysis are recorded on 35 mm film; it is general and strict practice to record a calibration specimen on each film. We employ a carbon grating replica (provided by E. F. Fullam, Schenectady, New York—No. 321) which has 21,600 lines/cm; the spacing of the lines is hence 0.463 μm or $4.63 \cdot 10^{-5}$ cm. If we call this line spacing g (for "grating") we can again assign it two values:

$g_t = 4.63 \cdot 10^{-5}$ cm is its *true* value;

g_a is its *apparent* value which can be measured on the micrograph, preferably expressed in cm.

Evidently, we find again

$$M = g_a/g_t = g_a/4.63 \cdot 10^{-5} \qquad (4.42)$$

It is now easy to see how the true value of d is calculated by substituting equation (4.42) into equation (4.41):

$$d_t = \frac{d_a}{M} = \frac{d_a \cdot g_t}{g_a} = \frac{d_a}{g_a} \cdot 4.63 \cdot 10^{-5} \text{ cm} \qquad (4.43)$$

In practice, one measures d_a and the apparent spacing of the lines of the calibration grating g_a as projected onto the test system; knowing the true spacing g_t one derives d_t. This completes the calibration.

4.2.7.   APPLICATION OF TEST SYSTEMS TO "MICROGRAPHS"

The term "micrograph" has been loosely used to mean "any microscopic image that can be superimposed on a test system". This can be a real image, such as a photographic print or the microscopic image projected onto a screen, or a virtual image as it is observed by looking into a light microscope. In either case a test system for stereological analysis must be applied to the "image" in such a way that both *image and test system lie exactly in the same plane*. Only in this way can a test point be precisely located in the material; if test system and image are in different planes parallax will render point location ambiguous, as the observer may shift the point with respect to the image by looking at it from different directions. The same holds evidently for test lines.

4.2.7.1.   *Eye-piece graticules*

In light microscopy test point and line systems may be used as graticules inserted into the back-focal plane of an eye-piece (ocular) provided with a back-lens which can be focused onto the graticule; this assures that the virtual microscopic image is seen in the same plane as the graticule, i.e. that they are parafocal. Whether this condition is fulfilled should be checked by slightly moving the head: if graticule and image are parafocal they move strictly in parallel.

Most microscope manufacturers carry so-called measuring eye-pieces which fulfill these conditions. Some also provide special graticules for stereological analysis, but many of them are not of the type of "coherent test systems" here propagated. Zeiss offers a large selection of test systems, together with a convenient measuring eye-piece with a revolving disc that accommodates six different graticules as well as two so-called "integrating eye-pieces" with test systems designed after A. Hennig; these test systems are not coherent however. Wild Heerbrugg markets two special graticules presented here, namely the multipurpose test system (M 42) and the curvilinear test system of Merz (L25). Leitz-Wetzlar also has some test systems that can be used. A large selection of graticules is marketed by Graticules Ltd. of London.

4.2.7.2.   *Drawing tubes*

Some microscopes can be fitted with a drawing tube attachment: a prism is inserted into the light path through which the image of a drawing board can be merged with the microscope image. A test system can be drawn on the drawing board and thus brought to coincide with the image of the section.

The advantage of this solution is great flexibility in the design of test systems; its disadvantage is poor image quality. This can be improved if the drawing board is replaced by a light box on which the test system is placed in the form of a photographic negative: the test system then appears as bright lines (cf. Elias and Botz, 1976).

4.2.7.3. *Projection screens for light microscopes*
In many microscopes the eye-piece can be replaced by a projection head which allows the image to be projected onto a fine ground-glass screen (Fig. 4.13). The screen is often supplied with a Fresnel-lens which projects the image onto the front screen. If the test system is intercalated in the form of a transparent foil between Fresnel-lens and screen it will be projected onto the screen together with the microscope image. This is the solution used by Wild Heerbrugg which provides a number of screens with coherent test systems for use in stereology.

Fɪɢ. 4.13: WILD sampling stage microscope with projection head carrying test grid.

4.2.7.4. *Projection screens for electron micrographs*
We are routinely recording electron micrographs destined for stereological evaluation on 35 mm film. Positive contact prints on film strips are easily produced on a long light box. They can be viewed at 10 × magnification in a back projector unit as shown in Fig. 4.14. The image is picked up by an

exchangeable screen which contains the test system. This screen is built of two optical-quality ground-glass plates between which the test system which is printed on high-contrast photographic sheet film is sandwiched; it is also projected onto the observation plane together with the image, thus assuring that both image and test systems are in the same plane.

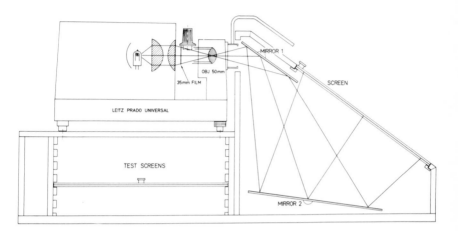

FIG. 4.14: Back projector unit for 35 mm (and 70 mm) film strips with exchangeable test screens. (top) Construction diagram, (bottom) photograph of unit.

4.2.7.5.   Overlay test systems for photographic prints

The simplest "technology" is to overlay a transparent sheet with the test system on a photographic print of the micrograph. Such test systems can either be drawn directly on the sheet with india ink—or even a pencil if a suitable acetate sheet is used; better quality is however obtained by reproducing a precisely drawn test system and photographically printing it on high contrast sheet film (e.g. Kodalith).

Various tricks can be used to ensure parallax-free superposition. Firstly, the film surface carrying the test lines should face the print. Secondly, a glass plate may be used to weigh the test film down onto the print. Paper clips may be the simplest way to prevent displacement of the test film during work.

4.2.7.6.   Simultaneous printing of micrograph and test system

It should finally be mentioned that some workers like to print the test system onto the paper print together with the micrograph. Thus Loud *et al.* (1965) propose the fitting of a fine wire screen to the frame of the enlarger easel which leaves white test lines on the micrograph. The same can be achieved by using transparent film with the test system printed on it (Fig. 4.15). Alternatively one could use double-exposure with a high-contrast negative test system to produce black test lines.

The advantages of this method are two-fold: (1) a permanent record of the combined micrograph–test system is produced; (2) the well-known distortion of paper prints affects both image and test system. Its disadvantage is that it does not allow for much flexibility and that a large volume of paper prints are—needlessly—produced.

4.3.   Calculation of V_V, S_V, J_V and K_V

The basic formulas for calculating V_V, S_V, J_V and K_V are simple: one has to count points, intersections, transections or tangents and divide by the appropriate test system parameter point number, line length or area. However, in practice it will be important to consider a number of conditions which must be satisfied in order to obtain a reliable result. If we consider the complex sampling schemes by which we have proceeded from an organ to micrographs and finally to point counts the question must be asked: at which point do we calculate the stereological parameter, and how do we estimate the statistical error? Furthermore, the point counts may be in error because of imperfect preparations: when and how do we correct for the effects of finite section thickness, or for the effect of section compression? These are the questions addressed in this section.

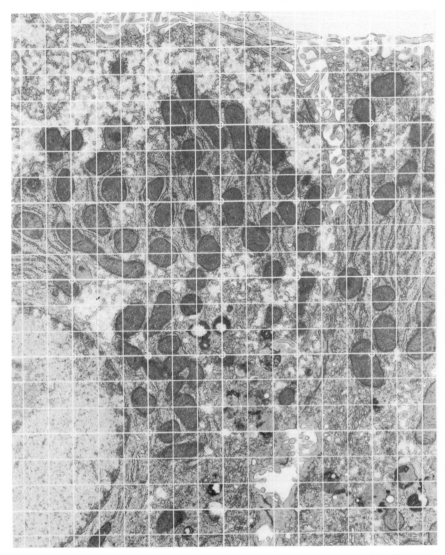

FIG. 4.15: Test lattice used as an overlay in printing easel produces white lines on photographic print. Coarse grid spacing 1.7 μm.

4.3.1. STEREOLOGICAL PARAMETERS AND RATIO ESTIMATE

The general formula for volume density of component a in the containing volume c is

$$V_{Va,c} = \frac{P_a}{P_c} \tag{4.44}$$

Note that we have replaced the symbol P_T for test point number by P_c, the number of test points falling on the profile of the containing space. P_c is equal to P_T if the entire test system is filled by the containing space. Very often, however, only part of the test field is covered by the reference or containing space so that $P_c < P_T$; in this case P_c may be variable from field to field.

Using a coherent test system the surface density of component a in c is

$$S_{Va,c} = 2 \cdot \frac{I_a}{L_c} = \left(\frac{2}{k_1 \cdot d} \right) \cdot \frac{I_a}{P_c} \qquad (4.45)$$

where the term in brackets is a constant depending exclusively on the test system used (see Section 4.2 and Appendix 3). Likewise the formulas for length and curvature density are, respectively

$$J_{Va,c} = 2 \cdot \frac{Q_a}{A_c} = \left(\frac{2}{k_2 \cdot d^2} \right) \cdot \frac{Q_a}{P_c} \qquad (4.46a)$$

$$K_{Va,c} = \pi \cdot \frac{T_{net}}{A_c} = \left(\frac{\pi}{k_2 \cdot d^2} \right) \cdot \frac{T_{net}}{P_c} \qquad (4.46b)$$

In each of these four formulas the parameter $Y_{Va,c}$ is hence calculated by the following general formula

$$Y_{Va,c} = \kappa \cdot \frac{X_a}{P_c} \qquad (4.47)$$

where κ is a constant and X_a ($= P_a$, I_a, or Q_a) and P_c are point counts. The only variable is hence the dimensionless ratio X_a/P_c whereby both X_a and P_c may vary more or less independently. The parameter $Y_{Va,c}$ accordingly depends on the estimate of this ratio and this implies that the statistical rules of ratio estimates, as outlined by Cochran (1953), are followed.

To avoid unnecessary complications we shall first demonstrate the computation scheme applying to ratio estimates on the example of volume density, and then generalize it by considering equation (4.47). Let us look at the simple and yet realistic example shown in Fig. 4.16: the containing space c is an irregular solid in which a certain number of elements of component a are irregularly distributed. In order to estimate $V_{Va,c}$ we cut n random sections and determine on each $P_a(i)$ and $P_c(i)$, where i is the number of the section. Fundamentally we can use two procedures to calculate $V_{Va,c}$:

(1) $V_{Va,c} = \frac{1}{n} \cdot \sum_{i=1}^{n} \frac{P_a(i)}{P_c(i)} = \frac{1}{n} \sum_{i=1}^{n} P_{Pa,c}(i)$ \qquad (4.48)

(2) $V_{Va,c} = \frac{\sum_{i=1}^{n} P_a(i)}{\sum_{i=1}^{n} P_c(i)}$ \qquad (4.49)

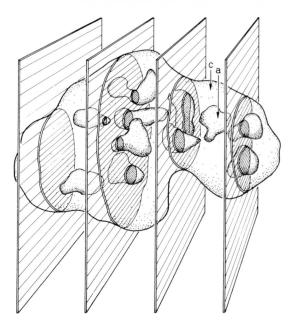

FIG. 4.16: Variable profile area of containing space produced by sectioning irregular solid.

We have shown in Section 3.7 that only equation 4.49 can give a consistent estimate of the ratio. To remind the reader of this important issue, let us look at formula (1), where the procedure is the following: on each section the ratio P_P as a direct estimator of V_V, or rather A_A, is determined, and these ratios are then simply averaged. But evidently, the different sections have a very different weight, one being very small, another one very large: the simple average is hence biased in favor of the small sections. The correct average would then be a *weighted* average, the weighting factor being the relative area in sample i:

$$w_i = \frac{A_c(i)}{\sum_{i=1}^n A_c(i)} = \frac{P_c(i)}{\sum_{i=1}^n P_c(i)} \tag{4.50}$$

If formula (1) is modified to calculate an area-weighted average we derive the following:

$$(1a) \ V_{V_{a,c}} = \sum_{i=1}^n w_i \cdot \frac{P_a(i)}{P_c(i)} = \sum_{i=1}^n \left(\frac{P_c(i)}{\Sigma P_c(i)} \right) \cdot \frac{P_a(i)}{P_c(i)} = \frac{\sum_{i=1}^n P_a(i)}{\sum_{i=1}^n P_c(i)} \tag{4.51}$$

Evidently, formula (1a) is identical with formula (2). This demonstrates again that *the consistent estimate of a ratio is the ratio of the sums of the two related variables.* Volume density must hence be calculated from the ratio of the sums of points counted on the various micrographs constituting one sample. Notice that calculating a weighted average is closely related to obtaining an A-weighted random sample discussed in Sections 3.3 and 3.4. In fact, the computation rule presented here is only able to eliminate bias if the sample is unbiased, i.e. if it was obtained according to the A-weighting sampling rules.

In the past this point has not been sufficiently stressed although it has been implicitly stated repeatedly. In work from our own laboratory we have used the term "pooling" of counts from several micrographs to form representative samples; this amounts to using formula (2). The point has been raised recently by Miles and Davy (1976) and by Mayhew and Cruz-Orive (1974) who were studying an "anomalous" situation, namely the estimation of the volume density of (single) nuclei within isolated cells. Under such circumstances the difference between formulas (1) and (2) become very important, whereas under more usual conditions the bias introduced by using formula (1) is often not very large: it depends on the degree of variation of $P_c(i)$. If $P_c(i) = P_T$ is constant it is evidently immaterial which formula is used because under these conditions they are identical.

If we now return to the general formula given in equation (4.47) we can immediately see that the consistent estimate is in all cases

$$Y_{Va,c} = \kappa \cdot \frac{\sum_{i=1}^{n} X_a(i)}{\sum_{i=1}^{n} P_c(i)} \tag{4.52}$$

The weighting factor that would have to be used in determining a weighted mean of the ratio is the same as that given by equation (4.50), irrespective of what type of point is meant by X_a. Explicitly we derive the following calculation procedures for surface, length and curvature density estimates:

$$S_{Va,c} = \left(\frac{2}{k_1 \cdot d}\right) \cdot \frac{\sum_{i=1}^{n} I_a(i)}{\sum_{i=1}^{n} P_c(i)} \tag{4.53}$$

$$J_{Va,c} = \left(\frac{2}{k_2 \cdot d^2}\right) \cdot \frac{\sum_{i=1}^{n} Q_a(i)}{\sum_{i=1}^{n} P_c(i)} \tag{4.54a}$$

$$K_{Va,c} = \left(\frac{\pi}{k_2 \cdot d^2}\right) \cdot \frac{\sum_{i=1}^{n} T_{net}(i)}{\sum_{i=1}^{n} P_c(i)} \tag{4.54b}$$

It should finally be mentioned that this calculation procedure also derives directly from the general stereological formulas of Miles (1972) which will be discussed in more detail in Volume 2.

4.3.2.   REPRESENTATIVE SAMPLE AND REVERSAL OF THE SAMPLING HIERARCHY

In Fig. 3.6 we defined a hierarchy of multiple stage sampling which leads stepwise from the population of animals to the point counts done on micrographs. In computing stereological parameters this sampling hierarchy must be reversed and we must indicate an appropriate computation procedure which goes from the counts to the population data through the following steps:

counts per field

↓

section data

↓

organism data

↓

population data.

In view of the discussion of ratio estimates we need to define at which point in this sequence point counts have to be summed, in other words, at which step the actual stereological parameters are calculated.

This question is intimately connected with the term *"representative sample"* which has been used repeatedly throughout this text without further definition. We call a sample representative if it is large enough to allow a "reliable" estimate of the parameter in which we are interested. The term "reliable" however must be qualified in statistical terms, i.e. by stating what kind of error we would accept with what probability. A representative sample must therefore yield an estimate whose expected standard error (or 95% confidence interval) is no more than a certain percentage of the mean, say $\pm 10\%$. This is the type of argument that we have used in order to determine the test point number required to yield an estimate with a certain precision, as it is derived through equation (3.29), (3.38) and (4.21), or from the nomograms of Figs. 4.7 and 4.9.

As was said in Section 4.2.2.4., the required number of test points thus derived must usually be distributed over several micrographs. In the example given in Table 4.2 we used 8 micrographs to accommodate the test points required; these 8 micrographs hence constitute the representative sample. They were all derived from one section, and we used 5 such sections (each yielding 8 micrographs) to constitute the sample per animal.

Under these conditions it would appear logical to compute the stereological parameters for each representative sample; i.e. we would sum the point counts of all 8 micrographs derived from the same section and compute the volume density of mitochondria in hepatocyte cytoplasm as

$$V_{Vmi,cy} = \frac{\sum_{i=1}^{8} P_{mi}(i)}{\sum_{i=1}^{8} P_{cy}(i)}$$ (4.55)

We would proceed likewise for the surface density estimates, etc.

From there on the parameters are averaged over the various section samples to yield organism means, as for example

$$\overline{V}_{Vmi,cy} = \frac{1}{m} \sum_{j=1}^{m} V_{Vmi,cy}(j)$$ (4.56)

where j is the sample number and m the total number of samples at this stage. We would then proceed likewise to compute the population mean from the organism estimates.

In this procedure we assume that the total number of test points on one representative sample, i.e. on all the micrographs derived from one section, is about the same for all sections. This should be checked carefully, for there may be cases where this condition is not fulfilled; this may occur, for example, if the cell type which serves as reference space is not uniformly distributed over the organ. In such instances it may be preferable to sum the counts over all sections and to calculate the stereological parameters by the ratio of sums directly for the organism. Alternatively, the number of micrographs recorded per section could be varied so as to produce a sample of the reference space of about equal size for all sections.

4.3.3. ESTIMATION OF STATISTICAL ERROR OF ESTIMATES

The average stereological parameters derived by the computation procedures outlined in the preceding sections are estimates of the true parameter existing in the specimen. How good is this estimate?

It is known from general statistical theory that the arithmetic mean of repeated sample estimates is an unbiased estimator of the true parameter (for further details consult any textbook on statistics, such as Snedecor, 1956; Dixon and Massey, 1957; Cochran, 1953; and others). But this estimate will be affected by a certain error which becomes smaller with larger numbers of sample estimates. In the following we shall call a "sample estimate" the parameter obtained on one representative sample. Such sample estimates show a distribution about the true mean; in general it is assumed, but by no means proven, that such sample estimates are normally distributed. The spread of this distribution is measured by the variance $\sigma^2(x)$ whereby x designates the parameter. This variance is estimated by the square of the standard deviation $s(x)$ or S.D.(x) by the following formula

$$\text{S.D. } (x) = \sqrt{\left(\frac{\Sigma(x - \bar{x})^2}{n - 1}\right)} = \sqrt{\left(\frac{\Sigma x^2 - \bar{x}\Sigma x}{n - 1}\right)} \qquad (4.57)$$

where n is the number of samples evaluated.

The mean $\bar{x}$ obtained from n samples is again normally distributed about the true mean, but evidently the spread will be smaller with larger numbers of samples used. The standard deviation of these means, $s(\bar{x})$, which is called the standard error of the mean, S.E. $(\bar{x})$, decreases with the square root of n, so that

$$\text{S.E. } (\bar{x}) = \frac{\text{S.D. } (x)}{\sqrt{n}} = \sqrt{\left(\frac{\Sigma x^2 - \bar{x}\Sigma x}{n(n - 1)}\right)} \qquad (4.58)$$

The standard error of the estimate thus derived is a measure of the precision realized with the sampling scheme chosen. From this we can derive the confidence interval within which the true mean will lie with a certain probability; for a probability of being "right" in 95% of the cases the confidence interval is

$$d = \bar{x} \pm t_{0.05} \cdot \text{S.E. } (\bar{x}) \qquad (4.59)$$

where $t_{0.05}$ is the t-variate for an error probability of 5% (0.05) and a degree of freedom $(n - 1)$. Some typical values of $t_{0.05}$ and $t_{0.01}$ (1% error probability) are presented in Table 4.3. In the range of degree of freedom usually prevailing in this type of analysis an appropriate rough value is hence $t_{0.05} \sim 2.4$ ot $t_{0.01} \sim 4$. Additional values can be found in any collection of mathematical tables.

Table 4.3

Values of t-variate for error probability of 5% and 1% and degree of freedom $(n - 1)$ of 4 to 15.

$n - 1$	$t_{0.05}$	$t_{0.01}$
4	2.776	4.604
5	2.571	4.032
6	2.447	3.707
7	2.356	3.500
8	2.306	3.355
9	2.262	3.250
10	2.228	3.169
11	2.201	3.106
12	2.179	3.055
13	2.160	3.012
14	2.145	2.977
15	2.132	2.947

The standard error of the mean also provides the data for a statistical comparison between two or more means which aim at establishing or rejecting the hypothesis that two means are estimates of the same population; if this hypothesis is rejected then the difference between the means is considered "statistically significant" whereby the error probability (e.g. $P < 0.05$) must be stated. The methods used for such a comparison are a matter of much debate. The popular t-test is based on a number of assumptions about the distribution of the differences which can usually not be proven. Statisticians therefore often prefer other so-called non-parametric tests, such as the Wilcoxon test based on a ranking of differences. This is not the place to expand on statistics; the reader should just be warned that he may be subject to attack by statisticians if he tries to blindly follow what others have done more or less successfully! The advice is to read on statistics and to look up an expert, if needed—but not to loose courage!

A special situation evidently occurs by the use of the ratio of sums as the primary estimator of stereological parameters. Here it is evidently not possible to apply the simple statistics outlined above. Cochran (1953) has developed the statistics applying to such ratio estimates. We shall present his formula for the variance of the ratio estimate by using the general formula given in equation (4.47) where the point count X_a can be either P_a, I_a, Q_a or T_{net} and κ assumes the values shown in equations (4.44) to (4.46).

If X_a and P_c are estimated over n fields then the estimate of the ratio is evidently

$$\hat{R} = \frac{\sum^n X_a}{\sum^n P_c} \qquad (4.60)$$

and the variance of this ratio depends on the covariance between X_a and P_c and is given by

$$\text{S.D.}^2\,(R) = \frac{n}{(n-1)\cdot(\Sigma P_c)^2}\{\Sigma X_a^2 + \hat{R}\Sigma P_c^2 - 2\hat{R}\Sigma X_a \cdot P_c\} \qquad (4.61)$$

From this the parameter and its variance are obtained by

$$Y_{Va,c} = \kappa \cdot \hat{R} \qquad (4.62)$$

$$\text{S.D.}^2\,(Y_{Va,c}) = \kappa^2 \cdot \text{S.D.}^2\,(R) \qquad (4.63)$$

The standard error of the ratio estimate or of the parameter respectively are obtained by dividing S.D. by $\sqrt{n}$. With these formulas at hand one can hence derive an estimate of the statistical error affecting a parameter calculated by the method of ratio estimates.

As outlined in Section 3.7, stereological estimates are obtained by a multiple stage sampling technique: (1) point counts to estimate $\hat{R}$ or a parameter density X_A per unit section area are performed on (2) a set of m planar sections from which we want to obtain an estimate of the parameter Y_V relating to the entire organ. Nicholson (1978) has emphasized the importance of properly accounting for this two-stage sampling approach by introducing multivariate statistics. It is particularly important to perform an analysis of variance by multivariate statistics if the statistical evaluation should serve the purpose of improving the analytical design of the study. It can help answer the question about where an increase in sample size is most likely to lead to an improved result: more points per micrograph, more micrographs per section, more sections per animal, or more animals? Such a thorough statistical analysis is particularly in place in the context of pilot studies for large experimental series.

It is beyond the scope of this text to present details and technicalities about multivariate statistics, for there is nothing specifically stereological in the approach. The reader is referred to the appropriate statistical literature, e.g. Mosteller and Rourke (1976), Anderson (1960), Gnanadesikan (1977), Morrison (1967), Rao (1968).

4.3.4. CORRECTION OF SYSTEMATIC ERRORS DUE TO FINITE SECTION THICKNESS

One of the major sources of error in applying stereological methods to real "thin" sections is their finite thickness. The basic stereological methods assume that the point counts are performed on true section planes; what we are in reality investigating is the projection onto an observation plane of the whole content of a relatively thick slice (Fig. 4.17). Evidently, the thicker the slice the greater the amount of enclosed components which is projected onto the observation plane, and accordingly the greater the error in estimating V_V or S_V. This effect of section thickness on stereological estimates was first recognized by Holmes in 1927; it is hence generally called the "Holmes effect".

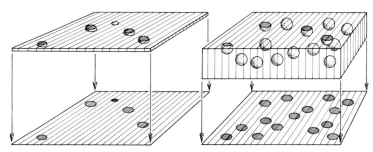

FIG. 4.17: Effect of slice thickness on profile number observed by transmission microscopy.

In Volume 2, Section 4.4, we shall present a set of formulae which would permit a correct estimation of stereological parameters from an analysis of thick sections (Miles, 1976). Since the conditions for direct application of these advanced methods are not yet worked out, we have exploited these formulae to derive correction factors which should allow an estimation of the magnitude of the Holmes effect, and a correction of the ensuing errors in the estimates of stereological parameters (Weibel and Paumgartner, 1978).

The rationale for the use of such correction factors is as follows. On a slice of thickness t we obtain a measurement of areal density of profiles, $A'_A = P'_P$, which, due to the Holmes effect, is larger than the true volume density, V_V; the latter is found by

$$V_V = A'_A \cdot K_t(V_V) = P'_P \cdot K_t(V_V) \qquad (4.64)$$

Likewise the apparent boundary length density of profiles, B'_A, or the apparent number of intersections per unit test line length, I'_L, observed on a thick slice are too large; the true surface density, S_V, is found by

$$S_V = \frac{\pi}{4} B'_A \cdot K_t(S_V) = 2 \cdot I'_L \cdot K_t(S_V) \qquad (4.65)$$

4.3.4.1.   *Correction factors for section thickness effect*
In practice, we do not know the true extent of overestimation; the values of K_t can however be estimated by theoretical model considerations (Volume 2, Chapter 4). In general terms, it is found that the correction factors depend on some model assumptions about the geometrical and statistical properties of the structure. With respect to the geometrical properties we have derived coefficients for structures made up of elements of the following size and shape characteristics (Fig. 4.18):

   (a) spheres of diameter $d = 2R$;
   (b) thin discs of thickness d and diameter $D = \delta \cdot d$;
   (c) long tubules of diameter d and length $L = \lambda \cdot d$.

Note that in each case d is the size characteristic, whereas δ and λ define the shape of the structural elements. The size characteristic d has been chosen in such a way that, for a given structural component, it is the "characteristic dimension", i.e. the one that shows least variation and should be measurable with some reliability. Thus for RER cisternae (discs) d is the cisternal width whereas for SER tubules it is their diameter: evidently, cisternal "diameter" and tubule length vary much more than their respective values of d; in either case d can be measured on "ideally" cut profiles, whereas δ and λ must be roughly estimated. In either case the shape coefficients are $\gg 1$, i.e. d is the smallest dimension of the structural elements.

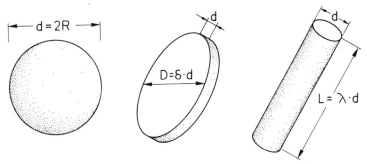

FIG. 4.18: Definition of dimensions of spheres, cylinders and discs as structural models for calculation of section thickness correction factors.

This choice of d as a "stable" size characteristic is also convenient because it allows the section thickness t to be expressed in relation to d. The correction coefficients K_t will not, in fact, contain d or t, but will be expressed in terms of the shape characteristics and of relative section thickness:

$$g = t/d \qquad (4.66)$$

The actual formulation of correction factors also depends on the statistical model which is assumed to describe the make-up of the structure. Essentially there are two models:

(1) a structure made of discrete convex particles, all clearly separated from each other (Cahn and Nutting, 1959);

(2) a structure made of convex elements assembled by a true random or Poisson process (Miles, 1976), which implies that some of the elements are not necessarily discrete but may interpenetrate; they will do so to an increasing degree the denser they are arranged.

The first model can, in fact, only be expected if the structural elements are very loosely arranged, i.e. if their V_V is very small; in this case the chance that profiles, generated by a thick slice, overlap is also small. The second model is, in a way, much more general: it allows for dense arrays of elements and for overlapping of profiles; but it depends essentially on the condition that the elements are assembled by a true random process, a condition often not fulfilled. These limitations must be known when the correction factors listed in the following are applied; nevertheless it will often be preferable to use correction factors based on an imperfect model, than to be misled by accepting uncorrected data at their face value.

The theoretical development leading to the derivation of such correction factors will be given in full detail in Volume 2, Chapter 4. The factors so derived are compiled in Tables 4.4 and 4.5 (equations 4.67 to 4.81); they relate to the model structures made of discrete particles (a) or interpenetrating elements (b), respectively. Note that the factors of Table 4.5 become identical to those of Table 4.4 if their density, measured by the parameter ζ becomes negligibly small, i.e. as $\xi \to 0$.

Table 4.4

Section thickness correction factors for V_V and S_V. Model structure made of discrete particles. (Relative section thickness $g = t/d$).

Spheres: diameter d

$$K_t(V_V) = \frac{1}{1 + \frac{3}{2}g} \tag{4.67}$$

$$K_t(S_V) = \frac{\pi}{\pi + 4g} \tag{4.68}$$

Spheres (truncated): diameter d; diameter smallest visible cap $d_0 = \rho \cdot d$

$$K_t(V_V) = \frac{2}{(2 + \rho^2)\sqrt{(1 - \rho^2)} + 3g} \tag{4.69}$$

$$K_t(S_V) = \frac{\pi}{2[\rho\sqrt{(1 - \rho^2)} + \sin^{-1}\sqrt{(1 - \rho^2)} + 2g]} \tag{4.70}$$

Discs: thickness d; diameter $D = \delta \cdot d$

$$K_t(V_V) = \frac{\delta}{\delta + g \cdot (\delta/2 + 1)} \tag{4.71}$$

$$K_t(S_V) = \frac{\delta(\delta/2 + 1)}{\delta(\delta/2 + 1) + g(\delta + 2/\pi)} \tag{4.72}$$

Tubules (closed): diameter d; length $L = \lambda \cdot d$

$$K_t(V_V) = \frac{\lambda}{\lambda + g(\lambda + \frac{1}{2})} \tag{4.73}$$

$$K_t(S_V) = \frac{\lambda + \frac{1}{2}}{\lambda + \frac{1}{2} + g(1 + 2\lambda/\pi)} \tag{4.74}$$

Tubules (open-ended):

$$K_t(S_V) = \frac{\lambda}{\lambda + g(1 + 2\lambda/\pi)} \tag{4.75}$$

It is interesting to note that for very large discs ($\delta \to \infty$) and very long tubules ($\lambda \to \infty$) occurring in very low number the correction factors become rather simple:

$$\text{Discs:} \quad K_t(V_V) = \frac{1}{1 + g/2} \tag{4.82}$$

$$K_t(S_V) = 1 \tag{4.83}$$

$$\text{Tubules:} \quad K_t(V_V) = \frac{1}{1 + g} \tag{4.84}$$

$$K_t(S_V) = \frac{\pi}{\pi + 2g} \tag{4.85}$$

Some of these factors were derived by Hennig (1969).

Table 4.5
Section thickness correction factors. Model structure made by Poisson process.
(Interpenetrating elements with dimensions as in Table 4.4.)

Volume density estimates:

$$\text{Spheres: } K_t(V_V) = \frac{1 - \exp\{-\xi \cdot \frac{2}{3}\ \}}{1 - \exp\{-\xi \cdot [\frac{2}{3} + g]\}} \tag{4.76}$$

$$\text{Discs: } \quad K_t(V_V) = \frac{1 - \exp\{-\xi\delta^2\}}{1 - \exp\{-\xi[\delta^2 + g \cdot \delta(\delta/2 + 1)]\}} \tag{4.77}$$

$$\text{Tubules: } K_t(V_V) = \frac{1 - \exp\{-\zeta\lambda\}}{1 - \exp\{-\xi[\lambda + g(\frac{1}{2} + \lambda)]\}} \tag{4.78}$$

Surface density estimates:

$$\text{Spheres: } K_t(S_V) = \frac{\pi \cdot \exp\{-\xi \cdot \frac{2}{3}\}}{[\pi + 4g] \cdot \exp\{-\xi \cdot [\frac{2}{3} + g]\}} \tag{4.79}$$

$$\text{Discs: } \quad K_t(S_V) = \frac{\delta(\delta/2 + 1)\exp\{-\xi\delta^2\}}{[\delta(\delta/2 + 1) + g(\delta + 2/\pi)]\exp\{-\xi[\delta^2 + g \cdot \delta(\delta/2 + 1)]\}} \tag{4.80}$$

$$\text{Tubules: } K_t(S_V) = \frac{\lambda \cdot \exp\{-\xi\lambda\}}{[\lambda + g(1 + 2\lambda/\pi)] \cdot \exp\{-\xi[\lambda + g(\lambda + \frac{1}{2})]\}} \tag{4.81}$$

N.B. Parameter ξ is a measure of the density of the fictitious elements; for most practical purposes its value can be taken to be of the order of 10^{-4} (see Volume 2, Chapter 4, and Weibel and Paumgartner, 1978).

4.3.4.2. *Practical use of correction factors*
Any of these factors can easily be calculated directly from the formulas; for convenience Figs. 4.19 to 4.21 present graphs from which K_t can be read as a function of section thickness.

Note that all these formulas contain only relative parameters, namely the relative section thickness $g = t/d$ and the relative shape or truncation factors. In practice it is hence important to decide which model shape is the most appropriate descriptor of the elements and then to attempt a careful estimate of d by repeated measurement on the sections. With respect to the disc model one must seek out the narrowest profiles, because these most likely represent perpendicular sections of the disc; similarly for tubules d is represented by the small diameter of tubule profiles. Note that for spherical particles one must measure the equatorial diameter, or calculate d from the

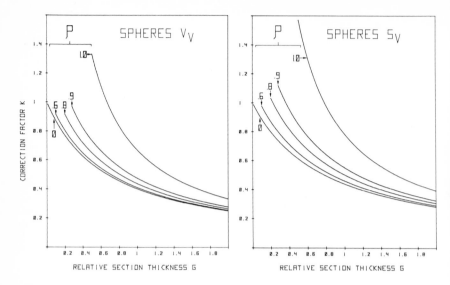

FIG. 4.19: Coefficients for correcting volume and surface density estimates for effect of section thickness t and truncation for sphere model (minimal detectable profile radius r_0). Curves for $\rho = 0$ represent Holmes correction with no truncation.

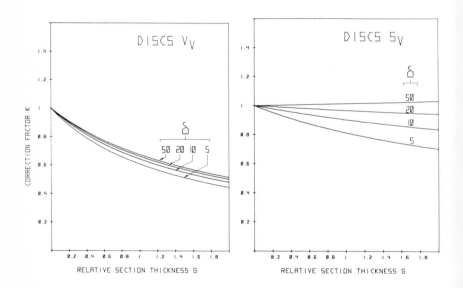

FIG. 4.20: Correction coefficients for section thickness effect on V_V and S_V estimates for disk model.

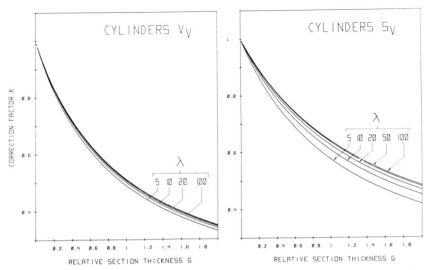

FIG. 4.21: Correction coefficients for section thickness effect on V_V and S_V estimates for tubule model.

diameters of section profiles (see Section 8.2), whereby in the present case d is the *sphere* diameter. By interpretation of the three-dimensional make-up of the structure one must then decide which model—discrete or true random —can best describe the structure and also attempt to judge the shape factors. Table 4.6 presents the results of an analysis of size and shape characteristics of subcellular organelles in liver cells (Weibel and Paumgartner, 1978); it also shows the Holmes correction factors for surface density that were calculated for the particular conditions prevailing in that study, assuming the true random model for the assembly of tubules and discs (equations 4.80 and 4.81), but the truncated discrete model for spherical vesicles (equation 4.67). This revealed that the surface density of RER was overestimated by no more than 2%, but that of SER by 30%.

One further example for applying these factors is the correction of estimates of volume and surface densities of small membrane vesicles as they were studied, for example, by Bolender (1974) in pancreatic acinar cells. The size of these vesicles is small, but truncation was evident, as very small caps cannot be recognized. Estimating the following parameters:

$$\text{section thickness} \quad t = 60\,\text{nm}$$
$$\text{vesicle diameter} \quad d = 100\,\text{nm}$$
$$\text{minimum cap diameter} \quad d_0 = 40\,\text{nm}$$

we find the following correction coefficients from equations (4.69) and (4.70):

$$K_t(V_V) = 0.53$$
$$K_t(S_V) = 0.58$$

It is thus apparent that in estimating volume and surface density of such small vesicular components a significant error is introduced. In estimating the volume and surface density of zymogen granules in the same study the error was evidently much smaller; since $\bar{D}_{zg} \sim 700\,\text{nm}$, the correction coefficients for both volume and surface density are of the order of 0.9. However, the

Table 4.6
Estimating section thickness correction factors for liver cell organelles.
(Modified from Weibel and Paumgartner, 1978.)

	RER cisternae	SER tubules	Mitochondria outer membrane	cristae
Model structure	Disk	Tubule	Sphere (trunc.)	Disk
Formulae	4.77; 4.80	4.78; 4.81	4.69; 4.70	4.77; 4.80
$\bar{d}$, nm ($\pm$ S.D.)	46 ± 13	66 ± 28	800	28 ± 8.5
D or L, nm	2,000	400	—	200
δ, λ or ρ	43	6	0.375	6.7
Section thickness $t = 36.7\,\text{nm}$ (± 4.6)				
$g = t/d$	0.804	0.561	0.046	1.321
Density estimate ξ	$2.7 \cdot 10^{-5}$	$5.1 \cdot 10^{-3}$	—	$2.2 \cdot 10^{-4}$
Correction factors $K_t(V_V)$	0.711	0.628	0.942	0.541
$K_t(S_V)$	0.986	0.702	0.966	0.824

size of zymogen granules may vary considerably so that more exact correction coefficients may have to be calculated following a rough determination of the moments of size distribution, using the corresponding formulae to be given in Volume 2, Chapter 4.

4.3.4.3. *Measurement of section thickness*

Successful application of any of the correction factors here derived evidently depends crucially on a reliable estimate of section thickness.

In light microscopy section thickness can be measured directly. With a high power objective ($60 \times$ to $100 \times$) and the numerical aperture of the condenser optimally adjusted the focal depth is no more than 0.1 to 0.2 μm; by bringing the bottom and top surfaces of a (paraffin) section a few microns thick sequentially into focus the section thickness can be read off the micrometer scales with which the fine focus knob of all good microscopes is fitted. The problem becomes more difficult with thin sections ($\leqslant 1\,\mu$) as they can be cut from plastic embedded blocks; the precision of this method, if possible at all, is inadequate. One can then resort to interferometry, which can also be used with "ultrathin" sections destined for electron microscopy.

The principle for section thickness measurement by interference microscopy can be summarized as follows. The monochromatic illuminating light beam of wave length λ is split in two beams, one of them (measuring beam M) traversing the section, the other one (reference beam R) the background. The difference in optical density of the background (support film, embedding medium etc.), n_0, and of the section, n, leads to an optical path difference

$$\delta = (n_1 - n_0) \cdot t \qquad (4.86)$$

which is hence dependent on section thickness. As the measuring and reference beams are recombined before they enter the microscope objective the optical path difference leads to interference contrast between section and background, which is quantitatively proportional to δ. There are several ways of estimating δ, which are exploited in the various interferometers that are on the market. One can either measure the displacement of interference fringes caused by δ, or use a polarized light beam; if the measuring beam passes through an object (section) of different refractive index than the background this results in elliptic polarization whose degree can be measured by use of a compensator. Such systems have been described for section thickness measurement by Gillis and Wibo (1971) and by Pluta (1969, 1971), for example.

It should be noted that this measurement of section thickness is particularly easy if the reference medium is air whose refractive index $n_0 = 1$.

A different approach is to use the optical or electron density of the section in conjunction with some standards. Casley-Smith and Crocker (1975) measured the differences in electron scattering between the section on the supporting film and the supporting film on the one hand, and Latex spheres as standards on the other hand. These measurements can be made directly in the electron microscope by measuring the differences in electron current on the viewing screen (exposure meter!). With suitable calibration fairly reproducible estimates of section thickness can be obtained, but one depends on measuring this thickness in plain plastic; the tissue-containing part of the section may perhaps be of different thickness.

Another method is proposed by Silverman et al. (1969) who embed an "electron density standard" in the block, along with the material. They chose relatively large anion exchange resin spheres which they had stained under standard conditions with phosphotungstic acid. Profiles of the sectioned Dowex spheres were micrographed under standard conditions, and the negatives evaluated by microdensitometry. Evidently, to calculate section thickness a number of parameters, such as the density of Dowex stained with PTA, had to be determined. For sections 30 to 100 nm thick the authors obtained fairly reproducible results which compared reasonably well with interferometric measurements.

In fact, most tissues contain a "built-in standard", namely erythrocytes, which can be used in a similar way. When working with well-standardized preparation techniques one should assume normal red blood cells from one species to always have very similar electron density. By recording the difference in electron current between plain supporting film and an erythrocyte profile at fairly high power one obtains an estimate of relative section thickness; with suitable calibration, for example by means of either an interferometric method, the fold method, or the re-sectioning method of Yang and Shea (1975), one may derive absolute section thickness.

We have found the fold method proposed by Small (1968) to be extremely practical, mainly because it is so very simple and reproducible, without the need of indirect standardization. Small proposes to look for "minimal" folds in the section which have a parallel outline over a certain distance (Fig. 4.22). At higher power one should be able to resolve a very thin cleft in the middle of the fold (Fig. 4.23), and, in general, one can still see some of the section

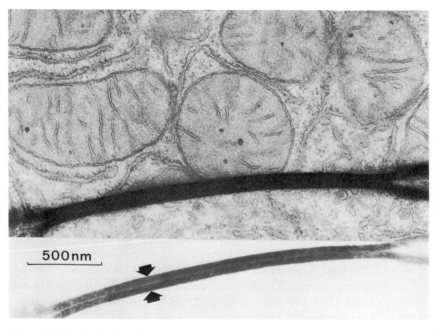

FIG. 4.22: Minimal fold ("Small fold") in thin section of liver cell comprises two section thicknesses. Magnification 46'900 ×.

content in the fold itself. If a membrane trace crosses the fold obliquely it will appear kinked. The nature of such a fold is shown diagrammatically in Fig. 4.24: the section is just slightly pinched so that the fold does not fall over: it stands upright and displays two pleats of the section sideways. We shall call such a fold a "Small-fold", evidently with a double meaning: it is indeed the smallest fold we can find, and we owe it to Dr. Small!

It is now quite evident that the width of a Small-fold is $2t$; it is easily measured. The estimates of section thickness thus obtained agree well with

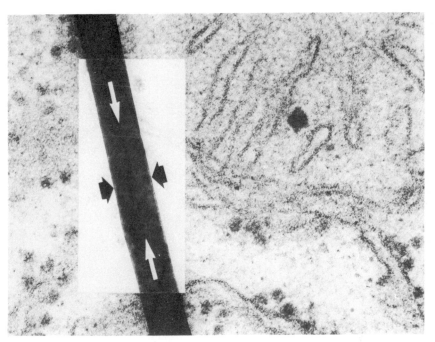

FIG. 4.23: Higher power of "Small fold" reveals fine cleft between the two apposed section thicknesses.

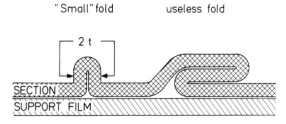

FIG. 4.24: Diagrammatic representation of "Small" fold.

interferometric measurements (Fig. 4.25). In our laboratory it has now become routine to record on each 35 mm film destined for stereological analysis one Small-fold at a higher magnification (such that the final magnification is at least 100,000 ×), and to overexpose the negative by about 50% in order to resolve the interior of the fold. As is seen in Fig. 4.23 a well focussed Small-fold will have a sharp edge, but it will also show, because of its height, an overfocus as well as an underfocus fringe; the exact width of the fold is the

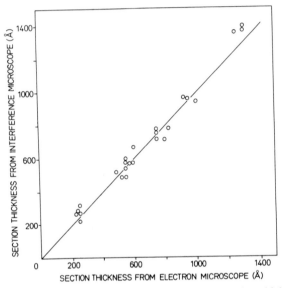

FIG. 4.25: Agreement between interferometric estimation of section thickness and fold method. (After Small, 1968.)

distance between the outer edge of the dark lines between the two light fringes. The main advantages of this method are (a) its simplicity, (b) the possibility of having a permanent record of section thickness, and (c) the possibility of measuring section thickness in the tissue part of the section rather than on clear plastic.

4.3.5. CORRECTION OF SYSTEMATIC ERROR DUE TO SECTION COMPRESSION

It is well known that most histological sections—be it paraffin sections for light microscopy or plastic sections for electron microscopy—are compressed upon sectioning. All sorts of tricks are used to re-expand them during floating, but rarely is there complete compensation for compression.

This compression tends to introduce some systematic error. If all components are about equally affected the volume density estimate should not be in error. However, all other estimates which are not dimensionless parameters will be in error: they will be overestimated.

What happens is shown diagrammatically in Fig. 4.25. After cutting, the section is shorter than the original block face. If we can measure the original

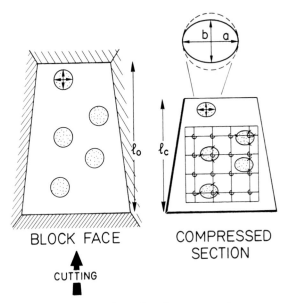

BLOCK FACE   COMPRESSED
                    SECTION

CUTTING

Fig. 4.26: Effect of section compression.

length of the block face in the direction of the cutting stroke, l_0, and the same compressed length on the section, l_c, then the section has been compressed by the compression factor

$$f_c = l_c/l_0 \qquad (4.87)$$

We assume that the width of the section has not been affected. The result is that the content of the section has been squeezed into an apparent area A_c which is

$$A_c = A_0 \cdot (l_c/l_0) \qquad (4.88)$$

where A_0 is the original area of the block face. This is presumably accompanied by a thickening of the section in proportion to l_c/l_0.

How does this affect the stereological data? Assume "ideal" conditions: all tissue components become compressed to the same extent (this may not always be the case, though). Under these conditions the profiles of the objects of interest and of the embedding matrix become reduced in area proportionately; the estimation of V_V from A_A or from a point count remains unaffected. This is not the case for the surface density: Loud *et al.* (1965) have argued that isotropic membrane systems, which can be represented by a sphere or by a circular profile on sections, become anisotropic by the amount of section compression; their profile is now represented by an ellipse. On this model Loud *et al.* (1965) have worked out the following correction formula

$$K_c(S_V) = \frac{S_V}{S_V'} = \frac{f_c}{\sqrt{\left(\frac{1 + f_c^2}{2}\right)}} \tag{4.89}$$

where S_V is the correct surface density and $S_V' = (4/\pi) \cdot B_A' = 2\Gamma_L$ is the one estimated on compressed sections. This model sounds plausible, but it is probably only a first approximation. The compression events are probably more complex and a more appropriate model would be the compression of a sphere into an oblate spheroid or triaxial ellipsoid of which one observes some zonal sections in projection. Factors based on such a model may be difficult to work out.

The correction factor applying to any counts per test area, whether transection or profile counts, is

$$K_c(N_A) = K_c(Q_a) = \frac{l_c}{l_0} \tag{4.90}$$

as can easily be seen; because these profiles just become pushed into a smaller area the test area A_T will be too large by l_0/l_c.

The compression factor (l_c/l_0) can be measured by comparing the dimensions of section and block face, but this is tricky because the face of a block cut in the form of a pyramid becomes larger as more sections are cut. Loud *et al.* (1965) have suggested estimating (l_c/l_0) directly on sections. If the material contains any spherical element, such as the nuclei in hepatocytes or lymphocytes, they are distorted by section compression to ellipses whose ratio of minor to major axis is equal to (l_c/l_0). This is a good procedure. An alternative could be to use a square lattice test system with one line direction oriented parallel to the cutting stroke (knife marks may help orientation); the number of intersections of an isotropic membrane system with the lines perpendicular to the cutting stroke divided by those with the parallel lines is also a direct estimate of (l_c/l_0).

4.3.6. CORRECTION OF ERRORS DUE TO TISSUE SHRINKAGE

The effect of tissue shrinkage is in some way similar to that of compression, only it acts not in one but in three dimensions, isotropic shrinkage granted.

The shrunken tissue block occupies a smaller volume than the original fresh tissue volume which has, supposedly, been measured, or to which the data should really be related. This means that the shrunken tissue volume at which the measurements were obtained, V_s, must be "re-swollen" to the original volume, V_0. If we can determine a linear shrinkage factor

$$f_s = \frac{l_s}{l_0} \qquad (4.91)$$

where l_s and l_0 are some characteristic lengths in the shrunken and the original tissue block, respectively, then the original volume is

$$V_0 = V_s \cdot f_s^{-3} \qquad (4.92)$$

Volume densities are again not affected by isotropic uniform shrinkage, but surface densities etc. are. The apparent surface density S_V' measured on the sections is related to V_s; true surface density must be related to V_0, which is obtained by

$$S_V = \frac{S}{V_0} = \frac{S}{V_s \cdot f_s^{-3}} = S_V' \cdot f_s^{-3} \qquad (4.93)$$

Length density and numerical density estimates are corrected the same way.

The amount of shrinkage occurring depends greatly on the method used. It is very dramatic with paraffin sections where $f_s \sim 0.74$ (Weibel, 1963), but it is much less important in electron microscopy using epoxy plastic embedding (Weibel and Knight, 1964). Shrinkage can be measured on free cells, small tissue blocks, or thin tissue flakes, which one follows through the different steps under a microscope.

4.3.7. THE EFFECT OF RESOLUTION ON STEREOLOGICAL MEASUREMENTS

We must finally consider a disturbing recent finding, namely that some of the measurements obtained may depend on the magnification or microscopic resolution used for the study.

If one measures the alveolar surface area of human lungs by light microscopy on paraffin sections one obtains values of the order of $80\,m^2$ (Weibel, 1963; Thurlbeck, 1967). Similar lungs studied by electron microscopy appeared to have surface areas of the order of $140\,m^2$ (Gehr et al., 1978). To clarify this point Keller et al. (1976) have measured alveolar surface area in one lung at magnifications ranging from $100\times$ to $10,000\times$. Figure 4.27 shows that the estimate of volume density of tissue septa achieved a stable value, whereas the estimate of surface density appeared to be steadily increasing.

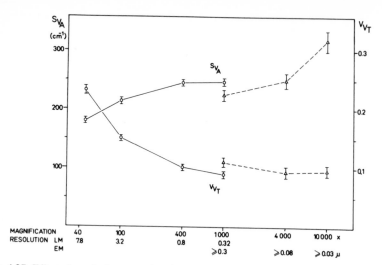

Fɪɢ. 4.27: Effect of resolution on estimation of surface density ("Coast of England effect"). (Reproduced by permission from Keller *et al.*, 1976.)

This effect has indeed been known to cartographers for a long time; they are constantly faced with the problem of simplifying the outline of coasts or of rivers etc. as they draw maps at smaller scale: thus, details are lost and the coast line becomes shorter. "How long is the coast of Britain?" was a question raised by Mandelbrot (1967). As one looks at the coast of Britain from decreasing distances one resolves an increasing amount of detail in the contour of the British Isles (Fig. 4.28). The same is the case when looking at the lung surface, or any other surface. It is hence to be expected that the estimate of surface density increases—indefinitely?—with increasing resolution.

Mandelbrot has analysed this question in great detail and has presented it in a fashion that is accessible to non-mathematicians in his recent book "Fractals: Form, Chance and Dimension" (1977), a fascinating and beautifully illustrated account of some "geometric fictions" which, however, may become of serious consequence for stereology.

Figure 4.29 shows one of the illustrations from Mandelbrot (1977): a quadric Koch Island, on which we can show the "coast of Britain effect". It starts out with a square. Each of its sides of length *d* is divided into four parts (hence "quadric"); one little square is cut out at position 3 and added as a "peninsula" at position 2 (Fig. 4.29): the coast line of the second island is twice as long as that of the first one, but its surface area is unchanged. We now take each of the sides of length *d*/4 and repeat the same operation: the coast line of the third island is again twice as long, i.e. 4 times that of the original square, but with unchanged area. This can evidently be carried out an infinite number of times, and the coast line eventually becomes infinitely long! Which is then the "real" coast line length?

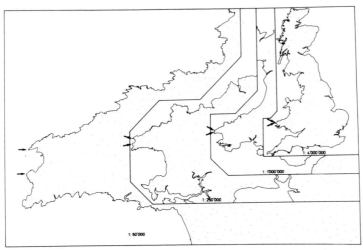

FIG. 4.28: Outline of the coast of Britain on maps of different scale shows varying details but similarity of contour complexity. Paired arrows point to Porth-mawr on the coast of Wales.

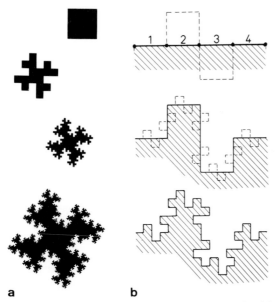

a   **b**

FIG. 4.29: (a) Quadric Koch island increases its boundary length without change of area. (Reproduced by permission from Mandelbrot, 1977.) (b) Standard polygon used to progressively modify Koch island of (a).

We cannot dwell on this fascinating problem much longer in the present context, although there is probably much in store to help stereology solve some of its paradoxical problems. Mandelbrot has proposed the characterization of such curves—and other forms as well—by what he calls its "fractal dimension" D_f. In the present case the operation consisted in dividing the side of the square etc. into four segments of length $r = \frac{1}{4}$ and then replacing it by a "standard polygon" of $N = 8$ segments of length r. The fractal dimension is obtained by

$$D_f = \log N / \log (1/r) \qquad (4.94)$$

which comes out at $D_f = 1.5$ for this case.

The essence of the structural process is that the coastline is modified at each step by the same "standard polygon"; Mandelbrot calls this type of curve "self-similar". If one therefore plots the length of the coastline logarithmically against the logarithm of r—which corresponds to the scale or to the resolution achieved—a straight line with slope D_f results. For the actual coast of Britain, where self-similarity seems to exist D_f is found empirically to be 1.25. When our data on the lung are plotted in the same way we find $D_f(\text{lung}) = 1.17$.

What then is the real surface density of the lung? Is it infinite? This does not make much sense. So we must ask: What is the *relevant detail of surface texture* that must be resolved? This question can evidently only be answered with the objective of a particular stereological analysis in mind. If we stay with the lung it is clear that fine texture details of the alveolar epithelial cells such as microvilli etc. play no role with respect to gas exchange because the epithelial surface is smoothed out by a fluid extracellular lining layer, resulting in a free or functional surface of alveolar air spaces which is smaller than the epithelial surface (Gil and Weibel, 1972; Weibel *et al.*, 1973; Gil *et al.*, 1978). On the other hand, when considering solute exchange between epithelial cells and the lining layer the actual membrane surface with all its detail texture becomes relevant. One hope we may have of solving this riddle is that a natural surface such as that of the lung may not be self-similar. Transitions from one pattern (standard polygon) to the next, i.e. patterns of "non-similarity", may be very important in trying to characterize such real surfaces. In fact, in the lung these "standard polygons" must be related to some structures, the size of erythrocytes and of microvilli for example, and it is likely that they will be different. As appears from Fig. 4.26 the curve in S_{V_a} seems to form a plateau in the resolution range $\{0.8 \rightarrow 0.08 \ \mu m\}$. This is possibly related to the fact that the resolution of detail of surface texture does not progress smoothly, i.e. not in a self-similar manner; if this is confirmed by further study, the plateaus thus formed may define certain levels of "relevant resolution", and thus facilitate the choice of appropriate magnification.

We should finally mention that this resolution effect is probably a widespread general phenomenon. It has now also been shown to affect estimates of mitochondrial and ER membrane surface area (Paumgartner et al., 1978); this may explain some of the discrepancies in stereological data found in the literature (Reith, 1976; Bolender et al., 1978). The choice of appropriate resolution must therefore receive particular attention in practical stereology.

4.4.   Calculation of N_V

In Section 2.5 we developed the stereological principles which permitted an estimation of the numerical density of discrete convex particles, N_V, from a count of their profiles per unit area of section, N_A (see also Volume 2, Chapter 5). The basic principle (Wicksell, 1925, 1926; De Hoff and Rhines, 1961) says that

$$N_V = N_A/\overline{D} \tag{4.95}$$

where $\overline{D}$ is the mean of the mean caliper diameter of the particles. Two independent stereological measurements must therefore be obtained: (a) a count of profiles, and (b) an estimation of particle size. We shall deal with the latter in the following sections. For the present we shall assume that $\overline{D}$ is known and discuss the basis for calculating N_V from N_A, and also discuss some alternative methods which avoid a precise knowledge of $\overline{D}$.

4.4.1.   CALCULATION OF N_V BY BASIC FORMULA

The counts of N_A are performed by means of the same test systems used for estimating the other stereological parameters. As was discussed in Section 4.2.3 the test area available for profile counting is given by the frame of the test system; in a coherent test system it is

$$A_T = P_T \cdot k_2 \cdot d^2 \tag{4.96}$$

where for a square lattice $k_2 = 1$. In analogy to the other counts the profiles can also be obtained with respect to a restricted containing space—or, in this case, area—which is then estimated by the number of test points, P_c, falling into this space, so that

$$A_c = P_c \cdot k_2 \cdot d^2 \tag{4.97}$$

Note that in counting profiles one has to take certain precautions with respect to those profiles which are intercepted by the frame of the test system, as has been outlined in Section 4.2 and in Fig. 4.12.

With the possibility of having a variable size of the test or reference area we have to consider some precautions about how to calculate

$$N_{Aa,c} = N_a/A_c \tag{4.98}$$

where N_a is the number of profiles of the particulate component a counted. From what has been extensively discussed in Section 4.3 it is evident that here too we are dealing with the estimate of a ratio; using equation (4.37) we find

$$N_{Aa,c} = \left(\frac{1}{k_2 d^2}\right)\frac{N_a}{P_c} \tag{4.99}$$

Since $\bar{D}$ is a property of the entire structure, independently derived, it can here be considered as a constant, so that the formula for numerical density of particles becomes

$$N_{Va,c} = \left(\frac{1}{k_2 \cdot d^2 \cdot \bar{D}}\right)\frac{N_a}{P_c} \tag{4.100}$$

It is hence of the same form as the general equation (4.47), with the only difference that the constant κ now includes a material property, $\bar{D}$. Nevertheless, N_V is basically estimated by the ratio of profile number per points on containing space. It is therefore evident that the ratio must be calculated by summing both profile counts and point counts over all n subsamples constituting one representative sample:

$$N_{Va,c} = \left(\frac{1}{k_2 \cdot d^2 \cdot \bar{D}}\right)\frac{\sum_{i=1}^{n} N_a(i)}{\sum_{i=1}^{n} P_c(i)} \tag{4.101}$$

4.4.2. Effect of finite section thickness

The basic formula for numerical density assumes that a true two-dimensional section plane intercepts the particles. We have shown that the probability that a particle is hit by a plane is determined by its mean caliper diameter (Section 2.1.3 and Volume 2, Chapter 2). If instead of a plane, a slice of thickness t cuts into the tissue, the probability of hitting a particle with the slice will be larger, in fact it will increase by t. This has been shown for spheres in Section 4.3.4.2; but it holds as well for any convex particle. The numerical density of profiles is then

$$N_A = N_V(\bar{D} + t) \tag{4.102}$$

as was shown by Abercrombie (1946), Hennig (1957), and Cahn and Nutting (1959). As was outlined for spheres, not all profiles may be detectable; if h_0 is the depth by which a particle has to penetrate into the slice before it is detected (see Fig. 2.34) the numerical profile density expected is, as shown by Floderus (1944) and Haug (1967a, c),

$$N_A = N_V(\overline{D} + t - 2h_0) \qquad (4.103)$$

The term in brackets defines the thickness of the "super-slice" within which we may observe profiles.

The terms t and h_0 are again constant properties, as long as the fields used to constitute a representative sample are taken from one section or at least from sections of equal thickness. Considering finite section thickness equation (4.101) becomes

$$N_{V_{a,c}} = \left[\frac{1}{k_2 \cdot d^2} \cdot \frac{1}{(\overline{D} + t - 2h_0)} \right] \cdot \frac{\sum_{i=1}^{n} N_a(i)}{\sum_{i=1}^{n} P_c(i)} \qquad (4.104)$$

If section thickness varies within one representative sample the second term within the bracket $1/(\overline{D} + t - 2h_0)$ has to be applied as multiplicating factor to each of the values of $P_c(i)$; this then introduces a weight for the section thickness effect in proportion to the test area sampled at each thickness.

We must finally mention again for the sake of completeness that the estimated numerical profile density may be an overestimate if the section is compressed. As was said in Section 4.3.5 the correction factor applying is

$$K_c(N_A) = \frac{l_c}{l_0} \qquad (4.105)$$

which must be introduced into the brackets of equation (4.104).

4.4.3.   CALCULATION OF N_V BY ALTERNATIVE METHODS

In Section 2.6.2 and Volume 2, Section 5.2, we have developed some alternative methods for N_V which attempt to avoid the determination of $\overline{D}$. This is achieved by combining other stereological parameters with a count of N_A, and by postulating a knowledge of particle shape.

Two of these formulas lend themselves particularly well to application. The method of Weibel and Gomez (1962) uses N_A together with V_V:

$$N_V = \left[\frac{K}{\beta} \right] \cdot \frac{N_A^{\frac{3}{2}}}{V_V^{\frac{1}{2}}} \qquad (4.106)$$

The shape coefficient β can be taken from Fig. 2.30 for ellipsoids and cylinders and from Table 2.2 for some other solids. As will be discussed in the next chapter the shape of real objects can often be approximated by some rather simple geometric solids. The size distribution coefficient K is merely related to the relative spread of the distribution measured by its coefficient of variation; its values can be read from Fig. 2.31. K is evidently the main point of uncertainty, because for its correct evaluation we would need to measure the size distribution of particles; but if we were to do this we could as well use the basic formula (equation 4.104). The use of the present formula represents a short-cut and we must be content with some approximations; so it may suffice to use some previously available information on size variation to estimate K. For example, Elias et al. (1961) measured the size distribution of glomeruli in the kidney and found a coefficient of variation of 10%, which gives $K = 1.014$. The diameter of hepatocyte nuclei was found to be $\bar{D} = 7.95\,\mu m$ with a standard deviation of $+1.12\,\mu m$ or 14% (Weibel et al., 1969), which yields $K = 1.027$. In an analysis of subcellular fractions Baudhuin (1968) estimated the size distribution of dense bodies (lysosomes) and peroxisomes; the coefficients of variation were 30% and 19% respectively, corresponding to values for K of 1.13 and 1.05. For most practical applications K will be in the range of 1.02 to 1.1; it may therefore often suffice to introduce an arbitrary value of K, but if small differences in the number of particles are to be interpreted one must remember that K will change if the size distribution changes. However, even quite drastic changes in the size distribution, e.g. doubling the standard deviation, will produce changes in K from, for example, 1.03 to 1.13, i.e. by 10% only (see Fig. 2.31).

The formula (equation 4.106) must again be used on data collected from a representative sample. What we need to count are the number of profiles N_a contained in the area $A_c = P_c \cdot k_2 \cdot d^2$ if we use a coherent test system, and the number of points falling onto the profiles, P_a. On this basis we can rewrite equation (4.106) as follows

$$N_{V_{a,c}} = \left[\frac{K}{\beta \cdot k^{\frac{3}{2}}_2 \cdot d^3} \right] \frac{[\Sigma N_a]^{\frac{3}{2}}}{[\Sigma P_c] \cdot [\Sigma P_a]^{\frac{1}{2}}} \tag{4.107}$$

where Σ means the sum of all counts over the representative sample.

De Hoff (1964) has proposed a system of alternative methods applicable to constant size particles. Of these the method developed in equation (2.81) is particularly attractive because it uses three different stereological estimates and avoids the use of powers, thus increasing the reliability. It says

$$N_V = \left[\frac{2\gamma_1 \cdot \gamma_2}{\gamma_3} \right] \frac{N_A \cdot I_L}{P_P} \tag{4.108}$$

The coefficients γ are shape factors some values of which are found in Table 2.2. If the counts are obtained by use of a coherent test system, where k_1 and k_2 are the coefficients linking test line length and frame area to the point spacing d, this becomes

$$N_{V_{a,c}} = \left[\frac{2\gamma_1\gamma_2}{\gamma_3 \cdot k_1 \cdot k_2 \cdot d^3} \right] \cdot \frac{[\Sigma N_a]}{[\Sigma P_a]} \frac{[\Sigma I_a]}{[\Sigma P_c]} \qquad (4.109)$$

where Σ means again summation of all counts over the representative sample.

In applying both these methods the necessary counts should be performed simultaneously to ensure that they relate to exactly the same sample. These two methods are convenient because they make use of stereological parameters that are easy to obtain and that form part of most protocols of morphometric analysis. If one determines V_V of a certain particulate component a simple count of the profiles allows an estimation of the particle number by equation (4.107). If intersections with the surface are also counted equation (4.109) may be used. De Hoff has also given a formula that requires only N_A and I_L (see Volume 2, Section 5.2.2).

CHAPTER 5

Particle size and shape

5.1. Approximation of particle shape

Very rarely will the shape of natural particles be accurately describable by one of the solids to be discussed in Volume 2, Section 5.5. Mostly we will have to find the shape which "fits best" by a lot of guess-work. In a practical sense it is therefore most important that we examine the range of errors in estimating the mean particle diameter if, in guessing, we miss the true structure.

Let us do this by estimating the error introduced by describing various solids by a sphere of equal volume; we shall call this sphere a "volume-equivalent sphere". This is a convenient procedure for various reasons: (a) the sphere has the smallest tangent diameter of all solids of equal volume, so that the error introduced in substituting a sphere for another solid will always be in the direction of underestimation; (b) various methods for practical estimation of D are based on the assumption that the particles can be adequately described by spheres. Wicksell (1926) has also arrived at the conclusion that this is in many ways the best procedure for comparing similar particles.

Assume a solid of known shape whose volume is v_s. Since the volume of a sphere is

$$v_0 = (4\pi/3) \cdot R^3 \tag{5.1}$$

and its diameter

$$D = 2R \tag{5.2}$$

we immediately find that the mean tangent diameter of the volume-equivalent sphere of the solid is:

$$\overline{D}_{\text{eq.sph.}} = 2 \cdot \left[\frac{3}{4\pi} \cdot v_s\right]^{\frac{1}{3}} = \left[\frac{6}{\pi} \cdot v_s\right]^{\frac{1}{3}} = 1.2407 \cdot v_s^{\frac{1}{3}} \qquad (5.3)$$

The formulas for calculating the mean tangent diameter of various solids will be discussed in Volume 2, Section 5.5.

Let us first look at some regular polyhedra (Fig. 5.1). If we take the simple case of a cube of side l we find its mean tangent diameter to be $1.5 \cdot l$, and that of its volume-equivalent sphere $1.2407 \cdot l$. By substituting a sphere of equal volume we would hence underestimate the mean tangent diameter of the cube by some 20%. Table 5.1 compares the true mean tangent diameter of various solids with that of the equivalent sphere; we observe that the error

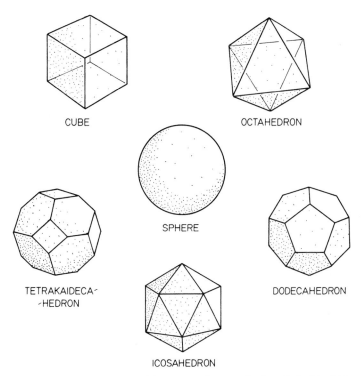

CUBE   OCTAHEDRON

SPHERE

TETRAKAIDECA-
-HEDRON   DODECAHEDRON

ICOSAHEDRON

Fig. 5.1: Regular polyhedra for which tangent diameter is given in Table 5.1 in relation to volume equivalent sphere.

introduced is of the order of 20% for cube and octahedron, whereas for the other polyhedra of higher order the error is only of the order of 8%. A remarkable phenomenon is the very close similarity of error for dodekahedron, ikosahedron and tetrakaidekahedron. This would indicate that $\overline{D}$ will be

Table 5.1

Comparison of mean tangent diameter $\bar{D}$ of various solids with $\bar{D}_{eq.sph}$ of sphere of equal volume (volume equivalent sphere); characteristic dimension of solid taken as unity

Solid	$\bar{D}_{solid}$	$\bar{D}_{eq.sph.}$	$\dfrac{\bar{D}_{solid}}{\bar{D}_{eq.sph.}}$
Cube	1.50	1.241	1.209
Octahedron	1.175	0.966	1.217
Dodecahedron	2.643	2.446	1.081
Icosahedron	1.742	1.609	1.083
Tetrakaidecahedron	3.0	2.785	1.077
Prolate ellipsoid ($a/b = 2$)	2.760	2.520	1.095
Oblate ellipsoid ($a/b = 2$)	3.418	3.175	1.076
Cylinder ($l/r = 2$)	2.571	2.289	1.123

very similar for any of these higher order polyhedra. This can be tested by calculating the mean tangent diameter for polyhedra of equal volume. Taking l as the edge length of the polyhedra we can write for its volume (see Volume 2, Section 5.5)

$$v = l^3 \cdot k_3 \qquad (5.4)$$

and for the mean tangent diameter

$$\bar{D} = l \cdot k_1 \qquad (5.5)$$

For polyhedra of volume $v = 1$ we hence find by equations (5.4) and (5.5)

$$\bar{D} = k_1/k_3^{\frac{1}{3}} \qquad (5.6)$$

The results of this calculation are shown in Table 5.2. We notice that the mean tangent diameters of these three polyhedra differ by no more than 0·5%! This is a result of great practical importance: *densely packed cells* will be

Table 5.2

Comparison of $\bar{D}$ for dodecahedron, icosahedron and tetrakaidecahedron of equal volume.

Solid	k_3	$k_3^{1/3}$	k_1	$\bar{D}$
Dodecahedron	7.6631	1.9715	2.6431	1.3406
Icosahedron	2.1817	1.2970	1.7421	1.3432
Tetrakeaidecahedron	11.3137	2.2449	3.0	1.3363

irregular polyhedra with a number of faces similar to the regular polyhedra here described (see also Hucher and Grolier, 1977). This result tells us that the actual shape of the polyhedron will have very little effect on $\overline{D}$, at least as long as the cells are not grossly distorted. In such cases $\overline{D}$ *will be about* 8% *larger than the diameter of a sphere of equal volume.*

The practical value of this result is as follows. If we count profiles of polyhedral particles on section, and if we have some idea of the volume of the particles, we may calculate their mean diameter D for a spherical model, in first approximation, and correct for the "excess" profiles generated by the corners of the polyhedron by multiplying with 1.08. The formula for estimating numerical density then is

$$N_V \approx N_A/(1.08 \cdot D) \qquad (5.7)$$

In many practical cases this will be as good as any other estimate.

Many particles can be conveniently described by ellipsoids, either prolate (egg-shaped) or oblate (pill-shaped). The prolate ellipsoid is a "drawn-out" sphere, whereas the oblate ellipsoid is a "compressed" sphere (Fig. 5.2). The

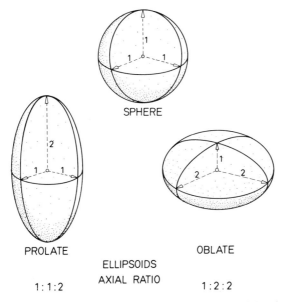

FIG. 5.2: Volume equivalent spheres and ellipsoids of axial ratio 1:2.

formulae for calculating $\overline{D}$ for ellipsoids of various axial ratios will be discussed in detail in Volume 2, Section 5.5. Here again it would be of interest to examine the error introduced if we were to substitute a volume-equivalent sphere. In Table 5.1 this is done for ellipsoids which have an axial ratio of

1:2; the error is of the order of 8 to 10%. This error evidently depends on the axial ratio of the ellipse from which the ellipsoid was generated by revolution. Figure 5.3 shows how this error increases as the ellipsoids become longer or flatter. Up to an axial ratio of 1.5:1 the error in substituting a sphere is no more than 3%. If the prolate ellipsoid is five times as long as it is thick the error will be 60%. It is noteworthy that the error increases less steeply for oblate ellipsoids.

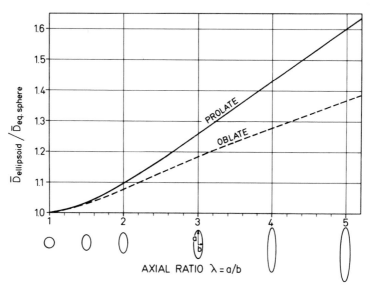

FIG. 5.3: Ratio of mean tangent diameters of ellipsoids and volume equivalent sphere.

The practical importance of this result is evident. In many cases—one example is the granules of leucocytes—a cell or tissue component produces elliptical profiles; we must hence conclude that the component is an ellipsoid. We can find out its approximate axial ratio by searching for the profile which is most elongated; if we find that it is, for example, 1.8 times as long as it is wide, we know that the axial ratio of the ellipsoid will not be greater than 1.8:1. We can then estimate from Fig. 5.3 the error introduced by substituting a spherical equivalent for estimating $\bar{D}$. To decide on whether we are dealing with an oblate or a prolate ellipsoid we search for the largest circular profile: if its diameter is of the size of the major diameter of the largest elliptic profile the particles are pill-shaped (oblate), if it is equal to the small diameter of the largest elliptic profile the particles are prolate. Also, in the case of prolate ellipsoids we expect the number of circular (or near-circular) profiles to be larger than those of elongated elliptic profiles, whereas with oblate ellipsoids circular profiles are comparatively rare.

5.2. Estimation of particle size from measurement of profile size distribution

If a population of (spherical) particles is cut by a random plane a population of profiles results whose largest diameter corresponds to the diameter of the largest particle, whereas the smallest profiles must always be vanishingly small, i.e. its diameter must approach zero. As we have shown in Section 2.7 and will pursue in more detail in Volume 2, Chapter 6, the size distribution pattern of the profiles is predictable if the size distribution of the particles is known.

The practical problem however is inverse: we can experimentally determine the profile size distribution and would then like to find information on the size distribution of the particles. One often encounters a great number of difficulties in this type of stereological analysis for a number of reasons: (1) the profile size distributions are incomplete because small profiles cannot be reliably measured, often not even recognized; (2) the profile frequencies are affected by errors due to section thickness; (3) the derivation of particle size often requires an assumption on the form of the particle size distributions; (4) a large number of measurements is required; (5) the analysis of the data is time-consuming and often involves elaborate computation procedures. It is therefore most important to evaluate whether particle size is *really* the most informative parameter, and whether the biological problem cannot be solved by using other morphometric parameters which may be obtained by one of the basic stereological procedures (see Section 4.1.3).

5.2.1. MEASUREMENT OF PROFILE SIZE

The size of a profile can be measured by three quantities: (a) by its area; (b) by its "diameter"; (c) by its intercept length distribution. The profile diameter is the most practical quantity, but it is also problematic, because it may be given different definitions as soon as the profile is not circular. The most common "diameters" of plane convex figures are defined in Fig. 5.4.

As we expressed the size of a particle by its caliper or tangent diameter (see Fig. 2.13) we may likewise determine *caliper diameters* of plane figures: they are defined as the distance of two parallel lines that are made tangent to the figure. Any such figure can be characterized by the *largest* (d_{max}) and *smallest* (d_{min}) *caliper diameter*, as shown in Fig. 5.4a; they are determined by rotating the "caliper" around the profile and noting its largest and smallest value. Maximal and minimal caliper diameters are in a way shape descriptors of the figure.

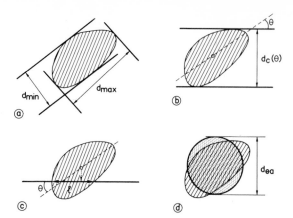

FIG. 5.4: Linear measures of profile size: (a) minimal and maximal caliper diameter; (b) Feret's or random caliper diameter; (c) random intercept or chord length; (d) diameter of area equivalent circle.

A profile measure of great importance is its *random* caliper diameter, d_c, also called "*Feret's diameter*" (Feret, 1931) (Fig. 5.4b): the caliper is kept in fixed position and the profile is allowed to assume any orientation θ. The mean caliper diameter (or mean Feret diameter) is the quantity which describes the probability that a random test line intercepts the profile; remember that analogously the probability of cutting a particle with a plane is proportional to its mean caliper diameter.

It has been shown (Tomkeieff, 1945; Walton, 1948) that the mean caliper diameter, $\bar{d}_c$, is equal to the diameter d_{eb} of a circle of equivalent circumference, b:

$$\bar{d}_c = d_{eb} = \frac{b}{\pi} \tag{5.8}$$

Another linear measurement of profile size is the *chord or intercept length l*, measured along a random test line (Fig. 5.4c). This quantity evidently depends (a) on the line's distance from the profile center, and (b) on the orientation of the profile with respect to the test line. At any fixed orientation θ the mean intercept length is directly related to the caliper diameter by

$$\bar{l}(\theta) = \frac{a}{d_c(\theta)} \tag{5.9}$$

where a is the profile area. In determining the overall average we must integrate over θ and note that the probability of measuring $\bar{l}(\theta)$ is proportional to $d_c(\theta)$. Consequently, the mean linear intercept length for a two-dimensional figure, l_2, is directly related to the mean caliper diameter by

$$\bar{l}_2 = a/\bar{d}_c \tag{5.10}$$

It is also related in a very general way (Crofton, 1885; Tomkeieff, 1945; Underwood, 1968) to area a and perimeter or boundary length b of the figure by

$$\bar{l}_2 = \pi \cdot \frac{a}{b} \tag{5.11}$$

This fundamental relationship can be easily derived by substituting equation (5.8) into equation (5.10).

A practical and very useful measure of profile size is the diameter of a circle of equal area called the *"area equivalent diameter"*, d_{ea} (Fig. 5.4d) which is

$$d_{ea} = 2 \cdot \sqrt{\left(\frac{a}{\pi}\right)} \tag{5.12}$$

Its usefulness derives from the fact that the human eye is extraordinarily adept in matching two areas of equal size, even if one is a circle and the other a complicated jagged figure (Fig. 5.5a). Miles and Davy (1977) comment on this approach of "fitting a control variate", that it is characterized by a large positive covariance, and may be a very efficient method for size estimation.

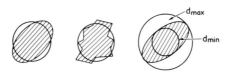

FIG. 5.5: Fitting of circles to obtain (a) diameter of area equivalent circle, and (b) maximal and minimal diameter.

Fischmeister (1968) has also discussed a number of other measures of profile size which are, however, of limited practical value.

Table 5.3 summarizes the principal size measures and their relations, as they apply for convex figures. In the special case of a circle mean caliper and area equivalent diameters are equal; but as the shape of the figure deviates from the circle we find $\bar{d}_c > d_{ea}$. For a hexagon the ratio $\bar{d}_c/d_{ea}$ is 1.05, for a square 1.13, and for an equilateral triangle 1.29. For ellipses of axial ratio $q > 1$ the ratio of these two diameters is

$$\frac{\bar{d}_c}{d_{ea}} \cong \sqrt{\left(\frac{1+q^2}{2q}\right)} \qquad (5.13)$$

which, for $q = 2$, for example, yields 1.118. Evidently, in each case the area equivalent diameter is smaller than the mean caliper diameter.

Table 5.3
Principal size measures of a profile of area a and perimeter b.

	Convex figures	Circle (r)
Mean caliper diameter:	$\bar{d}_c = \dfrac{b}{\pi} = \dfrac{a}{\bar{l}_2}$	$d_c = 2r$
Mean intercept (chord) length:	$\bar{l}_2 = \pi \cdot \dfrac{a}{b} = \dfrac{a}{\bar{d}_c}$	$\bar{l}_2 = \dfrac{\pi}{2}r$
Area equivalent diameter:	$d_{ea} = \sqrt{\left(\dfrac{4}{\pi} \cdot a\right)}$	$d_a = 2r$

The practical question is: which one of these measures should be used to size profiles? And the answer is that it depends on the use that is to be made of the measurement. One should note that only the area equivalent diameter d_{ea} can be obtained with only one reading for each profile: the integration over all orientations is performed automatically by our visual system! The estimation of *mean* caliper diameter for an individual profile requires multiple measurements in many directions, and so does the estimation of mean chord length. With respect to efficiency, it is hence evident that the *area equivalent diameter* is the parameter of choice if a large number of profiles need to be sized, particularly if automatic image analysis is not feasible. It is also justified to use this parameter if the subsequent evaluation of the data makes use of a spherical model for the derivation of particle size.

The practical procedure for estimating d_{ea} is very simple: one requires a set of test circles of increasing diameter. Convenient tools are transparent plastic stencils, as they are used for graphic work (Fig. 5.6a); note that the actual circle size needs to be calibrated as the diameter indicated often refers to that of the circle drawn by using a pen of a particular thickness! Alternatively test systems of concentric circles, or linear arrays of circles, as shown in Fig. 5.6b and c, can easily be made and reproduced photographically. In either case, about 15 to 25 circles will provide a convenient test system. Note that this type of test system can also be used to measure maximum and minimum diameter of the profile by determining the size of the smallest circumscribed and largest inscribed circle diameter, respectively (Fig. 5.5b).

In general, one will choose a linear sequence of circle diameters, each circle being larger than the preceding one by Δd (Fig. 5.6a–c); this then produces directly an estimate of the frequency distribution of profile size in

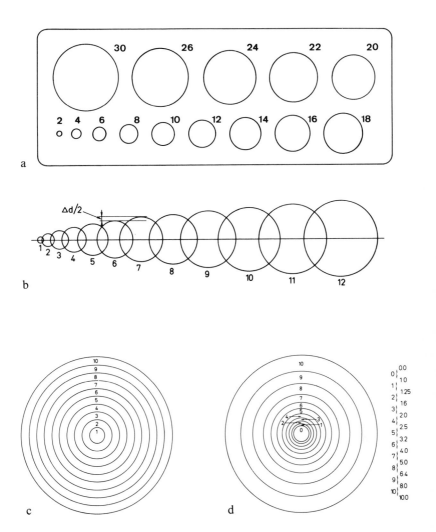

FIG. 5.6: Test circles for measuring profile size. (a) graphic stencil; (b) linear array of circles; (c) concentric circles on linear scale; (d) concentric circles on logarithmic scale.

classes of size Δd, as it will be used in constructing histograms. Alternatively, however, a logarithmic scale may prove useful in cases where there is a very large span in profile size and where small profiles are predominant; this then produces a frequency distribution on a logarithmic scale.

An excellent opto-mechanical device may be of great help in this type of analysis, that is the particle size analyser of Zeiss TGZ 3 in which a circular light disc is fitted to the profile by means of a hand-wheel and the area equivalent diameter is automatically read off and recorded in a counter (Endter and Gebauer, 1956), This instrument offers both linear and logarithmic scales, and can be connected to computing systems (Gahm, 1971).

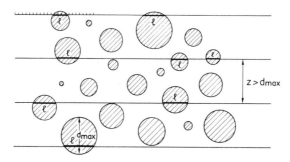

FIG. 5.7: Test lines for measuring intercept lengths l; their spacing z should be larger than the maximal caliper diameter of the largest profile to avoid multiple sampling of profiles.

For certain applications it may be required to measure the random linear intercept length of the profiles. For this purpose a set of (parallel) test lines is deposited on the micrograph, and the length of the chords within the profiles is measured (Fig. 5.7). The lines should be spaced at adequate intervals so that no more than one intercept is formed with the same profile. The reason is evident: if $z < d_{max}$ then all profiles with $d > z$ will form multiple intercepts which are then not independent as required by the theory. A classified count of the frequency distribution of intercept lengths can be easily obtained if the lines are replaced by "rows of points" conveniently spaced so that the largest chord is of the order of 15 to 25 points.

Automatic image analysers as well as some of the computer-assisted tracing devices (see Section 7.4) allow the estimation of chord lengths, caliper (Feret) diameter, largest and smallest diameter, shape factors, as well as of profile area and boundary. This evidently makes possible a more sophisticated approach to profile sizing.

A last point that requires attention relates to sampling of profiles for measurement. In this respect we must remember that the frequency distributions of profiles discussed in Section 2.7 (see also Volume 2, Chapter 6) were expressed in terms of N_A. It is therefore evident that the profile sample to be measured consists of all profiles contained in the test area of the test system. The same rules for accepting and rejecting profiles that intersect the frame of the test area apply as discussed above (see Fig. 4.12).

5.2.2. GRAPHIC REPRESENTATION OF DATA

The sizing of profiles with a set of graded test circles automatically results in a frequency distribution by profile size classes, which can be plotted in the form of a histogram (Fig. 5.8). The class width is Δd, the interval between test circle diameters. The method described above results in an estimate in the range

$$\{d_i \pm \tfrac{1}{2}\Delta d\}$$

i.e. the diameter of the test circle marks the center of the class (Fig. 5.8a). Note that in this case the smallest class extends between 0 and $\tfrac{1}{2}\Delta d$ with class mid-point $\tfrac{1}{4}\Delta d$. This is the method of plotting required for applying the transformation of Wicksell (1925) (see Section 5.2.9) or the method of Giger and Riedwyl (1970) (see Section 5.2.6).

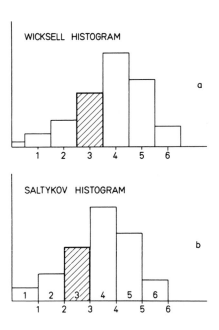

FIG. 5.8: Plotting histograms according to (a) Wicksell (1925) and (b) Saltykov (1958).

The transformation method of Saltykov (1958) (see Underwood, 1968, and Section 5.2.9) requires the data to be presented differently: the diameter characteristic for each class is the diameter corresponding to the *upper* class limit (Fig. 5.8b). In practice this is obtained by slightly modifying the measuring procedure: one determines which *two* test circles of the graded set enclose the "true" area equivalent circle, and then records the size of the *larger* of the two. In a plot of these data the smallest class extends from 0 to Δd, the next from Δd to $2\Delta d$, and so on (Fig. 5.8b).

5.2.3. COMPLETION OF HISTOGRAMS

All histograms of profile size distributions are deficient in the lower part of the plot because of the loss of small profiles (Fig. 5.9). This defect must be corrected by fitting in the missing profiles, for we know that the histogram must extend to zero. This completion must be performed by guess-work, but often one can check secondarily whether the completion has been adequate.

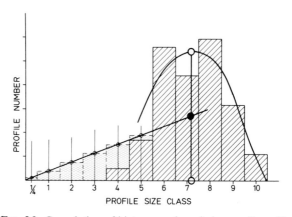

FIG. 5.9: Completion of histograms for missing small profiles.

The rule of thumb for this histogram completion is as follows (Fig. 5.9): (1) smooth your histogram by hand, eliminating excessive peaks, and class-to-class fluctuations (which are usually due to errors); (2) determine the mode of the distribution (peak of the smoothed curve) and mark a point at half the height of this peak; (3) extrapolate linearly from this point towards 0; (4) draw histogram bars at the intersections of the extrapolated line with the class mid-points. For Saltykov type histograms one proceeds similarly, using the intersection of the extrapolated line with the upper class limit to estimate the corrected frequency of small profiles.

All subsequent calculations are done with the data derived from the completed histograms; accordingly the original table of data must also be corrected.

Note that this completion is always necessary, because there is always an unknown number of small profiles missing. For that reason one may very well not make any special effort to try to measure even very small profiles. In fact, one can predict that the lower quarter to third of the histogram is almost certainly in error; one is however not helped in the correction procedure by whatever measurements one has obtained in these size classes, so one can just as well not record them to begin with.

These considerations have an important repercussion on number estimations. Because many of the small profiles are not recognized, N_A obtained by direct counting is too low and accordingly N_V estimates calculated by any of the methods presented above are going to be underestimates, unless the profile number has been corrected by completing a size distribution histogram! This effect is at least partly compensated in the methods of Weibel and Gomez (1962) and De Hoff (1964) because they use a ratio of two density measurements which are both affected by a similar error of this kind (see Sections 2.5.2 and 4.4.3).

5.2.4. EFFECT OF SECTION THICKNESS ON PROFILE SIZE DISTRIBUTION AND ITS CORRECTION

In Volume 2, Chapter 6.4 we will show that finite section thickness t increases the number or profiles in the largest profile size class derived from any particular particle. Referring to Fig. 2.40 we can easily see that the height of the largest profile size class will be increased in proportion to $t/(D + t)$, as shown in Fig. 5.10. If several particle size classes are present, then this effect will result in an excessive profile number in all size classes that correspond to the particle size distribution (Fig. 5.11). Whereas most methods for extracting from the profile size distribution information on particles require the particles to be cut by true section planes, most real profile size distributions will hence be "contaminated" by this section thickness effect in the larger size classes.

The question we now have to ask is whether one can correct the actually observed profile size distribution (completed for missing small profiles) for the overestimation of large profiles due to section thickness effect. This is rigorously only possible in the framework of an analytical solution, such as in connection with the methods of Bach (1959, 1967) or of Wicksell (1925), as described below. Here we offer an approximate correction which is based on the following reasoning.

G

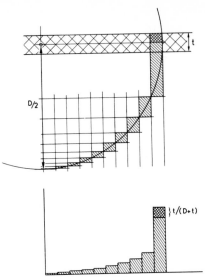

FIG. 5.10: Demonstration that the frequency of profiles in largest class alone is increased in proportion to t/D (compare with Fig. 2.40).

Let us designate by $F(D_j)$ the particle size distribution, by $f(d_i)$ the "true" profile size frequency, i.e. that which we should observe if the spheres were cut by a true section, and by $f'(d_i)$ the one we actually observe (Fig. 5.11). With this symbolism we can write (see Volume 2, Section 6.4)

$$f'(d_i) = f(d_i) + (t/D_j) \cdot F(D_j) \tag{5.15}$$

This is shown in Fig. 5.11. Solving equation (5.15) for the "true" frequency we obtain

$$f(d_i) = f'(d_i) - (t/D_j) \cdot F(D_j) \tag{5.16}$$

To correct the profile frequency we must hence subtract a value $(t/D_j) \cdot F(D_j)$.

The problem is that, evidently, we do not know $F(D_j)$, but we can approximate it. If $t \ll \overline{D}$ we could say that for all values of $D_j \geqslant \overline{D}$ we have $f(d_i) \gg (t/D_i) \cdot F(D_j)$ and we accept $f'(d_i) \approx f(d_i)$. In this case we find a first approximation of $F(D_j)$ by

$$F_1(D_j) \approx \left(1 - (t/D_j)\right) f'(d_i) \qquad \{\overline{D} \geqslant D_j \geqslant D_{max}\} \tag{5.17}$$

But this approximation holds only for large D, i.e. the range extending from $\overline{D}$ to D_{max}. If we can assume $F(D_j)$, the sphere size distribution, to be symmetrical about the mean, then $F_1(D_j)$ can be reflected symmetrically to the left of the mean; for this one should note that to some diameter class D_r on the right corresponds a diameter class D_l on the left (Fig. 5.11), such that

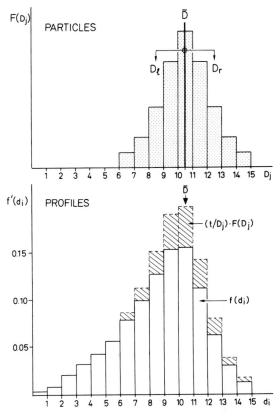

FIG. 5.11: Model distribution of spheres and profile size distribution generated by true section (open histogram) with addition due to section thickness $t = 2$ shown as hatched bars.

$$D_t = 2\overline{D} - D_r \qquad (5.18)$$

Symmetrical reflection of $F_1(D_r)$ implies that we accept

$$F_1(D_l) = F_1(D_r) \qquad (5.19)$$

or

$$F_1(D_l) \approx \left(1 - (t/D_r)\right)f'(d_r) \qquad (5.20)$$

It can be shown that this approximation works reasonably well at least for model distributions which are symmetrical.

The correction procedure now consists in substituting these approximations to $F(D_j)$ into equation (5.16). The true profile size distribution is then found by

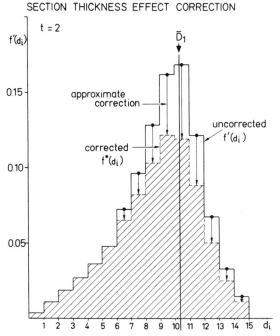

SECTION THICKNESS EFFECT CORRECTION

FIG. 5.12: Normalized profile size distribution from thick section obtained from Fig. 5.11; arrows indicate correction for section thickness derived by approximate procedure, resulting in corrected profile distribution (hatched histogram).

$$f(d_i) \approx f'(d_i) - \frac{t}{D_j}\left[F_1(D_j) \right]_r \qquad (5.21)$$

where the subscript r indicates that F_1 is estimated on the right hand side of the histogram. For the classes to the right of $\overline{D}$, including the one containing $\overline{D}$, i.e. in the range $\{\overline{D} \leqslant D_j \leqslant D_{max}\}$, this is

$$f(d_i) \approx f'(d_i) - \frac{t}{D_j}\left(1 - \frac{t}{D_j} \right)f'(d_i) \qquad (5.22)$$

whereas to the left of $\overline{D}$ it is

$$f(d_i) \approx f'(d_l) - \frac{t}{D_l}\left(1 - \frac{t}{D_r} \right)f'(d_r) \qquad (5.23)$$

This correction was performed on the histogram of Fig. 5.12 which is the same as that constructed in Fig. 5.11; the arrows indicate the estimated contribution of equatorial profiles seen because of finite section thickness. In Fig. 5.13 the resulting curve of $f^*(d_i)$ is compared to the original or correct profile

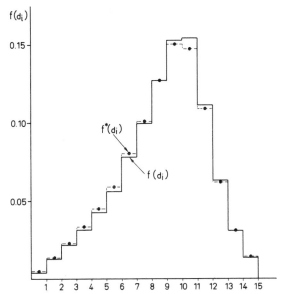

FIG. 5.13: Comparison of original $(f(d_i))$ and derived $(f^*(d_i))$ profile size distributions.

size distribution $f(d_i)$, if the particles were cut by a plane; the agreement is rather good.

The *practical procedure* is as follows (Fig. 5.12 and Table 5.4):

(1) Plot the histogram of observed profile numbers and correct for missing small profiles (Fig. 5.9). Calculate the frequencies $f'(d_i)$ by

$$f'(d_i) = N_A(d_i)/\sum_i N_A(d_i) \qquad (5.24)$$

(2) Estimate section thickness t (in the same units as the profile size d!).

(3) Calculate the mean profile diameter $\bar{d}$, from the corrected distribution. A first estimate of the mean particle diameter is then obtained by (see Section 5.2.5)

$$\bar{D}_1 \approx \frac{4}{\pi} \bar{d} - t/2 \qquad (5.25)$$

Half the section thickness is subtracted to roughly correct for the over-estimation of $\bar{d}$ due to section thickness.

(4) Plot $\bar{D}_1$ into the histogram (Fig. 5.12).

(5) Calculate from equation (5.17) the profile frequencies $F_1(D_r)$ for all classes $j = r$ to the right of the mean, and note these frequencies symmetrically for the classes to the left of the mean (Table 5.4). Note that $F_1(D_r)$

must be normalized, i.e. the values first obtained by equations (5.17) and (5.20) must be made to add up to 1 by dividing them by their sum.

(6) Calculate $f(d_i)$ by equations (5.22) and (5.23); this results in a set of relative frequencies which are shown in Fig. 5.12 by the dashed lines. Note that $f'(d_i) = f(d_i)$ for all profile classes $i < 7$, i.e. those that did not contain spheres (Fig. 5.11).

(7) The values of $f(d_i)$ thus obtained must again be normalized by dividing them by their sum which is < 1. These normalized frequencies $f^*(d_i)$ characterize the "true" profile size distribution corrected for section thickness effect.

Table 5.4

Numerical example for calculating section thickness correction on profile size distribution according to Section 5.2.4.

i	d_i (D_j)	Step (1) $f'(d_i)$		Step (5) $F_1(D_i)$	Step (6) $f(d_i)$	Step (7) $f^*(d_i)$
1	0.5	0.0036		—	(0.0036)	0.0045
2	1.5	0.0109		—	(0.0109)	0.0135
3	2.5	0.0185		—	(0.0185)	0.0229
4	3.5	0.0268		—	(0.0268)	0.0332
5	4.5	0.0362		—	(0.0362)	0.0449
6	5.5	0.0475		—	(0.0475)	0.0589
7	6.5	0.0722		(0.0122)	0.0651	0.0807
8	7.5	0.0957		(0.0277)	0.0817	0.1012
9	8.5	0.1277		(0.0564)	0.1026	0.1271
10	9.5	0.1613		(0.0999)	0.1251	0.1505
11	10.5	0.1680	$\leftarrow \bar{D}_1$	0.1360	0.1190	0.1474
12	11.5	0.1209		0.0999	0.0880	0.1090
13	12.5	0.0671		0.0564	0.0500	0.0620
14	13.5	0.0325		0.0277	0·0247	0.0306
15	14.5	0.0142		0.0122	0.0110	0.0136

Step (2): $t = 2$ $\uparrow$ Step (4)
Step (3): $\bar{d} = 9.007$
        $\bar{D}_1 = 10.47$

5.2.5. APPROXIMATE ESTIMATION OF MEAN DIAMETER OF SPHERICAL PARTICLES

If we have constructed a histogram of profile diameter distribution corrected for the loss of small profiles, and possibly also for section thickness effects, we can immediately calculate the mean profile diameter by

$$\bar{d} = \frac{\sum_{i=1}^{k} n_i \cdot d_i}{\sum_{i=1}^{k} n_i} = \sum_{i=1}^{k} f(i) \cdot d_i \qquad (5.26)$$

here d_i is the mid-point of the size class i and n_i the (corrected) profile umber in that class. One can evidently substitute frequencies $f(i)$ for n_i if 1ese have been calculated. It should be noted that for this purpose it does ot matter too much by which method the data have been classified (see ection 5.2.2, Fig. 5.8): with the first method after Wicksell (1925) the test ircle size corresponding to each class is directly equal to d_i, but then the first ze class has diameter $\Delta d/4$; with the second method after Saltykov (1958) 1e class midpoint has value

$$d_i' = d_i - \Delta d/2 \qquad (5.27)$$

ecause the recorded value of d_i corresponded to the upper class limit.

We should now recall that in Section 2.7.1 we found a simple relationship etween the sphere diameter D and the mean profile diameter $\bar{d}$ (see Fig. 2.39), hereby $\bar{d}$ was smaller than D by a factor of $\pi/4$. This applies strictly for a ngle size class of spheres, but it can also be used to derive, in first approxiation, the mean diameter for a population of spheres of varying diameter:

$$\overline{D_1} \approx \frac{4}{\pi} \cdot \bar{d} \qquad (5.28)$$

How good is this estimate of mean particle size? This evidently depends on he spread of the sphere size distribution, measured, for instance, by its tandard deviation S.D. The larger S.D. the greater the error will be, as we hall see in the next section; but for S.D. $\leqslant 20\%$ of D the error will be $< 5\%$, o that for many purposes this quick estimate of $\overline{D}$ is quite acceptable. The rror, by the way, is in the direction of an overestimation of $\overline{D}$, the reason eing that larger spheres contribute a larger number of profiles than smaller ones (remember that the probability of cutting a sphere is proportional to ts diameter). A word of caution should however be added: this procedure hould not be applied to profile size distributions uncorrected for the loss of small profiles because this would tend to exaggerate this overestimation of $\overline{D}$. Ve mentioned in Section 5.2.4 that section thickness also leads to an overstimation of $\bar{d}$ and this would hence contribute further to the tendency of verestimation of $\overline{D}$.

.2.6. THE METHOD OF GIGER AND RIEDWYL FOR ESTIMATING MEAN AND STANDARD DEVIATION OF PARTICLE DIAMETER

3iger and Riedwyl (1970) have developed a graphical method by which the correct mean particle diameter as well as the S.D. of the particle size distribution can be derived. This method assumes, however, that the particle size distribution is normal.

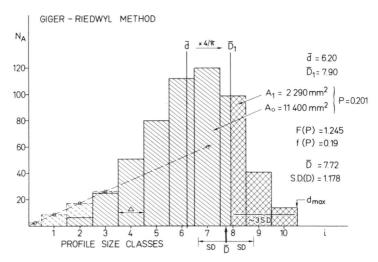

FIG. 5.14: Histogram of profile size distribution showing Giger–Riedwyl method.

The method starts with a histogram of profile size distribution obtained and plotted in the manner shown in Fig. 5.8a, i.e. according to the Wicksell type of classification. In Fig. 5.14 this original histogram of measured profile sizes is shown by the solid shaded bars. The principle is to complete and correct the histogram (according to Sections 5.2.3 and 4), and then to calculate $\bar{D}_1$ as indicated in the preceding section. It is evident from what was said in Section 2.7 that all profiles that are $> \bar{D}_1$ must have been derived from spheres whose diameter is $> \bar{D}_1$. Accordingly it is plausible that the relative number of profiles with $d > \bar{D}_1$ can yield an estimate of the relative number of spheres with $D > \bar{D}_1$, i.e. the relative number of spheres whose diameter deviates from the mean. That is however related to the variance or S.D. of the particle size distribution.

This type of reasoning is at the basis of the method of Giger and Riedwyl (1970). Without justifying this method mathematically (the interested reader is referred to the original paper) we shall present the instructions for the practical application of this very useful method.

5.2.6.1.   *Instructions for practical procedure*
1. *Determine* on independent random sections the *diameters d* of a large number *of profiles.* For the sections to be "independent" they must not allow

any particle to be sectioned more than once; this is evidently the case if only one very large section from the same block is used, but if several sections must be used they must be spaced by more than the diameter of the largest particle. Since the profile diameters d must be classified into no less than 10 and no more than 16 classes it will be convenient to do this measurement by means of a sequence of 10–16 graded circles (Fig. 5.6), whereby the largest circle should correspond to the largest profile observed. In practice one will first try to determine in a preliminary study the diameter of the largest profile and then choose a set of circles in which numbers 12 to 13 correspond to this estimated maximal profile size; if larger profiles should then occur the test system will still be able to account for them. Note that the data will have to be plotted on a Wicksell-type histogram; the profile size is thus recorded by the number of the circle which best fits to the profile and thus marks the class mid-point.

2. Plot a histogram according to the Wicksell-type (Fig. 5.8a). The circle size d_i marks the mid-point of class i which extends from

$$(d_i - \Delta d/2) < d_i < (d_i + \Delta d/2)$$

as shown in Fig. 5.14 by the solid outline of the histogram bars. Note that the lowest class should extend from 0 to $(\Delta d/2)$.

3. Very small profiles will usually be missing from the distributions because they are not recognized. The histogram must therefore be completed as discussed above (Fig. 5.9). The result of this procedure is shown in Fig. 5.14 by histogram bars with broken outlines. There is evidently some uncertainty attached to this procedure, but the possible error thus introduced can be tested secondarily (point 8 of this procedure).

4. If the section thickness is relatively large, i.e. if $(t/D) > 0.1$, it may be advisable to also correct for the section thickness effect (Section 5.2.4).

5. Calculate the mean profile diameter $\bar{d}$ from the *corrected* histogram and derive the first estimate of mean particle diameter by

$$\bar{D}_1 = \frac{4}{\pi}\bar{d} \qquad (5.29)$$

Draw a vertical line over the histogram at $\bar{D}_1$ (Fig. 5.14).

6. We had indicated that all profiles with $d > \bar{D}_1$ are derived from spheres of diameter $D > \bar{D}_1$. A very rough estimate of the standard deviation S.D. of the particle size distribution can now be obtained if one considers that the range 3 S.D. encompasses 99.73% of all particles. Since the largest profile diameter recorded, d_{max}, should have been derived from the largest particle

a first rough estimate of S.D. should be

$$\text{S.D.}_1 \approx (d_{max} - \overline{D_1})/3 \tag{5.30}$$

7. For more precision in estimating $\overline{D}$ as well as S.D. one has to consider the relative number of profiles with $d > \overline{D_1}$. To do this we measure by planimetry the area of the entire completed histogram, A_0, and the area of the histogram to the right of $\overline{D_1}$, A_1; the latter area is crosshatched in Fig. 5.14. From this a factor

$$P = A_1/A_0 \tag{5.31}$$

is derived. The graph of Fig. 5.15 now allows us to read off two coefficients, $F(P)$ and $f(P)$, which serve to determine the correct mean particle diameter, $\overline{D}$, and the standard deviation of the particle size distribution, S.D.(D), respectively by

$$\overline{D} = \overline{d} \cdot F(P) \tag{5.32}$$

$$\text{S.D.}(D) = \overline{d} \cdot f(P) \tag{5.33}$$

It should be noted that $\overline{D} \approx \overline{D_1}$ if $P < 0.1$ because in this range $F(P) \approx 4/\pi$.

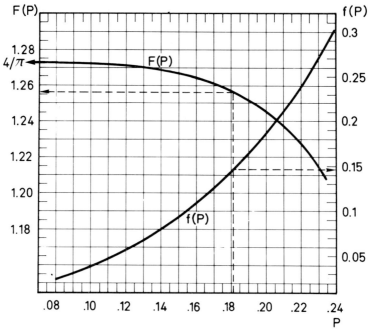

FIG. 5.15: Nomogram for deriving factors $F(P)$ and $f(P)$ in Giger-Riedwyl method. Reproduced by permission from Giger and Riedwyl (1970).

8. One can now test whether the completion of the histogram for the missing small profiles has been reasonable. If N is the total number of profiles in the completed histogram the numbers in the small classes should be

$$N_i^* \approx \frac{i \cdot (\Delta d)^2}{\overline{D}^2} \cdot N \qquad (i = \tfrac{1}{2}, 1, 2 \ldots) \qquad (5.34)$$

This test should be done for the two or three largest classes where corrections were done (e.g. classes 2 and 3 in the histogram of Fig. 5.15). If the empirically corrected profile numbers deviate appreciably, i.e. by more than 10–20% from the "true" number thus obtained, then the correction for missing profiles should be corrected accordingly, and the calculations repeated, starting at step 4.

5.2.6.2.   *Evaluation of this method*
The method of Giger and Riedwyl (1970) is very attractive because of its simplicity. Also, one very often only requires an estimate of $\overline{D}$, and in this case it appears unduly laborious to go through the elaborate reconstruction procedures according to Wicksell (1925) or Saltykov (1958) which also require larger numbers of profile measurements. The authors also claim that this graphical procedure avoids some of the anomalies one sometimes encounters with the reconstruction methods. It has also been shown that estimates of $\overline{D}$ and S.D. obtained by this method are very well comparable to those obtained by a Wicksell procedure (Bolender, 1974), and that they rather faithfully estimate the true population parameters (Table 5.7).

On the other hand the procedure suffers from some short-comings, particularly the requirement of a normal particle size distribution, and the inability to correct for section thickness effects, except of course by the approximate method proposed above. With these limitations in mind this simple method can be very useful in practical stereology.

5.2.7.   OTHER GRAPHICAL METHODS

The general approach to converting a profile size distribution into a particle size distribution consists in assuming some distribution function for the particle size distribution; by the process to be discussed in Volume 2, Section 6.2 one can then derive the corresponding theoretical profile size distribution and compare it to the empirically determined one. This comparison is mostly done by means of some parameters of the distribution.

Hennig and Elias (1970) have proposed an entirely visual approach to this problem. They present the user with a set of paired distribution curves for particles and profiles (Fig. 5.16); all he has to do is to find the profile curve that best fits his experimental distribution and he has also found the curve that describes his particle size distribution. But note that the profile size distribution considered must have been corrected for missing profiles and section thickness effects.

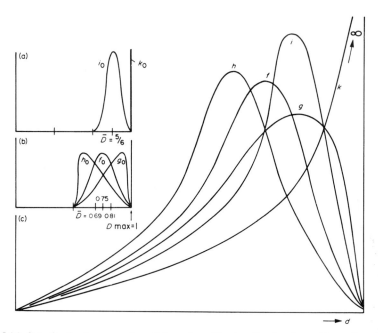

FIG. 5.16: Set of paired curves of particle and profile size distributions. (Reproduced by permission from Hennig and Elias, 1970.)

The method is indeed attractive because of its simplicity. There are a number of practical problems, though. One of them is that scaling of the empirical curve must be done with great care because this will evidently influence the visual judgement of matching curves. Another one is that this method yields a distribution *pattern* but does not directly lead to an estimate of distribution parameters which are usually more important than the pattern.

5.2.8. ESTIMATION OF MOMENTS OF THE PARTICLE SIZE DISTRIBUTION FROM THOSE OF THE PROFILE SIZE DISTRIBUTION

Before discussing the practical methods for reconstructing a particle size distribution we should point out that very often the result we seek is really some parameter related to a moment of that distribution, whereas the full particle size distribution is of secondary importance. The k-th moment of a particle size distribution can be written as

$$M_k = \left[\sum_{j=1}^{m} R_j^k \cdot N_V(j)\right]\Bigg/\left[\sum_{j=1}^{m} N_V(j)\right] = \sum_{j=1}^{m} R_j^k \cdot f(j) \qquad (5.35)$$

The k-th moment is thus evidently the expected or mean value of the size parameter raised to the power k: the 0-th moment $M_0 = 1$, the first moment, M_1, is the mean particle radius by

$$\bar{R} = M_1 = [\Sigma R_j \cdot N_V(j)]/[\Sigma N_V(j)] \qquad (5.36)$$

whereas the mean surface and volume are

$$\bar{s} = 4\pi \cdot M_2 = [4\pi\Sigma R_j^2 \cdot N_V(j)]/[\Sigma N_V(j)] \qquad (5.37)$$

$$\bar{v} = \tfrac{4}{3}\pi \cdot M_3 = [\tfrac{4}{3}\pi\Sigma R_j^3 \cdot N_V(j)]/[\Sigma N_V(j)] \qquad (5.38)$$

There is hence nothing magic about this often unfamiliar but very useful term.

In Volume 2, Section 6.5 we will show that a simple relation exists, due to Bach (1967), between the moments of the particle size distribution, M_k, and those of the profile size distribution, m_k. The final formulas there derived are, for profile size distributions obtained on sections of thickness t:

numerical density:

$$N_V = N_A/(2\bar{R} + t) \qquad (5.39)$$

or

$$N_A = N_V \cdot (2\bar{R} + t) \qquad (5.40)$$

mean radius:

$$\bar{R} = \tfrac{1}{2}\left(\frac{N_A}{N_V} - t\right) \qquad (5.41)$$

standard deviation of R:

$$\text{S.D.}(R) = \left[\frac{2}{\pi}\left(\frac{N_A}{N_V}\cdot\bar{r} - t\cdot\bar{R}\right) - (\bar{R})^2\right]^{\frac{1}{2}} \qquad (5.42)$$

mean particle surface;

$$\bar{s} = \frac{2}{\pi}\left(\frac{N_A}{N_V}\cdot\bar{r} - t\cdot\bar{R}\right) \qquad (5.43)$$

mean particle volume:

$$\bar{v} = 2\cdot t^2 \cdot \bar{R} + (2\bar{R} + t)\,[\pi(\sigma^2(r) + \bar{r}^2) - 2\cdot t\cdot\bar{r}] \qquad (5.44)$$

where $\sigma^2(r)$ is the variance of the profile radius distribution.

Practical procedure: To derive any of these parameters from the profile size distribution we need to calculate the following:

(1) the numerical density of profiles:

$$N_A = \sum_{i=1}^{m} N_A(i) \qquad (5.45)$$

(2) the numerical density of particles; it is obtained, according to Bach (1967), by a summation of the numerical densities of profiles in all size classes weighted by a factor $g(r_i/t)$ which can be read from Table 5.5

$$N_V = \frac{1}{t} \sqrt{\left(\frac{2}{\pi}\right)} \cdot \sum_{i=1}^{m} N_A(i) \cdot g(r_i/t) \qquad (5.46)$$

this then allows us to calculate $\overline{R}$ by equation (5.41).

(3) the mean profile radius, evidently by

$$\bar{r} = \frac{\sum_{i=1}^{m} r_i \cdot N_A(i)}{\sum_{i=1}^{m} N_A(i)} \qquad (5.47)$$

Table 5.5
Coefficients $g(r_i/t)$ for calculating N_V by equation (5.45) according to method of Bach (1967).

$r_i < t$		$r_i \geqslant t$			
r_i/t	$g(r_i/t)$	r_i/t	$g(r_i/t)$	r_i/t	$g(r_i/t)$
0.00	1.2533	1.0	0.3535	3.0	0.1307
0.05	1.1372	1.1	0.3270	3.2	0.12228
0.10	1.0373	1.2	0.3039	3.4	0.1158
0.15	0.9509	1.3	0.2837	3.6	0.1095
0.20	0.8756	1.4	0.2659	3.8	0.1039
0.25	0.8097	1.5	0.2502	4.0	0.0988
0.30	0.7517	1.6	0.2361	5	0.0793
0.35	0.7004	1.7	0.2234	6	0.0662
0.40	0.6548	1.8	0.2120	7	0.0568
0.45	0.6140	1.9	0.2017	8	0.0497
0.50	0.5775	2.0	0.1923	9	0.0442
0.5	0.5446	2.1	0.1838	10	0.0398
0.60	0.5149	2.2	0.1759	11	0.0362
0.65	0.4880	2.3	0.1686	12	0.0332
0.70	0.4634	2.4	0.1620		
0.75	0.4410	2.5	0.1558	for larger values:	
0.80	0.4205	2.6	0.1501		
0.85	0.4017	2.7	0.1447		
0.90	0.3844	2.8	0.1397		
0.95	0.3683	2.9	0.1351	$g(r_i/t) \approx t/(r_i \cdot \sqrt{2\pi})$	

(4) the variance of the profile radius distribution, $s^2(r)$, by

$$s^2(r) = \frac{\sum_{i=1}^{m} r_i^2 \cdot N_A(i)}{\sum_{i=1}^{m} N_A(i)} - \bar{r}^2 \qquad (5.48)$$

One must evidently also have an estimate of section thickness t (see Section 4.3.4.1.). It must be noted that these calculations must be done with profile size distributions that have been completed for missing profiles (Section 5.2.4), but that no correction for section thickness effect is required.

This approach allows an estimate of the mean and standard deviation of the particle diameters, as well as of their mean surface and volume. For many applications this will be entirely satisfactory. By going back to the original relationship between moments (equation 5.35) one may evidently also derive other parameters.

5.2.9. RECONSTRUCTION OF THE FREQUENCY DISTRIBUTION OF PARTICLE SIZE FROM THAT OF PROFILES

5.2.9.1. *Introduction to the procedure*
The procedures for reconstructing a particle size distribution by "unfolding" a measured profile size distribution are rather involved and difficult to fully understand. This chapter, which eventually presents in detail the practical steps to be taken, should therefore be introduced by a brief account of the main results of the detailed description of the methods to be given in Volume 2, Section 6.3, combined with a simple illustration taken from Underwood (1968).

We learn to understand the procedure if we first trace the steps backwards, i.e. if we remember (see Fig. 2.42) (1) that profiles of a certain size class i may be derived from all spheres of size classes $j \geqslant i$, and (2) that a given sphere of size class j contributes to all profile size classes $i \leqslant j$ in proportion to the probability of obtaining this profile size, which decreases as the profile becomes smaller. Referring to the Saltykov procedure (Section 2.8.2) we found the following number of profiles of size class i derived from particles of size class j (equation 2.120):

$$N_A(i,j) = N_V(j) \cdot \Delta[\sqrt{(j^2 - (i-1)^2)} - \sqrt{(j^2 - i^2)}] \qquad (5.49)$$

where Δ is the class interval, measured in real dimensions, such that $j \cdot \Delta = 2R$ and $i \cdot \Delta = 2r$. The term in brackets is a coefficient k_{ij} which evidently only depends on the class numbers for spheres and profiles.

We now introduce the simple example used by Underwood (1968) to illustrate the procedure: it comprises three classes of spheres ($j = 1, 2, 3$), with numerical densities $N_V(1)$, $N_V(2)$ and $N_V(3)$. Evidently there are also three profile size classes ($i = 1, 2, 3$) with numerical densities $N_A(1)$, $N_A(2)$, and $N_A(3)$ which are the sums of the profile numbers contributed by the sphere size classes:

$$N_V(1) \qquad N_V(2) \qquad N_V(3)$$
$$\downarrow \qquad\qquad \downarrow \qquad\qquad \downarrow$$

$$N_A(1) = N_A(1, 1) + N_A(1, 2) + N_A(1, 3) \qquad (5.50)$$

$$N_A(2) = \qquad\qquad N_A(2, 2) + N_A(2, 3) \qquad (5.51)$$

$$N_A(3) = \qquad\qquad\qquad\qquad N_A(3, 3) \qquad (5.52)$$

Through equation (5.49) we obtain the following values for $N_A(i, j)$:

For $j = 1, i = 1$:

$$N_A(1, 1) = N_V(1)\Delta$$

For $j = 2, i = 1$:

$$N_A(1, 2) = N_V(2) \cdot \Delta(2 - \sqrt{3}) = 0.26795 \cdot N_V(2)\Delta$$

For $j = 2, i = 2$:

$$N_A(2, 2) = N_V(2) \cdot \Delta(\sqrt{3}) \qquad = 1.73205 \cdot N_V(2)\Delta$$

For $j = 3, i = 1$:

$$N_A(1, 3) = N_V(3) \cdot \Delta(3 - \sqrt{8}) = 0.17157 \quad N_V(3)\Delta$$

For $j = 3, i = 2$:

$$N_A(2, 3) = N_V(3) \cdot \Delta(\sqrt{8} - \sqrt{5}) = 0.59236 \cdot N_V(3)\Delta$$

For $j = 3, i = 3$:

$$N_A(3, 3) = N_V(3) \cdot \Delta(\sqrt{5}) \qquad = 2.23607 \cdot N_V(3)\Delta$$

You can observe that the profiles of spheres in class 3 fall for the major part into class 3, with decreasing frequency in the lower classes (compare with Fig. 2.41) and that the numerical density of spheres influences in a linear fashion the profile number in each class.

Following the general formula that the numerical density of profiles in class i is

$$N_A(i) = \sum_{j=1}^{m} N_A(i,j) = N_A(i,i) + N_A(i,i + 1) + N_A(i,i + 2) + \ldots + N_A(i,m)$$

we find in this example the set of equations (5.50 to 5.52), and by substituting from above:

$$N_A(1) = 1.0N_V(1)\Delta + 0.26795N_V(2)\Delta + 0.17157N_V(3)\Delta$$

$$N_A(2) = \qquad\qquad 1.73205N_V(2)\Delta + 0.59236N_V(3)\Delta$$

$$N_A(3) = \qquad\qquad\qquad\qquad 2.23607N_V(3)\Delta$$

These are three linear equations with three unknowns [$N_V(1)$, $N_V(2)$, $N_V(3)$]

and three experimentally measured quantities $[N_A(1), N_A(2), N_A(3)]$ of the type

$$N_A(i) = \Delta \sum_{j=1}^{m} k_{ij} \cdot N_V(j) \tag{5.53}$$

$$= \Delta(k_{ii} \cdot N_V(i) + k_{i,i+1} \cdot N_V(i+1) + \ldots + k_{im} \cdot N_V(m))$$

where the coefficients k_{ij} are simply given by the square bracket in equation (5.49). These equations can be solved for the unknowns which requires a calculation of new coefficients by matrix inversion. The result is a set of equations by which one derives $N_V(j)$; for the particular example:

$$N_V(1) = 1.0 \cdot \frac{N_A(1)}{\Delta} - 0.1547 \cdot \frac{N_A(2)}{\Delta} - 0.0360 \cdot \frac{N_A(3)}{\Delta}$$

$$N_V(2) = \qquad\qquad 0.5774 \cdot \frac{N_A(2)}{\Delta} - 0.1529 \frac{N_A(3)}{\Delta}$$

$$N_V(3) = \qquad\qquad\qquad\qquad 0.4472 \frac{N_A(3)}{\Delta}$$

What this amounts to is that one must subtract from the profiles in class j those which are derived from spheres larger than j. In the general case one would write this as

$$N_V(j) = \frac{1}{\Delta}\left[\sum_{i=j}^{m} \alpha_{ij} N_A(i) \right] \tag{5.54}$$

The coefficients α_{ij} can be calculated once and for all because they only depend on the size class numbers j and i, whereas the size calibration factor Δ does not affect them. Note however that they can be either positive or negative.

The coefficients α_{ij} however depend critically on the way the sphere and profile diameters, D_j and d_i, are grouped into size classes. Equation (5.49) was based on the procedure of Saltykov (1958) (see Volume 2, Section 6.3.2) who proposed that we consider the upper limit of each size class as corresponding to the particle and profile size measure (Fig. 5.8b); accordingly the coefficients α_{ij} are calculated from a matrix of coefficients

$$k_{ij} \text{ (Saltykov)} = [\sqrt{(j^2 - (i-1)^2)} - \sqrt{(j^2 - i^2)}] \tag{5.55}$$

We have discussed above (Section 5.2.2) that Wicksell (1925) who proposed the first method of this kind (see Volume 2, Section 6.3.3) groups the profile diameters around the mid-point of each histogram class (Fig. 5.8a), so that α_{ij} is calculated from a matrix of coefficients

$$k_{ij} \text{ (Wicksell)} = [\sqrt{(j^2 - (i-\tfrac{1}{2})^2)} - \sqrt{(j^2 - (i+\tfrac{1}{2})^2)}] \tag{5.56}$$

Cruz-Orive (1978) has shown that Saltykov's approach leads to a poor estimate of the size distribution of particles due to large errors affecting the small and large size classes. He proposes to group the data according to a Saltykov-type histogram with class limits $i \cdot \Delta$ (Fig. 5.8b), but to take as class mid-point $x_i = (i - \frac{1}{2})\Delta$. The coefficient α_{ij} are then calculated from

$$k_{ij} \text{ (Cruz-Orive)} = [\sqrt{((j - \frac{1}{2})^2 - (i - 1)^2)} - \sqrt{((j - \frac{1}{2})^2 - i^2)}] \quad (5.57)$$

for $i = 1, 2, 3 \ldots (j - 1)$
and

$$k_{ii} = \sqrt{(j - \frac{3}{4})}$$

for $i = j$.

These sets of coefficients were tested on a synthetic distribution of spheres (Table 5.6): it turns out that the modified procedure of Cruz-Orive (1978) permits a rather faithful unfolding of the profile size distribution whereas Saltikov's method shows larger errors, particularly in the small and large size classes. Table 5.7 also shows that this modified procedure gives better estimates of mean particle diameter and its variance, as well as of mean particle volume and surface, and of numerical density.

Table 5.6

Comparison of unfolding methods of Saltykov (1958) and Cruz-Orive (1978) on synthetic example of size distribution of spheres. Courtesy L. M. Cruz-Orive

Granule diameter (cm or μm) $D(j)$	Numerical densities, $N_V(j) \cdot 10^{-3} \text{mm}^{-3}$		
	True data	Unfolding after Saltykov	Unfolding after Cruz-Orive
(0, 0.2)	341	901	247
(0.2, 0.4)	1703	3268	1405
(0.4, 0.6)	5827	8548	5787
(0.6, 0.8)	13647	16327	14345
(0.8, 1.0)	21893	22307	22958
(1.0, 1.2)	24063	21630	24848
(1.2, 1.4)	18122	14504	17829
(1.4, 1.6)	9349	6841	8858
(1.6, 1.8)	3303	2161	2888
(1.8, 2.0)	799	459	658
N_V (total)	99047	96946	99823

Table 5.8 gives the modified values of the set of coefficients α_{ij} as calculated by Cruz-Orive (1978) from a matrix of coefficients according to equation

(5.57). They can now be used to calculate $N_V(j)$ from the measured values $N_A(i)$ according to equation (5.54). For the practical procedure see below (Section 5.2.9.3).

Table 5.7
Comparison of particle parameters for synthetic size distribution of Table 5.6 obtained by three different unfolding procedures. Courtesy L. M. Cruz-Orive.

	True value	By Saltykov unfolding	By Cruz-Orive unfolding	By Giger–Riedwyl method
Mean Diameter $\bar{D}\,\mu m$	1.05	1.0728	1.0418	1.047
Diameter variance $S_x^2\,\mu m^2$	0.1024	0.1109	0.0971	0.1018
Mean num. density $N_V \cdot 10^{-3}\,mm^{-3}$	99047	96946	99823	99331
Granule volume density V_V	0.0768	0.0631	0.0750	—
Granule surface density $S_V\,\mu m^{-1}$	0.3749	0.3220	0.3709	—

5.2.9.2. Method of reconstruction of particle size distribution allowing for finite section thickness

The method of Bach (1967) allows this type of reconstruction—also called "Wicksell transformation"—to be done with profile size distributions recorded from sections of finite thickness t. This has been exploited practically by Baudhuin (1968), who also developed a computer program.

There is nothing fundamentally different from the original Wicksell or Saltykov methods in this procedure, except that, as we remember, the profile size class (i, i), that is the one containing the sphere size, will be augmented by projected great circles in proportion to $t/2R_i$ (Section 5.2.4, and Volume 2, Section 6.4). So this will increase the first term in equation (5.53) only. Looking at the numerical coefficients k_{ij} in the example (eqs 5.50 to 5.52) we note that always the first one on each line, i.e. the coefficients on the diagonal of the matrix, will be modified by adding $t/2R_i$.

From this verbal description it becomes clear that equation (5.53) must be slightly modified to account for section thickness. In the general form one writes

$$N_A(i) = \Delta \cdot \sum_{j=i}^{m} k_{ij}^* \cdot N_V(j) \qquad (5.58)$$

Table 5.8

Matrix of coefficients α_{ij} used in calculation of $N_V(j)$ by equation (5.54), as obtained by the method of Cruz-Orive (1978). Note that some coefficients are negative. The bottom line serves the calculation of the overall particle number per unit volume.

Courtesy Dr. L. M. Cruz-Orive.

Particle size class (j)	Profile size class (i)														
	$N_A(1)$	$N_A(2)$	$N_A(3)$	$N_A(4)$	$N_A(5)$	$N_A(6)$	$N_A(7)$	$N_A(8)$	$N_A(9)$	$N_A(10)$	$N_A(11)$	$N_A(12)$	$N_A(13)$	$N_A(14)$	$N_A(15)$
$N_V(1)$	+2.00000	-0.68328	+0.08217	-0.02799	-0.00048	-0.00298	-0.00126	-0.00094	-0.00062	-0.00045	-0.00034	-0.00026	-0.00020	-0.00016	-0.00013
$N_V(2)$		+0.89443	-0.47183	+0.04087	-0.02528	-0.00393	-0.00427	-0.00234	-0.00167	-0.00117	-0.00087	-0.00066	-0.00051	-0.00040	-0.00033
$N_V(3)$			+0.66667	-0.39550	+0.02903	-0.02416	-0.00516	-0.00488	-0.00288	-0.00208	-0.00150	-0.00113	-0.00087	-0.00069	-0.00055
$N_V(4)$				+0.55470	-0.34778	+0.02308	-0.02293	-0.00565	-0.00512	-0.00315	-0.00230	-0.00169	-0.00129	-0.00101	-0.00080
$N_V(5)$					+0.48507	-0.31409	+0.01947	-0.02176	-0.00582	-0.00518	-0.00327	-0.00242	-0.00180	-0.00139	-0.00109
$N_V(6)$						+0.43644	-0.28863	+0.01704	-0.02071	-0.00584	-0.00516	-0.00332	-0.00248	-0.00186	-0.00145
$N_V(7)$							+0.40000	-0.26850	+0.01527	-0.01977	-0.00579	-0.00509	-0.00332	-0.00250	-0.00189
$N_V(8)$								+0.37139	-0.25208	+0.01393	-0.01893	-0.00571	-0.00501	-0.00330	-0.00250
$N_V(9)$									+0.34816	-0.23834	+0.01287	-0.01818	-0.00560	-0.00491	-0.00327
$N_V(10)$										+0.32880	-0.22663	+0.01200	-0.01751	-0.00549	-0.00481
$N_V(11)$											+0.31235	-0.21649	+0.01128	-0.01691	-0.00538
$N_V(12)$												+0.29814	-0.20761	+0.01067	-0.01636
$N_V(13)$													+0.28571	-0.19973	+0.01015
$N_V(14)$														+0.27472	-0.19269
$N_V(15)$															+0.26491
N_V	+2.00000	+0.21115	+0.27700	+0.17208	+0.14056	+0.11435	+0.09722	+0.08436	+0.07453	+0.06674	+0.06042	+0.05519	+0.05080	+0.04705	+0.04382

where

$$k_{ij}^* = k_{ij} + \frac{t}{2\Delta} \cdot \delta_{ij}$$

and

$$\delta_{ij} = 1 \quad \text{if } j = i$$
$$\delta_{ij} = 0 \quad \text{if } j \neq i$$

Alternatively this becomes

$$N_A(i) = \Delta\left[\left(k_{ii} + \frac{t}{2\Delta}\right) \cdot N_V(i) + \sum_{j=i+1}^{m} k_{ij} \cdot N_V(j)\right] \qquad (5.59)$$

$$= \Delta\left[\left(k_{ii} + \frac{t}{2\Delta}\right) \cdot N_V(i) + k_{i,i+1} \cdot N_V(i+1) + \right.$$

$$\left. k_{i,i+2} \cdot N_V(i+2) + \ldots k_{im} \cdot N_V(m)\right]$$

This then defines the set of linear equations that can be solved for $N_V(j)$, as done for equation (5.53). The trouble is that the coefficients k_{ij}^* are not constant but contain a variable, $t/2\Delta$. It is therefore not possible to calculate an inverse matrix of coefficients similar to Table 5.8 which would allow straightforward calculation of $N_V(j)$ by equation (5.54). However, by using a computer this inversion is easily done (Baudhuin, 1968).

5.2.9.3. Remarks on practical procedures
The practical procedure is essentially the same for both methods.
(1) Record a size distribution of profiles by the methods described in Section 5.2.1. Note that you must decide here which type of analysis you wish to perform. In the Wicksell-type you record the diameter class of the circle that best fits the profile, whereas in the Saltykov-type you must determine between which two circles the profile fits and record the diameter of the larger one; this is the same for the modified procedure of Cruz-Orive here advocated. With about 12–15 classes a few hundred profiles must be measured to give adequate precision in the analysis. Note that the profile numbers should be recorded as numerical densities, i.e. that they must be sampled with respect to the test area applied, observing the counting rules given in Section 4.2 (Fig. 4.12).
(2) Calculate numerical densities $N_A(i)$ and plot them as a histogram (Section 5.2.2); correct for missing small profiles (Section 5.2.3) and

amend the table accordingly. The table is best arranged horizontally for convenience in the subsequent calculation.

(3) Estimate section thickness and decide whether it must be taken into consideration. If no, go to (5), if yes go to (4).

(4) If section thickness must be considered one must first set up the matrix of coefficients $k_{ij}*$ by adding $(t/2\Delta)$ to each k_{ii} in Table 5.9. By inversion of this matrix a set of specific coefficients $\alpha_{ij}(t)$ is obtained with which one can now proceed to step (5). (Alternatively one could also correct for section thickness effect according to Section 5.2.4 and proceed to step (5) using the standard coefficients α_{ij}).

(5) Use the table of coefficients α_{ij} given in Table 5.8 if section thickness correction is not required (Tables for the Wicksell or Saltykov methods will be found in Volume 2, Section 6.3), or alternatively the specific set of coefficients calculated in step (4), and perform the calculation of $N_V(j)$ according to equation (5.54). Note that, particularly in the smaller size classes, you may end up with some negative values of $N_V(j)$; obviously, this does not make sense, but Bach (1967) has shown that one may set such negative values equal to 0.

(6) Once all $N_V(j)$ are calculated one can check on whether the correction for missing small profiles (Section 5.2.2) has been appropriate. One calculates by equation (5.53) $N'_A(i)$ resulting from the sphere size distribution so obtained; it is evidently sufficient to do that for 2 or 3 of the small size classes where many profiles have been found missing. If $N'_A(i)$ thus calculated deviates considerably (e.g. $>20\%$) from the graphically corrected value then one may have to base the correction on $N'_A(i)$ and redo the calculations of $N_V(j)$.

Fig. 5.17 shows the result of a practical application of this approach to the sizing of hepatocyte nuclei (Weibel et al., 1969). The top panel shows the original frequency distribution of observed profiles (cross-hatched) and the profile distribution after correction for missed small profiles and for section thickness effect by the method of Baudhuin (1968). The bottom panel then reports the result of unfolding this profile size distribution according to the Wicksell procedure.

5.3.   Measurement of intercept length distribution and estimation of particle size

In Volume 2, Chapter 7, we shall show that one can derive estimates of the size distribution of particles by measuring, instead of the profile diameter,

Table 5.9

Coefficients k_{ij} by equation (5.57) after Cruz-Orive (1978).

Profile size class i	\multicolumn{15}{c}{Particle size class j}

i / j	1	2	3	4	5	6	7	8	9	10	11	12	13	14	15
1	0.5000	0.3820	0.2087	0.1459	0.1125	0.0917	0.0774	0.0670	0.0590	0.0528	0.0477	0.0436	0.0401	0.0371	0.0345
2		1.1180	0.7913	0.4818	0.3564	0.2849	0.2380	0.2046	0.1796	0.1601	0.1445	0.1317	0.1210	0.1119	0.1041
3			1.5000	1.0695	0.6770	0.5137	0.4184	0.3546	0.3084	0.2732	0.2455	0.2229	0.2043	0.1886	0.1751
4				1.8028	1.2925	0.8349	0.6428	0.5296	0.4530	0.3970	0.3541	0.3199	0.2919	0.2686	0.2489
5					2.0616	1.4836	0.9702	0.7541	0.6261	0.5391	0.4752	0.4258	0.3863	0.3539	0.3267
6						2.2913	1.6533	1.0902	0.8531	0.7123	0.6162	0.5454	0.4906	0.4465	0.4103
7							2.5000	1.8074	1.1990	0.9428	0.7906	0.6866	0.6097	0.5500	0.5020
8								2.6925	1.9495	1.2991	1.0255	0.8628	0.7515	0.6691	0.6050
9									2.8723	2.0821	1.3924	1.1024	0.9300	0.8120	0.7246
10										3.0414	2.2068	1.4800	1.1747	0.9931	0.8688
11											3.2016	2.3248	1.5628	1.2429	1.0528
12												3.3541	2.4372	1.6416	1.3078
13													3.5000	2.5446	1.7168
14														3.6401	2.6477
15															3.7749

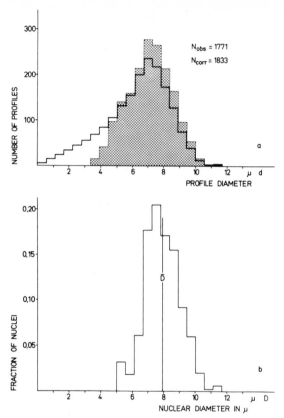

FIG. 5.17: Estimation of the size distribution of hepatocyte nuclei from a section profile distribution by Wicksell transformation according to Baudhuin (1968). (Reproduced by permission from Weibel *et al.*, 1969.)

the length of intercepts formed by random test lines with the profile. We have discussed above (Section 5.2.1) the meaning of the (mean) intercept length or chord as a measure of profile size; if the profiles are obtained by cutting particles with independent random planes, then the intercepts are also measures of particle size (Volume 2, Chapter 7).

The most convenient test system for estimating intercept length distributions is a set of parallel straight test lines. Their spacing should be large enough so that none of the profiles is hit by more than one line. This rule may appear peculiar but it is easily understood by considering a profile population where the largest profile measures 3 cm (Fig. 5.18): if one now uses a test system where the lines are spaced at 2 cm, then evidently profiles which are > 2 cm may be hit twice and yield two intercepts, all those < 2 cm will yield only one. The two intercepts obtained from large profiles are however not

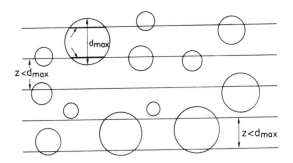

FIG. 5.18: Too closely spaced test lines may yield more than one intercept.

independent, but rather depend on the line spacing. So in order for each profile to have a chance to be hit simply in proportion to its diameter and for the intercept lengths generated to be independent of each other the spacing of test lines should be such that

$$z > d_{max}$$

where d_{max} is the largest profile (or particle) diameter (Fig. 5.19a). Alternatively, if one wants to better exploit the profile population the spacing should be chosen very dense (Fig. 5.19b), namely

$$z \leqslant \Delta$$

the interval between the size classes in particle diameter that one chooses for recording the size distribution; this way each profile yields a number of intercepts in proportion to its diameter. Evidently, this is only a practical procedure in automatic instruments; for manual measurement the wider spacing is to be preferred because it allows a much greater number of profiles to be sampled. Please note that no compromise between these two extreme line densities is possible.

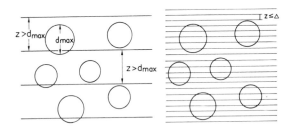

FIG. 5.19: Loose or very dense spacing of test lines for intercept length measurement.

It is evident that the total test line length applied to the micrograph must be known, because the size distribution is recorded in terms of numerical density per unit test line length, $N_L(l)$. This also defines the sampling procedure: one follows the set of test lines and measures all intercepts one encounters. Of the profiles hit by the end of the test line, one measures those hit by the left-hand end and rejects those hit by the right-hand end.

The actual measurement of l is simply performed with a ruler. Since one is in general recording the data in size classes of interval Δl one can use a special ruler, as shown in Fig. 5.20, on which a starting point and the class limits are marked. The starting point "0" is placed at the left end of the intercept; one then records the size by noting the class limit that is just outside the profile, i.e. "8" in the example of Fig. 5.20. This corresponds to the Saltykov type of classification. The mean intercept length for each size class i is then

$$l_i = \Delta l(i - \tfrac{1}{2}) \qquad (5.60)$$

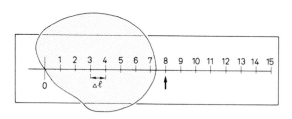

FIG. 5.20: Ruler to obtain intercept lengths distributed into size classes of spacing Δl.

In Fig. 5.21 the recorded numerical densities $N_L(l_i)$ have been plotted as a histogram; as observed for the profile diameter distributions the lowest size class is empty so that we have evidently missed the shortest intercepts because we have not seen the smallest profiles. In this case correction for the missing profiles is easy. We noted in Figs 2.43 and 2. 4 that the frequency of intercepts obtained from one sphere size class forms a straight line through the origin. All we now need to do is to draw a line through the top midpoint of the histogram columns and the origin. Starting from the right or from the largest size class the next histogram column must be above the preceding line as long as there are spheres in that size class (see Fig. 2.44); if there are no spheres in that class then the top of the histogram column must fall on the preceding line, as is the case for class 3. It is noted that in Fig. 5.21 the histogram columns of classes 2 and 1 are below the top line; this means that we must have missed some of the profiles in these two classes. It is now easy to correct the histogram by drawing the columns up to the last line, as indicated by arrows in Fig. 5.21. Note that this correction is required if one chooses to derive the particle size distribution analytically or for calculating moments;

it is particularly important if one wants to estimate particle number from the number of shortest intercepts (eq. 5.62). For the graphical solution it suffices to remember that no particles can be present in a certain size class if the histogram column is not higher than the line extrapolated from the column to the right.

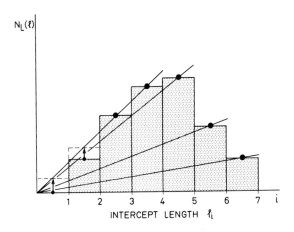

FIG. 5.21: Histogram of intercept length frequencies. Note correction for missing chords (arrows).

Strictly speaking, this correction is applicable only to a population of spherical particles; in first approximation it should however also work for ellipsoids, because DeHoff and Bousquet (1970) also found $N_L(l)$ to be proportional to l for ellipsoids. It would probably not hold however for polyhedra, because these would cause a greater number of very short intercepts to occur.

The easiest way to proceed with these data is to perform the graphical analysis of Lord and Willis (1951) as will be described in detail and justified theoretically in Volume 2, Section 7.2.1. Figure 5.22 demonstrates the practical procedure which is as follows:

(1) Measure the intercept lengths by means of a ruler (Fig. 5.20) and along a set of (parallel) test lines of known total length L_T (Fig. 5.19a).

(2) Plot the intercept length distribution as a histogram, with class interval Δl, and using $N_L(l_i)$ as a measure for frequency.

(3) Draw a vertical auxiliary line A at some convenient distance from the histogram.

(4) Draw straight lines from the origin through the top mid-point (circle) of each histogram column to the auxiliary line. Start with the largest inter-

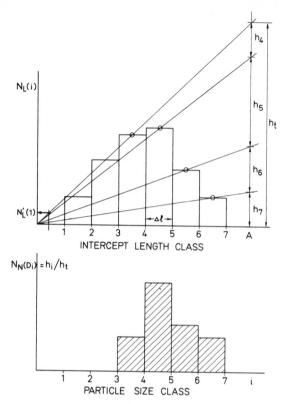

FIG. 5.22: Graphical unfolding procedure of Lord and Willis (1951) using auxiliary lines at A.

cept length class, and stop drawing when the column belonging to the next smaller size class lies on or below the last line (classes 3 and 2 in Fig. 5.22).

(5) Measure the lengths h_i on the auxiliary line and obtain the frequency distribution for spheres in size classes i by

$$F(i) = h_i/h_t = h_i/\sum_i h_i = N_N(D_i) \qquad (5.61)$$

which is equal to the relative number $N_N(D_i)$ of spheres of diameter D_i.

(6) To obtain the total number of particles in the unit volume read from the histogram the number of intercepts per unit test line length, $N'_L(1)$, by carrying the intersection between the uppermost line through the origin and the mid-line of class 1 to the ordinate (Fig. 5.22). By the method of Hilliard (1968) we find

$$N_V = \frac{4}{\pi}\left[\frac{N'_L(1)}{L_T \cdot (\Delta l/2)^2}\right] \qquad (5.62)$$

(7) One can now find the particle size distribution in terms of numerical densities by

$$N_V(D_i) = N_V \cdot N_N(D_i) \qquad (5.63)$$

For the calculation of the moments or for the analytical solution of the size distribution according to Volume 2, Chapters 7.2.2 and 8.4 one should work with data which have been corrected for missing chords. Also the number of classes and the sample size should not be too small because the analytical solution requires the estimation of slopes of the distribution curve.

Cruz-Orive (1978) has shown that the graphical solution of Lord and Willis (1951) is thwarted by the same kind of error as that discussed above in relation to the method of Saltykov (1958): the choice of the upper class limit as the value representing sphere size leads to appreciable errors, particularly with respect to the estimated frequencies of small and large size classes. Using class mid-points instead leads to improved estimates. However, this procedure does no longer lend itself to a simple graphical procedure, but rather requires an analytical solution using a table of coefficients. This will be presented in Volume 2, Section 7.2.2. Likewise, the possibilities in analysing size distributions of non-spherical particles are dealt with in Volume 2, Chapter 8.

CHAPTER 6

Various methods

6.1. Measurement of barrier thickness

A barrier is a layer of tissue separating two spaces; the example given in Fig. 6.1 is the tissue barrier separating air in alveoli and blood in capillaries in the lung, but we could as well have chosen the glomerular basement membrane, the capillary endothelium, the intestinal epithelium, etc. The main characteristics of a "barrier" is that it is thin compared to its extension in the other two dimensions, and that it should, essentially, separate the two spaces completely.

Two measures of barrier thickness are relevant and accessible to estimation by stereological methods:

(1) the arithmetic mean barrier thickness, $\bar{\tau}$, which estimates the tissue mass making the barrier

(2) the harmonic mean thickness, τ_h, which is a measure for the diffusion resistance of the barrier.

In addition one can also characterize the thickness distribution of the barrier (Volume 2, Chapter 9). The question which of these fundamentally different measures of barrier thickness is relevant to a particular study must be answered from considering the experimental situation (see Section 8.10 for a discussion of the functional meaning of these two parameters).

In general, it is not possible to measure τ directly on micrographs: because a random section hits the barrier in any direction the apparent thickness on the section will most of the time be greater than the true section. This will be dealt with in detail in Volume 2, Section 9.2. Simple stereological methods are however available to measure both arithmetic and harmonic mean barrier thickness. These are presented here in their practical aspect.

204

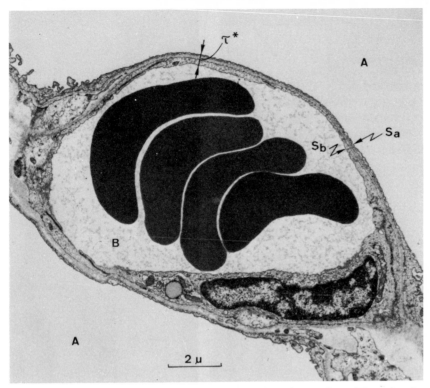

FIG. 6.1: Electron micrograph of alveolar capillary from human lung. Tissue barrier between air and blood, whose thickness τ is to be found, is bounded by two surfaces, one (S_a) towards the alveolar space (A), the other (S_b) towards the capillary blood (B). The apparent thickness τ^* is probably larger than the true thickness because the section does not necessarily cut the barrier at right angles.

Two methods are available for measuring the *arithmetic mean barrier thickness* $\bar{\tau}$.

(1) The first method (Weibel and Knight, 1964) estimates $\bar{\tau}$ from a point count (Volume 2, Section 9.4). We define $\bar{\tau}$ as the volume of the barrier V_b divided by one of its two bounding surfaces, S_b (Fig. 6.2). Using a coherent test system of lines and points (see Section 4.2.5) we can easily find:

$$\bar{\tau} = V_b/S_b = (l_T/2) \cdot (P_b/I_b) \qquad (6.1)$$

where P_b are test points falling onto barrier profiles and I_b are intersections with the trace of the *reference* surface; l_T is the test line length associated with each test point of the point lattice: for a square lattice it is $2d$, for the multi-purpose test system (e.g. M 168) it is $d/2$. Sometimes it will be more appropriate to relate the barrier volume to the average of both its bounding surfaces,

S_a and S_b (Fig. 6.1), rather than to one of them only; in this case the formula is:

$$\bar{\tau} = l_T \cdot P_b/(I_a + I_b) \tag{6.2}$$

The practical procedure is simple. One chooses a convenient coherent test system and applies it to random micrographs (Fig. 6.3). In every respect the rules are those for all point counting procedures (Section 4.2); in particular one should sum the point counts over all fields which make up a representative sample, and enter these sums into equation (6.1) or (6.2), so that, by equation (6.2)

$$\bar{\tau} = l_T \cdot (\Sigma P_b)/(\Sigma I_a + \Sigma I_b) \tag{6.3}$$

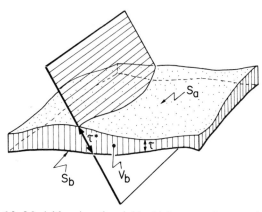

FIG. 6.2: Model barrier of variable thickness cut by a section.

(2) The second method for estimating $\bar{\tau}$ exploits the distribution of random intercept lengths l measured on random test lines (Gundersen et al., 1978). The mean barrier thickness $\bar{\tau}$ is obtained by

$$\bar{\tau} = \bar{l}/2 \tag{6.4}$$

as will be shown in Volume 2, Section 9.3. To estimate $\bar{l}$, a set of parallel random lines is desposited on the micrograph (Fig. 6.4), and l is measured with a scale, such as that shown in Fig. 5.20.

The method for estimating *harmonic mean barrier thickness* τ_h also works with random intercept lengths; Weibel and Knight (1964) have shown that (Volume 2, Section 9.3)

$$\tau_h = \tfrac{2}{3} l_h \tag{6.5}$$

where l_h is the harmonic mean intercept length.

The practical procedure for obtaining intercept length distributions and for using them to calculate $\bar{\tau}$ and τ_h is as follows.

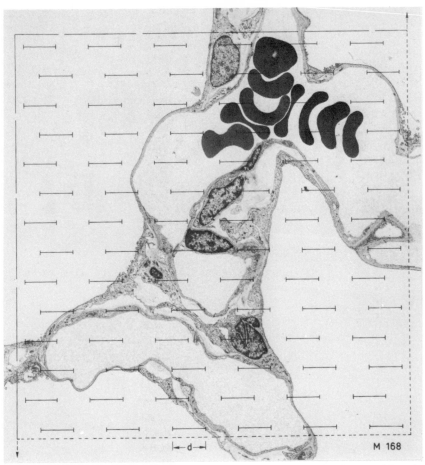

FIG. 6.3: Electron micrograph of lung capillaries with test system to estimate arithmetic mean barrier thickness from point and intersection count. $d = 4.6\,\mu$m.

(1) Select a test system of parallel straight lines (Fig. 6.4). The line spacing is not very important in this case. However, we have found it useful to use a simple square lattice (e.g. A 100 shown in Appendix 3) and to measure intercepts in both directions; this compensates partly for the high degree of local anisotropy which is always present in such barriers; in fact it is part of the very nature of such barriers!

(2) Identify intercepts. Figure 6.4 shows that there are two kinds of intercepts: those which enter the barrier at surface (a) and leave it at surface (b), in that example the alveolar and capillary surface respectively, are "true" intercepts because they span the entire barrier; other intercepts

H

Fig. 6.4: Estimating intercept length distribution for barrier in lung using square line test system. Note "false intercepts" indicated by wavy line.

which enter at one surface and leave again at the same surface are "false" intercepts. The "false" intercepts are due to the curvature of the barrier and should not be measured because they do not extend across the barrier; they are a source of error with this method because curvature was not considered in the theoretical model to be presented in Volume 2, Chapter 9, but it has been shown experimentally that this error is very small and can be disregarded (Weibel and Knight, 1964).

(3) Measure the intercept lengths. We can use the same type of scale as shown in Figure 5.20 and record the intercept length by fitting the 0 point to one end of the intercept and noting the class by the class limit that is just outside the barrier at the other end. This gives an intercept

length distribution classified on a linear scale, which is appropriate for estimating $\bar{\tau}$ or the barrier thickness distribution (Volume 2, Section 9.3). This scale is however not optimal for estimating l_h; here it is better to use a scale which is linear in $1/l$. Figure 6.5 shows such a ruler; the procedure of recording the intercept length is the same as for the linear ruler: fit the 0 point to one end of the intercept and record the class by noting the rule which is just outside the barrier. Figure 6.5 also indicates the values of the class midpoints, l_i, which are to be used in the subsequent calculations. Note that in this case there is no need of correction for missing small intercepts, because one should be able to identify all intercepts with a barrier.

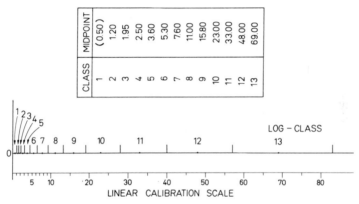

FIG. 6.5: Ruler with logarithmic scale for recording intercept length distribution in terms of $(1/l)$.

(4) Calculate the means. For estimating the arithmetic mean $\bar{\tau}$ we must average the intercept lengths l_i from the linear measurements (Fig. 5.20) by

$$\bar{l} = \left\{ \sum_{i=1}^{m} N_L(i) \cdot l_i \right\} \bigg/ \left\{ \sum_{i=1}^{m} N_L(i) \right\} \qquad (6.6)$$

where $\qquad l_i = \Delta \cdot (i - \tfrac{1}{2})$

Δ being the class interval or the distance between the units on the ruler. Whilst this is the correct formulation it is evidently not necessary to calculate $N_L(i)$, the number of intercepts per unit test line length, if we only require the mean; one can just use the raw relative numbers recorded for each class.

For estimating the harmonic mean l_h one uses the values of l_i indicated in Fig. 6.5 (or any appropriate values if other test systems are used) and computes l_h by

$$1/l_h = \left\{ \sum_{i=1}^{m} N_L(i) \cdot 1/l_i \right\} \bigg/ \left\{ \sum_{i=1}^{m} N_L(i) \right\} \qquad (6.7)$$

or

$$l_h = \left\{ \sum_{i=1}^{m} N_L(i) \right\} \bigg/ \left\{ \sum_{i=1}^{m} N_L(i) \cdot 1/l_i \right\} \qquad (6.8)$$

Note that here too the raw numbers can be used instead of $N_L(i)$.

(5) If the barrier thickness distribution should be required, histograms of l_i or $1/l_i$ must be plotted. In Volume 2, Section 9.3 the method of Gundersen *et al.* (1978) for graphically reconstructing the distribution in $1/\tau_i$ will be explained in detail.

6.2.   Combination parameters

It has been mentioned repeatedly that some of the basic stereological measurements can be combined to derive very useful additional parameters. One example, dealt with in Section 6.1, is the definition of mean barrier thickness as the ratio of volume to surface. In setting up this parameter we have set

$$\bar{\tau} = V_{Vb}/S_{Vb} \qquad (6.9)$$

where V_{Vb} stands for volume density of the barrier tissue and S_{Vb} for the surface density of the "reference surface" of the barrier; both densities are defined with respect to the same containing volume. It is now very simple to substitute $V_{Vb} = P_{Pb}$ and $S_{Vb} = 2 \cdot I_{Lb}$ and to solve as follows

$$\bar{\tau} = P_{Pb}/(2I_{Lb}) = (P_b \cdot L_T)/(2 . I_b \cdot P_T) \qquad (6.10)$$

and using a coherent test system where each test point is associated with l_T test line length this yields

$$\bar{\tau} = (P_b/I_b) \cdot (P_T \cdot l_T)/(2P_T) = P_b \cdot l_T/(2I_b) \qquad (6.11)$$

Another useful parameter is the surface-to-volume ratio (s/v) of particles or its inverse the volume-to-surface ratio (v/s) which is evidently similar to $\bar{\tau}$ for the barrier. It is easily found by

$$(v/s) = V_{Va}/S_{Va} = P_{Pa}/(2I_{La}) = (P_a/I_a) \cdot (l_T/2) \qquad (6.12)$$

Figure 6.6 shows how a coherent test system immediately yields the data required for this calculation. This parameter, first introduced by Chalkley *et al.* (1949), is useful because it is directly related to the mean caliper diameter of particles, $\bar{H}$, through a shape dependent constant, such that

$$\bar{H} = k \cdot (v/s) \qquad (6.13)$$

Some values of k are given in Table 6.1. This may often yield a sufficiently good estimate of $\overline{D}$ for calculating N_V from N_A. Furthermore, (v/s) is directly related to the mean linear intercept $\overline{l}$ of the particles, provided the latter are convex:

$$\overline{l} = 4 \cdot (v/s) \tag{6.14}$$

This indicates that $\overline{l}$ of a set of convex particles a can actually also be derived from a combination count, namely by

$$\overline{l}_a = 2 \cdot l_T \cdot P_a / I_a \tag{6.15}$$

whereby we evidently have to use a coherent test system.

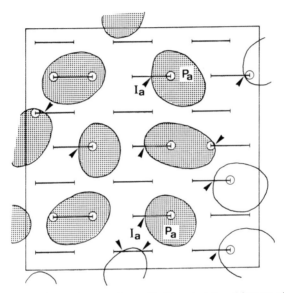

FIG. 6.6: Estimating volume-surface ratio from point and intersection count.

A useful parameter of structure is the "mean free distance" λ between particles, as will be discussed in Volume 2, Section 9.5. Whilst it can be estimated from the distribution of intercept lengths in the matrix it is also directly accessible through a point count. In fact, in relation to Fig. 6.6, λ_a is directly related to the volume-to-surface ratio of the "matrix" in which the particles are embedded:

$$\lambda_a = 4(1 - V_{Va})/S_{Va} \tag{6.16}$$

from where we derive directly

$$\lambda_a = 2 \cdot l_T \cdot (P_T - P_a)/I_a \tag{6.17}$$

Table 6.1
Relationship between mean caliper diameter $\bar{H}$ and
volume-to-surface ratio (v/s) for various solids.

Solid	$\bar{H} = k \cdot (v/s)$ $k =$
Sphere	6.0
Prolate ellipsoid $(a/b = 2)$	7.118
Oblate ellipsoid $(a/b = 2)$	9.026
Cube	9.0
Octahedron	8.638
Dodecahedron	7.128
Icosahedron	6.915
Tetrakaidecahedron	7.102

Evidently, there are many more such combination parameters that one can derive. However, such derivations must be done with great care and a good feeling for the basic rules of stereology! Furthermore one must note that often such parameters are not unbiased if they are derived on a population of objects of varying size; this is, for example, the case with respect to (v/s). But such considerations need to be referred to the theoretical part (Volume 2, Chapter 9).

6.3.   Measurement of surface curvature

In Section 2.5 we discussed the relationship between the density of integral mean curvature, K_V, and the number of tangents, T_A, which the surface forms, per unit section area, with a sweeping test line (equation 2.69). One of the physically important parameters of structure is the average mean curvature (or "mean mean" curvature) $\bar{K}$ because it is directly related to surface tension: by the formula of Gibbs (see Gil and Weibel, 1972) the pressure generated at a closed surface due to surface tension γ is

$$p = 2 \cdot \gamma \cdot \bar{K} \qquad (6.18)$$

Since K_V is the integral mean curvature, integrated over the entire surface of the component under consideration, S_V, it is plausible that $\bar{K}$ is obtained by averaging K_V over S_V. On this basis Cahn (1967) and De Hoff (1967) have developed a stereological method for estimating $\bar{K}$ on random sections which will be further discussed in Volume 2, Section 10.2. The practical method is as follows.

It is required to apply two test systems (Figs. 6.7 and 6.8): (1) a fixed test line system with which one estimates the area of the surface by counting its intersection density on the test lines, I_L; (2) a test area combined with a sweeping test line, i.e. a moveable test line that is swept in a parallel course across the entire test area, with which one estimates the number of positive and negative tangents, T_+ and T_-, formed between the sweeping line and the surface. The formula is

$$\overline{K} = \frac{\pi}{2} \cdot T_A(net)/I_L = \frac{\pi}{2}(T_+ - T_-)/(A_T \cdot I_L) \qquad (6.19)$$

where T_A (net) is the difference between positive and negative tangents per unit test area (equations 2.67 and 2.68).

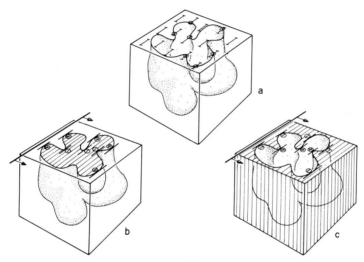

FIG. 6.7: Estimating mean surface curvature $\overline{K}$ from intersections (a) and tangent counts (b and c).

The estimation of the number of tangents is a new measurement, and it might need some additional explanation. Let us first define what is meant by a positive or negative tangent. A positive tangent is formed if the sweeping test line meets a *convex* region of the surface trace, a negative tangent if it meets a *concave* region of the trace (see Fig. 2.27). The terms "convex" and "concave" can however be ambiguous: in a space-filling system the "surface" is, in general, an interface between two compartments or phases, and each tangent point is convex with respect to one and concave with respect to the other phase. It will therefore be essential to define with respect to which phase the surface curvature is to be determined. Figure 6.7 shows the consequences of this decision: in (b) the curvature of the enclosed object is to be determined, and the net tangent count is $T_{net} = +2$; in (c) however we con-

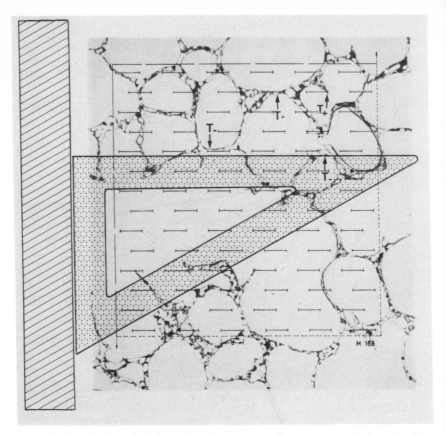

FIG. 6.8: Practical method for estimating mean surface curvature in lung fixed by vascular perfusion. Sliding ruler is used for tangent count.

sider the enclosed space as a void and consequently estimate the curvature of the enveloping surface; the net tangent count is $T_{net} = -2$. The intersection count is identical in both cases so that, in the final result, we will estimate a *positive* average mean curvature in case (b), but a *negative* average mean curvature in case (c). That this is a logical result is seen if we consider the use that can be made of $\bar{K}$ by calculating the pressure generated by surface forces (equation 6.18): in case (b) the pressure vector will act onto the "solid" phase or tend to compress it and hence be positive, whereas in case (c) it will act towards the void or "pull" on the solid phase and hence be negative with respect to the solid phase.

By equation (6.19) $\bar{K}$ is estimated as a combination parameter of two independent stereological measurements: T_A estimates the integral mean curvature, K_V, (equation 2.69), whereas I_L estimates S_V (equation 2.32). It is evident that this is best performed by using a coherent test system where L_T and A_T are linked (Section 4.2.1):

$$L_T = P_T \cdot k_1 \cdot d$$

$$A_T = P_T \cdot k_2 \cdot d^2$$

This allows equation (6.19) to be brought into the following practical form:

$$\overline{K} = \frac{\pi}{2} \frac{T_+ - T_-}{A_T} \cdot \frac{L_T}{I} = \frac{T_+ - T_-}{I} \cdot \frac{k_1 \cdot \pi}{k_2 \cdot d \cdot 2} \tag{6.20}$$

Which of the coherent test systems is chosen is predominantly a matter of taste and convenience. We have found the multi-purpose test system (Appendix 3: M42 or M168) to be convenient because its short test lines loosely arranged in rows on the one hand do not interfere with the tangent count and on the other even facilitate it. In this particular instance equation (6.20) becomes

$$\overline{K} = \frac{T_+ - T_-}{I} \cdot \left[\frac{\pi}{2 \cdot d \cdot \sqrt{3}} \right] \tag{6.21}$$

Figure 6.8 shows the practical approach for the case of lung tissue where we want to determine the mean curvature of the alveolar surface in view of calculating the surface pressure generated (Gil and Weibel, 1972). The screen of the projection head of a light microscope is fitted with a test system M168. For counting tangents a ruler is attached vertically and a clear plastic triangle is slid along this ruler considering its horizontal edge as the sweeping test lines. The use made of this information is discussed in Section 8.10.

The sampling scheme is no different than for any other stereological measurements. The tangent and intersection counts are summed over an appropriate number of fields which are considered to constitute a representative sample (Section 3.7), and $\overline{K}$ is then calculated by introducing these sums into equation (6.20) or, specifically, into equation (6.21).

6.4. Stereological study of anisotropic tissue

One of the practical problems the stereologist meets is that the component objects making up the structure show preferential orientation, an often functionally important result of biological organization (Weibel, 1978). Well-known examples are the myofibrillar organization of striated muscle cells (see Sections 8.4 and 8.5), the lamellar organization of bone (see Section 8.8), the parallel tubules in the renal medulla, or the parallel microvilli making up the brushborder of intestinal cells. In all these cases, a single section will not be representative of the structure, because the section image will depend on the angle θ at which the components are cut. In the case of parallel cylindrical tubules, a section at right angles to their axis yields circular profiles, a longitudinal section yields straight bands, whereas an oblique section produces elliptic profiles of axial ratio depending on the sectioning angle (Fig. 6.9). It is intuitively plausible that stereological measurements performed on such sections through anisotropic structures may depend on the orientation of the section plane to the structure and that special precautions are required to yield correct estimates of structure parameters. The problem will be dealt with in depth in Volume 2, Chapter 10. In the present context we restrict ourselves to a few practical considerations.

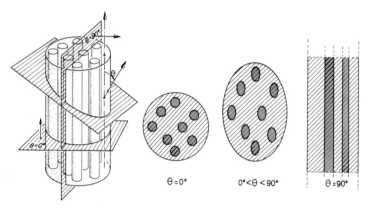

$\theta = 0°$ $0° < \theta < 90°$ $\theta = 90°$

FIG. 6.9: Cutting anisotropic fasciculated structure yields different section images, depending on sectioning angle θ.

6.4.1. PATTERNS AND DEGREE OF ANISOTROPY

The typical isotropic object is a sphere. Anisotropic objects are obtained if the sphere is deformed in a given direction (Fig. 6.10): (a) by "stretching" it (to infinite length) in the z-direction a circular cylinder is formed with marked anisotropy in the z-direction whereas in the x, y-plane the cross-section is isotropic; (b) by "squashing" it, for example in the y-direction, a thin disc or flake is formed with anisotropy in the x- and z-directions. This defines two possible models of totally anisotropic structures (Fig. 6.11): (a) *fasciculated structures* made of bundles of parallel cylindrical tubules, and (b) *lamellar structures* made of parallel discs. Evidently, intermediate structures modelled by aligned (c) cigar- or (d) pill-shaped ellipsoids are also possible.

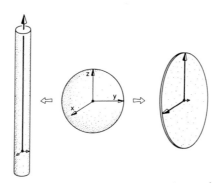

FIG. 6.10: Forming anisotropic solids from sphere.

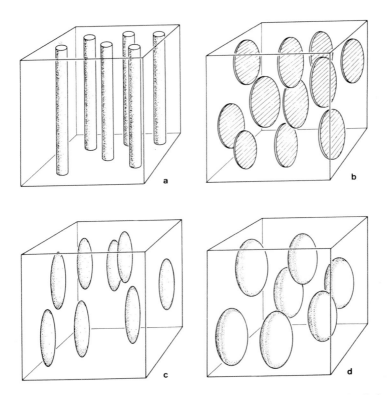

FIG. 6.11: Four models of anisotropic structures: (a) fasciculated parallel tubules or rods; (b) lamellar parallel discs of flakes; (c) aligned cigar-shaped ellipsoids; (d) aligned pill-shaped ellipsoids.

The degree of anisotropy is measured by the orientation distribution of the normals onto the surface of the objects. For spheres the normals are uniformly distributed over the sphere of orientation (see Fig. 2.10); for anisotropic surfaces some orientations will be preferred: in the case of lamellar structures all normals point in the same direction and for the cylinder they pass through the equator of the orientation sphere. For cigar- or pill-shaped ellipsoids the normals occupy all regions of the orientation sphere but they are inhomogeneously distributed with some preferred orientations. The result of this is that a test line shot at random into the structure will hit an object with a probability which varies with the angle between the test line and the orientation normal of the objects; this follows directly from Section 2.1.3 (see also Volume 2, Sections 2.3 and 10). The degree of anisotropy can therefore be estimated by counting intercepts of objects (or intersections with their surface) on test lines of varying orientation, and plotting the intercept number on polar coordinates as a function of the angle of orientation. This is extensively discussed by Underwood (1968), and further details will be given in Volume 2, Chapter 10.

6.4.2. ESTIMATING VOLUME DENSITY IN ANISOTROPIC STRUCTURES

By the principle of Delesse (1847) the volume density is obtained from the areal density of profiles (A_A) within sections of the containing space (A_c):

$$V_V = A_A = A_a/A_c \qquad (6.22)$$

If an oriented structure is cut at a certain angle θ this will affect both the profile area of the anisotropic objects and of the containing space to the same degree (Volume 2, Chapter 10) so that

$$A_a(\theta)/A_c(\theta) = A_A = V_V \qquad (6.23)$$

This is easy to see for the case of fasciculated structures made of circular cylinders within a containing space in the form of a cylinder, but it also holds for any other anisotropic structure.

The conclusion is that a stereological estimate of V_V for anisotropic structures is independent of the sectioning angle and may therefore be performed on any arbitrary section, e.g. by means of a point count (Sections 2.2.3; 4.3), without any further precautions, except that the error introduced by section thickness (Section 4.3.4) may be affected by the sectioning angle.

6.4.3. ESTIMATING SURFACE DENSITY IN ANISOTROPIC STRUCTURES

For isotropic structures we found (Section 2.3.3) that the boundary length density, B_A, is related to surface density, S_V, by

$$B_A = B_a/A_c = (\pi/4) \cdot S_V \tag{6.24}$$

In contrast to the ratio of two areas discussed above the ratio of boundary length to area of the containing space does depend on the sectioning angle θ, so that we find, in general,

$$B_A(\theta) = k(\theta) \cdot S_V \tag{6.25}$$

The factor $k(\theta)$ depends, however, on the shape of the anisotropic elements and it must therefore be specified for certain model assumptions. We shall do this for the two basic patterns of fasciculated tubular and lamellar structures (Fig. 6.11a and b).

Taking first the case of *fasciculated tubules*: we define the sectioning angle θ as the angle between the direction of the tubules and the normal to the section plane. We find that the profile area increases more rapidly with θ than the boundary of the elliptic tubule profiles (Volume 2, Chapter 10); it is found that

$$B_A(\theta) = S_V \cdot [\sqrt{((1 + \cos^2 \theta)/2)}] \tag{6.26}$$

so that

$$k(\theta) = \sqrt{((1 + \cos^2\theta)/2)} \tag{6.27}$$

From this we derive some interesting results: if the fasciculated structure is cut at right angles ($\theta = 0°$) we find

$$B_A(\theta = 0°) = S_V \tag{6.28}$$

but for a longitudinal section ($\theta = 90°$)

$$B_A(\theta = 90°) = S_V \cdot \sqrt{\tfrac{1}{2}} \tag{6.29}$$

[Note that the angle θ is measured between the direction of the tubules and the normal to the section plane (Fig. 6.9) rather than the section plane, itself, as was the case in Weibel (1972) and Eisenberg et al. (1974). This introduces some slight differences in the formulation of $k(\theta)$.]

Looking next at the case of *lamellar structures* we need to define the sectioning angle differently. The orientation of the lamellae is unambiguously defined by the direction of the normals to their surface (the y-direction in Fig. 6.10); the sectioning angle ψ is therefore the angle between this direction and the normal to the section plane (Fig. 6.12). As will be developed in Volume 2, Chapter 10 we find

$$B_A(\psi) = S_V \cdot [\sin \psi] \tag{6.30}$$

so that

$$k(\psi) = \sin \psi \tag{6.31}$$

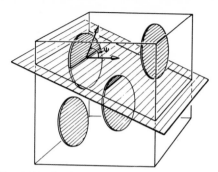

FIG. 6.12: Cutting anisotropic lamellar structure.

A cross-section of the lamellar structure is obtained if $\psi = 90°$; it is easily seen that in this case again

$$B_A(\psi = 90°) = S_V \qquad (6.32)$$

Notice that there was one unique cross-section for the fasciculated structures, but that there are many section planes which cut the lamellar structures perpendicularly. The section plane with $\psi = 0°$ is a plane parallel to the discs and is therefore not useful because B_A vanishes as $\psi \rightarrow 0°$.

The practical conclusion to be drawn from this study of two distinct models of anisotropic structure is that there must be two basic ways to approach the problem of stereological analysis of anisotropic structures:

(1) to cut the section at an arbitrary angle and to estimate $k(\theta)$ or $k(\psi)$ in order to derive S_V from the estimate $B_A(\theta$ or $\psi)$ obtained on this section (equations 6.26 and 6.30);

(2) to cut cross-sections and obtain S_V directly from the estimate of B_A; it is fortunate that neither $k(\theta)$ nor $k(\psi)$ deviate appreciably from 1 if the sectioning angle deviates by $\pm 10\text{--}15°$ from a true cross-section.

In either of these two cases B_A is estimated by an intersection count using a convenient test system (see Section 4.2.3). In the case of anisotropic structures it is advisable to use an *isotropic test system*, such as that of Merz (1967) made of a sequence of semicircles (see Fig. 4.11; Appendix 3, L 100). By equation (2.42) we have

$$B_A = \frac{\pi}{2} \cdot I_L \qquad (6.33)$$

so that we find

$$S_V(\theta \text{ or } \psi) = k^{-1}(\theta \text{ or } \psi) \cdot \frac{\pi}{2} \cdot I_L \qquad (6.34)$$

For the curvilinear test system with one semicircular arc per each of the test points spaced at a distance d (Section 4.2.3.3) this is explicitly

$$S_V(\theta \text{ or } \psi) = k^{-1}(\theta \text{ or } \psi) \cdot \frac{I_a}{P_c \cdot d} \qquad (6.35)$$

where I_a are the intersections with the surface trace of a and P_c are the test points falling onto the containing space. Evidently I_a and P_c are the sums of intersections and point hits obtained over a representative sample (Section 3.7).

Fasciculated structures offer a third possibility which has been advocated repeatedly (Sitte, 1967; Eisenberg et al., 1974; see also Section 8.4): the study of longitudinal sections. Here again B_A can be estimated by an intersection count with an isotropic test system; S_V is then obtained by equations (6.34) and (6.35), i.e. using $k^{-1}(\theta) = \sqrt{2}$.

Sitte (1967) has proposed the estimation of I_L with a quadratic test system. In this case I_L depends on the angle α between test lines and direction of anisotropy; the "correct" or average value of I_L is obtained if the test lines are oriented at an "optimal" angle $\alpha = 19°$ (and $71°$) to the direction of anisotropy (Fig. 6.13). The rationale for this arrangement will be discussed in Volume 2, Chapter 10; it is related to the fact that with a square lattice the number of intersections is proportional to $(\sin \alpha + \cos \alpha)$ and that $\alpha = 19°$ gives the mean value of $4/\pi$ that is required.

Finally we should mention the method proposed by Hilliard (1967b): if a test line in the form of that shown in Fig. 6.14 is laid down on a longitudinal section with the arrow pointing in the direction of anisotropy, an intersection count I_L can be used directly to calculate S_V from

$$S_V = 2 \cdot I_L$$

It should finally be noted that the methods proposed for longitudinal sections of fasciculated structures work also for cross-sections or longitudinal sections of lamellar structures.

We should not end this section without a strong word of warning. It has become clear that the methods proposed are all related to a particular model structure: fasciculated or lamellar structures. This clearly demonstrates that the stereological evaluation of anisotropic structure is a tricky matter and must be preceded by a careful analysis of the type and degree of anisotropy prevailing. Eisenberg et al. (1974) have shown, for example, that the sarco-tubular reticulum of skeletal muscle is only partially oriented; this can be obtained by recording intersection counts on a longitudinal section as a function of the angle α and comparing $I_L(\alpha)$ with the theoretically expected curve. A plot of this kind is shown by Eisenberg et al. (1974); Underwood (1968) extensively discusses the "rose of orientation" plots where the inter-

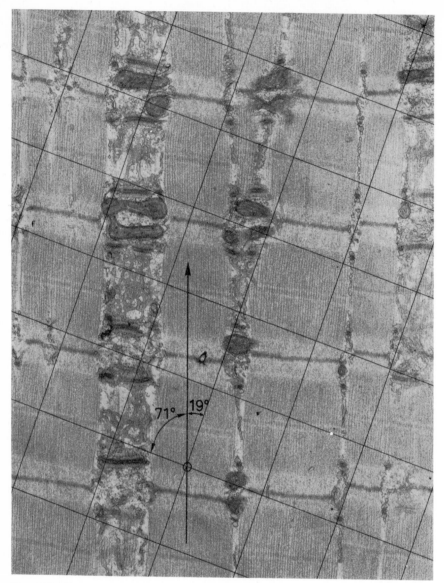

FIG. 6.13: Electron micrograph of part of skeletal muscle fiber cut longitudinally with square lattice test system oriented at "optimal angle" of 19° and 71° to long axis.

section counts are plotted against polar coordinates for the angle. Once the pattern and degree of anisotropy are determined a proper stereological approach can only be set up by working through an appropriate stereological

model (see Volume 2, Chapter 10). Further parameters characterizing anisotropic structures will be discussed in Volume 2, Chapter 10.

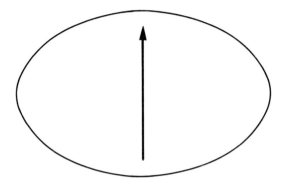

FIG. 6.14: Test figure of Hilliard (1967b) for obtaining intersection counts on longitudinal section.

6.5.   Stereological methods in quantitative autoradiography

In autoradiography a radiation source—usually a radio-actively labeled
precursor of some cell constituents—is introduced into the tissue. The
distribution of this source to the various tissue components is then estimated
from the distribution of silver grains as they are developed, after suitable
exposure, from a thin layer of photographic emulsion overlayed over the
section. To make this approach quantitative meets with a number of problems which can partly be solved by means of stereological methods.

The first problem is that a given component is the more likely to contribute
radiation sources the more frequent it is. In order to interpret autoradiographs it is hence essential to dispose of estimates of morphometric parameters such as the volume density of the various components or the surface
density of different membrane types; this information is evidently easily
obtained by stereological methods. Ross and Benditt (1962) were the first to
recognize this and to use point counting for estimating the relative volumes
of cell organelles in autoradiographs.

The second problem stems from the fact that radiation spreads from the
source and randomly hits a silver grain at a variable distance from the
source, i.e. the radiation source is usually not "exactly" under the developed
grain. Much careful work has been done to work out the probability distribution of grains around a radiation source (see Salpeter and Bachmann,
1972); the general result is that 95% of the grains should be located within a
circle of 200–300nm diameter around the source. It is thus evident that

localization of the potential radiation source may be ambiguous in high resolution autoradiography, as it is practiced in electron microscopy, because many cell constituents are of the same size or even smaller than this 95% probability circle. Indeed, all components within this circle around the grain are more or less equally likely to have contained the radiation source; this concept has been exploited by Whur et al. (1969). To be more precise, one could say that the probability that the radiation source was located in one or the other cell component depends on its relative volume within the section in the region marked by the 95% probability circle around the grain; here again stereology could be of use in estimating this probability.

This is not the place to discuss autoradiography in all its aspects; rather, we shall limit ourselves to a few suggestions on improving the quantitative evaluation of autoradiographs using stereological methods.

The general aim of quantitative autoradiography should be to estimate, as objectively as possible, a *specific radiation label density* for the various cell components, i.e. the (relative) amount of label contained in the unit volume of a particular organelle or compartment. This requires two steps: (1) estimation from autoradiographs of the radiation label density, R_{Vi}, associated with the different compartments, expressed as the relative number of grains on a compartment per unit volume of cell (or tissue); (2) estimation of the volume density, V_{Vi}, of that compartment in the cell (or tissue). The specific radiation label density is then obtained as

$$R_{Vi}^* = R_{Vi}/V_{Vi} = R_i/V_i \qquad (6.36)$$

As mentioned above, one of the major problems of autoradiography is the fact that the location of a developed silver grain indicates the location of the radiation source in terms of a certain probability (Caro, 1962; Salpeter and Bachmann, 1964, 1972; Bachmann et al., 1968; Whur et al., 1969). Although opinions differ regarding resolution (see Salpeter and Bachmann, 1972), there is general agreement that the grains are distributed around the radiation source in a probability pattern as shown in Fig. 6.15a. Inverting the argument, we note that the probability for source location decreases according to a bell-shaped function constructed around the grain. In analogy to the work of Bachmann et al. (1968) one may, therefore, draw three concentric circles around the grain in such a way that they delimit the fields which contain the radiation source with a 30, 60, and 90% probability (Fig. 6.15a, b); each of the three shaded areas in Fig. 6.15b contains the radiation source with 30% probability, although their areas increase towards the periphery. In order to estimate the probability with which the radiation source lies within the various cell compartments we would need to estimate the composition of each of these areas and weigh it according to decreasing probability for containing the radiation source. Composition is easily estimated by performing a point count on each of the areas; a weighted estimate is obtained directly, if the same number of test points is applied to each area. Figs. 6.15b–d show three arbitrary examples of the kind of test point lattices one could construct in that way; note that one or two additional points are placed outside the outermost circle to account for the 10% probability that the radiation source lies beyond that circle.

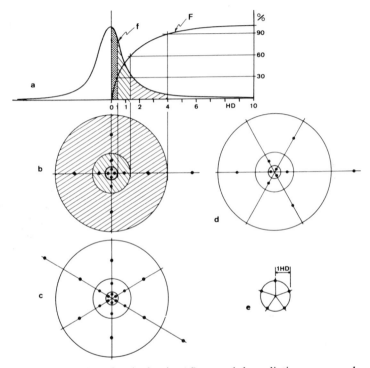

FIG. 6.15: (a) Distribution of grain density (f) around the radiation source and probability (F) that the radiation source lies within a circle around the developed grain, expressed as half-distances (HD); drawn according to Bachmann *et al.* (1968). (b)–(d) Test point systems for estimating the probability of location of the radiation source. (e) Simplified test system of five points on a circle having a half-distance radius. (Reproduced by permission from Weibel and Bolender, 1973.)

Any number of similar test systems can evidently be constructed by the same principle. In practice one will find that complex test point sets of this kind are difficult to use. Indeed, they may estimate the probability of source location with too much precision. After all, one is dealing with a complex statistical process where a number of independent probabilistic events lead to the formation of a silver grain. The significance of the estimation of source location probability is hence influenced both by the precision with which the source is located for an individual grain and by the number of grains evaluated; but the total number of grains evaluated has a much greater effect on the over-all precision of the estimate. If a large number of grains is available for analysis it appears therefore preferable to simplify the test point set placed around the grain. One example is a set of 5 points placed on a circle at about the "half-distance" around the grain, i.e. the circle which contains the radiation source with 50% probability (Fig. 6.15e).

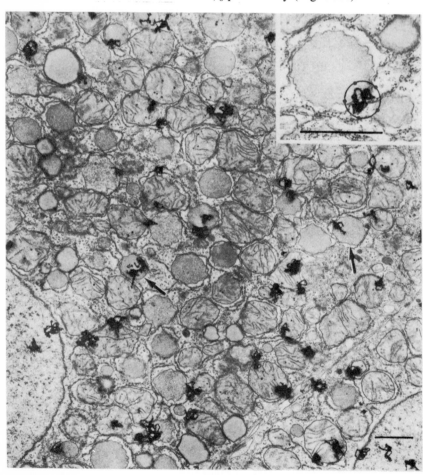

FIG. 6.16: Electron microscope autoradiograph of hepatocyte three days after injection of tritiated arginine to test for localization of newly synthesized protein. Inset shows silver grain lying partly over a peroxisome (arrows); five test points are arranged around the grain, in this case on the 95% probability circle, to obtain the "grain associated point count". Bar 1 μm. (Reproduced by permission from Stäubli et al., 1977.)

The practical procedure is now to center such a test point set around the individual silver grains (Fig. 6.16) and to count the number of points, G_i, falling onto compartment i. If this is done for a total number R_T of grains the total number of points counted is

$$G_T = R_T \cdot g \tag{6.37}$$

where g is the number of points in the point set, e.g. $g = 5$ in Fig. 6.16. If G_i points fall on component i the probable relative radiation label density on this component is

$$R_{Ri} = G_i/G_T \tag{6.38}$$

Note that G_i is the sum of all the points falling on i for all the R_T silver grains, and that the sum of R_{Ri} over all compartments is equal to 1.

In order to estimate from this grain-associated point count the "true" label density R_{Vi} we need to consider the volume of tissue within which the source may have been contained; for this we need to know the section thickness t (see Section 4.3.4.5) and the section area occupied by the cell, A_c, or by the profiles of compartment i, A_i. These profile areas can be obtained by point counting, using, for example, a square lattice with point spacing d (see Section 4.2.2); with P_c and P_i points falling onto profiles of the cell or compartment i we find

$$A_c = P_c \cdot d^2$$

$$A_i = P_i \cdot d^2$$

It is easily seen that the probable label density for compartment i in the entire cell is

$$R_{Vi} = G_i/(g \cdot A_c \cdot t) = G_i/(P_c \cdot g \cdot d^2 \cdot t) \tag{6.39}$$

If we now relate the grain-associated point counts, G_i, to the volume of compartment i contained in the section we obtain immediately the probable specific label density

$$R_{Vi}{}^* = G_i/(g \cdot A_i \cdot t) = G_i/(P_i \cdot g \cdot d^2 \cdot t) \tag{6.40}$$

It is readily apparent that the formula for specific label density corresponds to

$$R_{Vi}{}^* = R_{Vi}/V_{Vi} \tag{6.41}$$

the ratio of radiation label density to volume density for compartment i. It is also evident that the specific label density is essentially the ratio of a "grain-associated" point count to a random point count, G_i/P_i, with some calibration factor related to the characteristics of the test system and to section thickness.

This approach differs from the simpler methods for quantitative autoradiography in that it makes the location of the radiation source more objective. The location of the grain around the source is the result of a probabilistic event; it is hence logical to give this aspect full consideration in the process of evaluation of the autoradiographs. Whur *et al.* (1969) have achieved this partially by attributing each grain fractionally to the various components contained in a 95% probability circle; the approach suggested here simply extends this notion by exploiting the possibilities offered by stereology.

This method has been used in a study of the proliferative response of hepatic peroxisomes to some drug treatment (Stäubli *et al.*, 1977). To estimate specific label densities was particularly important because drug treatment induced a massive change in cell composition, the over-all volume of peroxisomes being increased by a factor of 7 at the peak of drug effect, with corresponding changes in the relative volumes of other organelles. It was thus possible to work out a time course for peak specific radiation label densities from endoplasmic reticulum through Golgi to peroxisomes; an interpretation of this kind would have been difficult on the basis of simpler radiation label density estimates because the effect of changes in V_{Vi} would have remained undetermined.

6.6.   Stereological methods in histochemistry and cytochemistry

Histochemistry and cytochemistry have become important tools for the functional characterization of tissue and cell constituents: their chemical nature or biochemical function are revealed by the deposition of some specific reaction product (Pearse, 1960; Burstone, 1962). This approach allows a functional differentiation of constituents which are similar on purely morphological grounds.

The possible interaction between cytochemistry and stereology is twofold: (1) stereological methods could be used to quantify the qualitative results of a cytochemical reaction; (2) cytochemical methods could be used to tag cell (or tissue) constituents for a specific function in view of a stereological analysis. The two objectives are similar but yet different in some respects.

The quantification of a cyto- or histochemical reaction is best performed by either densitometry or scanning photometry; this is however only feasible if the reaction is quantitatively controlled, i.e. if the amount of reaction product ("stain") deposited is strictly proportional to the amount of chemical constituent or enzyme present. This is the case for the Feulgen reaction for DNA and for some similar reactions; it fails however for many other methods, such as most of the enzyme reactions, because the reaction conditions (substrate concentration, pH, temperature, etc.) are not as easily controlled in the tissue as in the test tube. Partial quantification can be achieved in such cases by using point counting techniques: this allows determination of the relative volume of the tissue constituents that have reacted, irrespective of the intensity of the reaction and its local variations, but it evidently relies on having some reaction product deposited in the right place!

Cell membranes are structurally very similar although they perform widely differing functions; whilst they can be differentiated by context and configuration in intact cells these criteria fail in subcellular fractions obtained from homogenates. In such instances one may resort to a tagging of the various membrane fragments by means of a cytochemical reaction for specific marker enzymes. Another very promising technique is that of immunocytochemistry (Kraehenbühl and Jamieson, 1972; Papermaster *et al.*, 1978): specific membrane constituents are identified by having them labelled by an antibody tagged with ferritin or microperoxidase, for example. In either case of membrane tagging the basic measurement required from a stereological approach is the relative membrane area marked by the tag, S_{Sa}. It is obtained by a simple intersection count using any convenient test line system and recording intersections with marked (I_a) and unmarked (I_0) parts of the membrane trace. The marked surface fraction is

$$S_{Sa} = I_a/(I_a + I_0) \tag{6.42}$$

where, evidently, the sums of intersections obtained over a representative sample of micrographs must be entered (see Section 3.7).

In the case of ferritin labelled antibody techniques the ultimate aim is to estimate the number of antibodies bound to the unit area of membrane. It should be possible to obtain this estimate by a stereological approach, but the situation is a bit tricky because it requires factors such as section thickness and inclination of the membrane profile to the section plane to be taken into consideration. There are some first attempts in this direction but the concepts are not yet sufficiently advanced to expand on this topic any further at this point.

6.7. Stereological methods in freeze-fracture electron microscopy

The method of freeze-fracturing (Moor, 1969) has become a tool of increasing importance in analysing cells and, particularly, the specialization of their membranes. The main virtue of this method is that it splits membranes along a mid-plane and thus reveals their population of intramembranous particles, which is to some extent characteristic of the membrane type (Losa *et al.*, 1978). But this virtue is evidently also a major handicap for the possible use of stereological methods in the quantitative evaluation of freeze-fracture preparations: the fracture faces—whose texture is revealed by shadow casting with heavy metals—are, in general, non-flat, and they are heavily biased towards the membranes. Standard stereological techniques are therefore not strictly applicable, except under very special circumstances.

If the special events of "fracturing" and "shadow-casting" are properly considered it appears however well possible to develop quantitative methods of analysis based on the principles of stereology, but these methods will have to be designed for specific purposes and will therefore have limited applicability.

One such method will be briefly explained (Weibel et al., 1976). It was designed to allow, in microsomal fractions of liver tissue, a quantitative estimation of the relative membrane area of vesicles derived from the various membrane classes of liver cells (Losa et al., 1978). It was found that the various membrane classes could be characterized by adquately separated ranges of particle density $\{\alpha\}$, where α is the number of intramembranous particles per μm^2 of membrane. The general strategy was therefore to classify each vesicle according to its particle density α and to obtain the relative number $N_N\{\alpha\}$ of vesicles falling into a specified range $\{\alpha\}$. By stereological considerations a relationship between $N_N\{\alpha\}$ and the relative surface, $S_S\{\alpha\}$, represented by these vesicles in the fraction was to be found. These two relative parameters are defined as follows:

$$N_N\{\alpha\} = N_V\{\alpha\}/N_V(T) \qquad (6.43)$$

$$S_S\{\alpha\} = S_V\{\alpha\}/S_V(T) \qquad (6.44)$$

where $N_V(T)$ and $S_V(T)$ are the total number and membrane surface of all vesicles in the unit volume of the microsomal fraction. Because of the asymmetry of the membranes all particle counts were to be obtained on the outer membrane leaflet of the vesicles, i.e. on concave vesicle profiles (Weibel et al., 1976).

The validity of this strategy evidently depends on whether the following postulates are satisfactorily fulfilled:

(1) The various membrane classes (endoplasmic reticulum, plasma membrane, mitochondrial membranes, etc.) are unambiguously identifiable by a range of particle density $\{\alpha\}$ on concave vesicle profiles.

(2) The membranes vesiculate consistently upon homogenization; e.g. the outer leaflet of an ER cisterna consistently becomes the outer leaflet of a microsomal vesicle, etc.

(3) The freeze-fracture procedure and the selection of cast-shadow-free concave profiles yields an unbiased sample of the vesicle population.

(4) A stereological method exists by which the relative surface of membranes with a certain range of particle densities can be derived from the numerical frequency of vesicles showing this particle density.

It has been shown that postulates (1) and (2) are satisfactorily fulfilled (Losa *et al.*, 1978); postulates (3) and (4) are stereological in nature and need some comment (see Weibel *et al.*, 1976 for details).

The stereological considerations are facilitated by the special situation that in the microsomal fraction the membranes occur in the form of spherical vesicles of varying size. As the cleavage plane fractures the ice block it will hit those vesicles for which the distance of the center from the cleavage plane is less than the radius (Fig. 6.17). The membrane of these vesicles splits in half; if the center lies below the plane a convex vesicle profile is formed, if the center lies above a concave profile results. This is revealed by shadow-casting the fracture surface with heavy metal (Fig. 6.18).

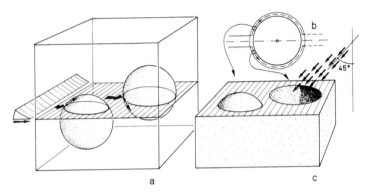

FIG. 6.17: Cleavage of ice block with membrane vesicles (a) produces fracture surface (c) with convex or concave profiles depending on whether fracture plane hits vesicle above or below equator (b). Cloud of small arrows in (c) symbolizes shadow casting at 45°. (Reproduced by permission from Weibel *et al.*, 1976.)

The first conclusion from stereology (see Section 2.6) is that the numbers of concave and convex vesicle profiles should be equal, and this is indeed found to be the case. The second conclusion is that the diameter of the profiles depends (1) on the diameter of the vesicle, and (2) on the distance between the center and the fraction plane (see Figs 2.36–38).

The purpose of the analysis was to determine the density of intramembranous particles by counting within a small test circle applied to the center of concave profiles as shown in Fig. 6.19 (Losa *et al.*, 1978). This is only possible with profiles whose surface texture is not obscured by the cast shadow of the profile edge (Figs 6.18 and 19). Such profiles are formed if the cleavage plane hits the vesicle below the lower 45° tangent point, assuming that the heavy metal is shot at the fracture surface at an angle of 45° (Fig. 6.17). Looking at Fig. 6.20 and considering what was said in Sections 2.1.2

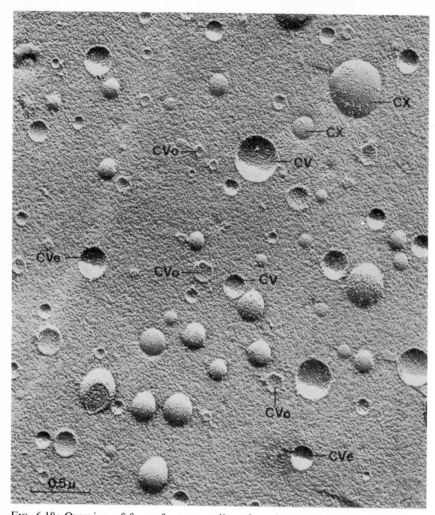

Fig. 6.18: Overview of freeze-fracture replica of a microsomal fraction. Vesicles of various origin form convex (*CX*) or concave (*CV*) caps. Concave caps without cast shadow (*CVo*) are used for analysis. Note "equatorial" profiles (*CVe*) with half the cavity covered by cast shadow. Bar, 0.5 μm. Magnification 32'000 ×. (Reproduced by permission from Losa *et al.*, 1978.)

and 2.6.1 we can immediately derive that the number of such profiles per unit area of the fracture plane is

$$N_A^* = N_V \cdot X$$
$$= N_V \cdot (1 - \sqrt{\tfrac{1}{2}}) \cdot R \tag{6.45}$$

And since the total number of profiles is (equation 2.77)

$$N_A = N_V \cdot 2R \qquad (6.46)$$

the relative number of concave profiles free of cast shadow is

$$N_N{}^* = (1 - \sqrt{\tfrac{1}{2}})/2 \approx 14.65\% \qquad (6.47)$$

It can be shown that this is independent of the size distribution of vesicles; it is also confirmed by actual counting that about 15% of all profiles are "concave and free of cast shadow". Stereological considerations have therefore shown that the sample of vesicle profiles available for this anslysis is unbiased; postulate (3) appears fulfilled.

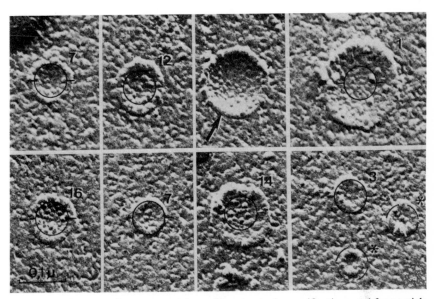

FIG. 6.19: Montage of concave vesicle profiles at actual magnification used for particle counting. Test circles are superimposed and the particle numbers counted indicated. Arrow indicates profile not evaluated because of cast shadow. Asterisks mark profiles lost because they are smaller than the test circle. Bar, 0.1 μm. Magnification 125'000 × . (Reproduced by permission from Losa *et al.*, 1978.)

A certain bias is however introduced in the actual analysis by applying a test circle of finite size to these profiles, for this eliminates a number of very small profiles from the sample, namely those with diameter smaller than the test circle (Fig. 6.19); this corresponds to the loss of cap sections in counting and sizing of profiles as discussed in Section 4.4.2 and 5.2.3, and given in more detail in Volume 2, Sections 4.4 and 5.3. A biased sample results because the population of small vesicles is more seriously affected than that of large vesicles (Weibel *et al.*, 1976). An additional size-dependent error is introduced by the finite test circle size because the particles counted within the circle are contained in a curved membrane surface which is the larger the smaller the vesicle size; in practice, this error is however small and can in general be disregarded (Weibel *et al.*, 1976).

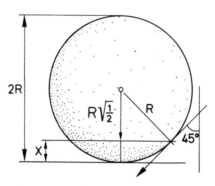

FIG. 6.20: Sampling of concave profiles without cast shadow. (Reproduced by permission from Weibel *et al.*, 1976.)

It is however evident that the sample of profiles for which α can be determined is biased; a stereologically valid method for estimating $S_S\{\alpha\}$ from $N_N\{\alpha\}$ needs to account and correct for this bias. It was found that this depends on the size distribution of the vesicles whose radius R is expressed as a multiple q of the (constant) test circle radius g:

$$R = q \cdot g \qquad (6.48)$$

We shall not go into the details of the development (see Weibel *et al.*, 1976), but just state the final result. Designating by E_α ($\cdot$) and E_T ($\cdot$) the expected (mean) values of the size parameter ($\cdot$) in the $\{\alpha\}$ vesicle population and the total (T) population respectively, the relative membrane surface area of $\{\alpha\}$ vesicles, $S_S\{\alpha\}$, can be estimated from the observed relative number of vesicles with particle density $\{\alpha\}$, $N_N\{\alpha\}$, by

$$S_S\{\alpha\} = N_N\{\alpha\} \cdot \frac{E_\alpha(q^2)/\{E_\alpha(\sqrt{(q^2-1)}) - [\sqrt{\tfrac{1}{2}}]E_\alpha(q)\}}{E_T(q^2)/\{E_T(\sqrt{(q^2-1)}) - [\sqrt{\tfrac{1}{2}}]E_T(q)\}} = N_N\{\alpha\} \cdot \frac{\lambda_\alpha}{\lambda T} \quad (6.49)$$

The relative membrane area of α vesicles can hence be derived from the estimated numerical frequency $N_{N\alpha}$ if the size distribution dependent coefficient

$$\lambda = \frac{E(q^2)}{\{E(\sqrt{(q^2-1)}) - [\sqrt{\tfrac{1}{2}}]E(q)\}} \qquad (6.50)$$

can be estimated for the total vesicle population and for $\{\alpha\}$ vesicles.

It is evident that the correction factor λ_a/λ_T depends on the degree by which the size distribution of $\{\alpha\}$ vesicles deviates from that of the total vesicle population; its value may be >1 if the mean size of $\{\alpha\}$ vesicles is smaller than the population mean, or <1 in the reverse case (Weibel et al., 1976). In the practical case of estimating the proportion of ER-derived vesicles in liver microsomes we found λ_{er}/λ_T to be about 0.98, but it is different for other membrane types (Losa et al., 1978).

The application of this method to liver microsomes has shown it to give quite reproducible results. It revealed some most intriguing findings, namely that a smaller fraction of the microsomal membranes than expected was derived from endoplasmic reticulum (Losa et al., 1978).

The method here described applies to a very special case and cannot claim broad validity. It is presented as an example of how stereological reasoning can help to set up methods of morphometric evaluation of membrane systems in conjunction with the valuable method of freeze-fracturing which, at first sight, appears to preclude a stereological approach.

6.8. Stereological methods in scanning electron microscopy

Stereological methods have not been used to any appreciable extent in conjunction with scanning electron microscopy, at least not in biology. The reason is that the basic operations of scanning electron microscopy (SEM) and stereology appear to be mutually exclusive: in SEM one looks at "natural" surfaces of the objects whereas in stereology objects need to be sectioned.

One could imagine some applications of stereology in SEM. Firstly, on sections SEM can differentiate components of material by their atomic composition because the secondary electrons emitted and recorded by the detector depend on the atomic weight; this is being exploited in the materials sciences, e.g. in the study of metal alloys, and one could imagine some contributions of this effect in biology as well. Secondly, it should appear possible to use a stereological approach in attempting to quantify some surface characteristics, e.g. the density of microvilli on a cell surface, or of antibodies tagged with a large marker binding to a surface. The main problem in developing such methods is that parts of the specimen surface are usually obliterated and that the surface projection is distorted due to undetermined inclination of the surface to the electron beam. One will have to devise special sampling methods for special cases, similar to the approach outlined in Section 6.7 for freeze-fracture preparations.

In porous structures, such as trabecular bone (Whitehouse, 1974a) or lung, SEM can afford the possibility of looking beyond the section plane and this can, for certain purposes, be a valuable aid in identifying structural components. Stereology is then done on the sectioned surface (Whitehouse, 1974b), but one may still have to cope with the problems of distortion (Hilliard, 1972).

CHAPTER 7

Performing a stereological study

To add a stereological analysis to a simple morphological study means to add cost—in general important costs in terms of labor, but also investment in some equipment. Stereology adds precision to the study, but that also means that the study becomes more sensitive to "irrelevant" effects such as artefacts, differences in fixation and tissue processing etc. Very careful planning and good control of all steps is therefore an essential prerequisite for any study in which it is intended to use stereological methods. It extends from setting up a strategy and operational plan over proper consideration of specimen preparation to the choice of appropriate instrumentation. This chapter should therefore offer a few guidelines in regard to this effort in planning.

7.1. Strategy and operational plan

To work with microscopic images offers so much esthetic satisfaction that morphologists often tend to assume in their research more the attitude of an artist or of an explorer than that of an analytical scientist; intuition is their main guide. Morphometry by stereological methods is sober business, to a large extent boring, exciting only in the results that come out after lengthy calculations; the work must follow a carefully elaborated plan.

When setting up this operational plan we must begin at the end, i.e. by asking what kind of result we expect. In other words, before engaging in the stereological study we must formulate the *hypothesis* which we would like to test. To be able to do so we must already have done extensive explorative work, we must already have "discovered"—or believe ourselves to have done so—the structural changes we are interested in or the relevant structure-function relationship. Morphometry *per se* rarely leads to discoveries, but it yields "hard data" on structure that may be tested for their relevance by statistics, and that may eventually lead to new insights.

With the hypothesis formulated—if possible in the form of mathematical functions—one should be able to decide on what are the *most informative stereological parameters*, and what parameters may yield additional valuable information. In other words, one should list the possible morphometric descriptors in order of priority with respect to testing the hypothesis. Typical questions to be answered are: do simple bulk measurements on aggregates, such as volume and surface density, yield adequate information or are particle statistics such as particle number and size distribution essential? (See Section 4.1 and Weibel, 1974).

The second planning step is to decide on the *microscopic methods* that are required (light or electron microscopy?), and to set up the *sampling scheme* accordingly. Can we proceed by means of a single sampling stage or do we need multiple stage sampling? To answer this question it is in general necessary to *set up the computation formulas* that lead from primary measurements or counts to the final parameters, and this should consider the statistics to be used.

Careful consideration must then be given to the *preparation* of the specimens, fixation and processing, and to the *background information* that is required. Do we need to do some physiological measurements before the organ is fixed? Do we need to determine and record such important base line data as body weight, sex, age, organ size, ancillary observations on other organs, etc.? Experience shows that one never goes wrong in recording as many of these base line data as possible, because they may become important in the subsequent explanation of unexpected results. In this respect it is advisable to prepare data sheets which allow these data to be recorded in an orderly fashion and which serve as check lists during the experiments. A small detail: one should also note down who has been directly involved in a particular experiment and at what step (animal caretaker, technician, experimenter, etc.).

Last but not least, *economic considerations* are important. How much time may be invested? How can we arrive at the result with least cost in terms of time of investigators and technical personnel, of animals and material? What equipment is available or required?

When all this planning is completed it is most important to perform a *preliminary experiment* in order to test whether the procedure selected works satisfactorily. This will allow final decisions to be made on the sampling scheme and on the choice of magnification and test system. This is important because one should avoid changing the procedures in the midst of an experiment.

7.2.   Organ size and body weight

It need not be emphasized that recording the size of the organ under consideration is one of the most important bits of background information. Most stereological estimates come out as densities per unit volume; knowing the the organ volume, V_0, one can then obtain estimates of the absolute size of the parameter investigated. For example, estimating surface density, S_{Vi}, of component i, one can calculate its total surface in absolute terms by:

$$S_i = V_0 \cdot S_{Vi} \qquad (7.1)$$

The simplest method for measuring the organ volume is that of water displacement (Fig. 7.1). A suitable graduated cylinder is filled with water or buffer solution to a certain level (V_1); the organ is then totally submerged and the volume V_0 is read off as the difference between the new fluid volume V_2 and the original volume V_1:

$$V_0 = V_2 - V_1 \qquad (7.2)$$

An alternative method, particularly suited for large organs, is that of overflow: fill a cylindrical vessel to the brim, until the first drop flows over the edge; then submerse the organ and remove it; fill the cylinder again and record the volume of fluid required until the first drop flows over the edge: this volume is V_0.

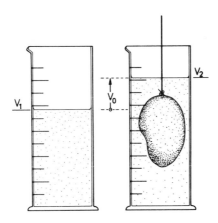

FIG. 7.1: Measuring organ volume by fluid displacement.

I

More precision is achieved by using the submersion method of Scherle (1970) in which the water displacement due to organ volume is recorded by weighing (Fig. 7.2). A sufficiently large container filled to about 2/3 with buffer or isotonic saline is placed on a balance and the initial weight W_1 is read off. The organ is now totally submerged (see below), whereby care must be taken that it touches neither the vessel wall nor its bottom; the resulting weight W_2 is read off. The weight of the fluid displaced by the organ is found by

$$W_0 = W_2 - W_1 \tag{7.3}$$

If the specific gravity g of the fluid is known then the organ volume is obtained by

$$V_0 = W_0/g \tag{7.4}$$

As the specific gravity of isotonic saline is 1.0048 one can, without undue error, disregard g and set

$$V_0 \approx W_0 = W_2 - W_1 \tag{7.5}$$

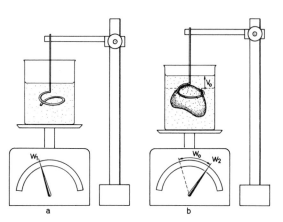

FIG. 7.2: Measuring organ volume by weighing displaced fluid according to method of Scherle (1970).

An important condition must be met for this method to work: the difference in specific gravity between the organ and the submersion fluid must be compensated by outside force. If the organ is heavier than the fluid—as is the case for liver and many solid organs—it must be suspended on a stand by a thin thread; if it is lighter—as with the lung—the buoyancy must be compensated by carefully pushing it down by means of a clamp or a wire spiral or the like (Fig. 7.2). If such a device is used its contribution to water displacement must be eliminated; this is best done by recording W_1 with the device lowered into the fluid to the same level as is required for submersing the organ (Fig. 7.2a).

It need not be stressed that with any of these methods it is important that the organ be carefully freed of attached tissue such as fat, etc., and that it should be blotted free of excess fluid before submersion.

In the normal animal the organ size is usually found to be in consistent relation to body weight, W. This allows the compensation of variations in body size within the population of experimental animals by calculating "specific organ volumes", i.e. the organ volume per unit body weight

$$V_W = V_0/W \qquad (7.6)$$

or "specific parameters", such as a specific surface

$$S_{Wi} = S_i/W = S_{Vi} \cdot V_0/W \qquad (7.7)$$

This procedure of normalizing data with respect to body size must, however, be handled with caution, particularly if body weight varies over a large range. The relations to body weight are not always simple; differences in species, strain, and sex may influence it. One must also consider scaling through allometric relations where a parameter Y is related to W through a power function of the type

$$Y = aW^b \qquad (7.8)$$

In this case a proper normalization procedure would require the absolute morphometric parameter to be divided by W^b rather than by W, as was done in equations (7.6 and 7.7).

7.3.   Specimen preparation

The procedures of fixation and processing to be used in the framework of a stereological study depend very much on the specific conditions of the study. But there is one basic rule that must be observed: the dimensions of cells and other tissue constituents should be preserved, or, if changes are unavoidable, the amount of dimensional change (shrinkage or swelling) should be controlled or at least assessed.

Processing for light microscopy almost inevitably results in shrinkage, particularly with paraffin embedding (see Section 4.3.6). The situation is considerably improved with embedding in resins; embedding in epoxy resins, as practiced for electron microscopy, induces only a small percentage of shrinkage (Weibel and Knight, 1964).

Dimensional changes as they occur during fixation are sometimes more difficult to assess and control. With buffered solutions of fixatives (glutaraldehyde or osmium tetroxide) one attempts to reduce such artefacts by adjusting the osmolarity to the osmotic pressures prevailing in the cells.

Some debate has recently arisen as to whether the total fixative solution or the buffer in which the fixative is dissolved should be made isotonic. Some authors have suggested that glutaraldehyde *per se* contributes little or nothing to the osmotic pressure of a buffered glutaraldehyde fixative (Arborgh *et al.*, 1976); on the other hand it has been claimed that the osmolarity of glutaraldehyde cannot be disregarded altogether. Barnard (1976) arrived empirically at the conclusion that the "effective osmotic pressure" (EOP) of a fixative was the sum of the osmolarity of the buffer (M_B) plus half the osmolarity of glutaraldehyde (M_G):

$$EOP \sim M_B + \tfrac{1}{2}M_G \qquad (7.9)$$

The experimental analysis of this problem is rendered difficult by the fact that the "true" dimensions of cells and their constituents are usually unknown. Mathieu *et al.* (1978) have therefore attempted to study the relative osmotic effects of glutaraldehyde and buffer solutions by using the rat lung as a "biological osmometer": by instilling the fixative solution into the airways of a collapsed lung it is brought into immediate contact with the air–blood barrier which separates it from the capillary blood. The fixative reaches the blood after traversing two thin cell layers, each bounded by two plasma membranes and about 0.2 μm thick. The osmotic effect of the fixative, as it acts on tissue and capillary blood, can be ascertained (1) by shape changes of erythrocytes, and (2) by the thickness of the air–blood barrier, the latter being measured by its harmonic mean.

Let us look at the dimensional changes of barrier thickness, as the fixative osmolarity is systematically varied. In Fig. 7.3 total fixative osmolarity, M_T, is varied from 145 to 1080 mOsm, using two glutaraldehyde concentrations (1 % and 2.5 %). The data, plotted on a double-logarithmic scale, show that the barrier thickness decreases as total osmolarity increases. This was to be expected; indeed, if the barrier behaves like a good osmometer, then the barrier thickness τ_h should vary inversely proportionally to the "effective osmolarity", or

$$\tau_h = a \cdot M^{-1} \qquad (7.10)$$

Figure 7.3 shows that this is nearly the case for total osmolarity ($\tau_h = a \cdot M_T^{-0.83}$), but that the glutaraldehyde concentration plays an additional role. Looking at solutions with a total osmolarity of 360 mOsm, measured by freezing point depression, we find that the barrier is thicker if 2.5 % glutaraldehyde has been used than with 1 % glutaraldehyde. This should be interpreted to mean that the osmotic effect of glutaraldehyde is smaller than

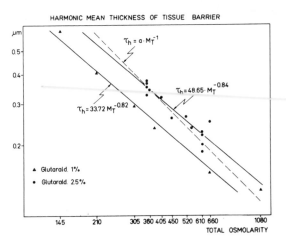

HARMONIC MEAN THICKNESS OF TISSUE BARRIER

FIG. 7.3: Effect of total osmolarity of fixative and concentration of glutaraldehyde on thickness of air-blood barrier in lung. (Reproduced by permission from Mathieu et al., 1978.)

one would anticipate from its osmolarity. This is hence in line with Barnard's (1976) general conclusion. Testing some 21 solutions with varying buffer and glutaraldehyde concentrations Mathieu et al. (1978) found that the effective osmolarity could be satisfactorily calculated by

$$M_{eff} = 1056 \cdot \frac{M_T^{(1.3 \cdot 10^{-4} \cdot M_G + 0.81)}}{M_G + 239} \tag{7.11}$$

This relationship considers that an effective osmolarity of 300 mOsm is required to maintain red cell shape, i.e. that this osmolarity is "physiological". This equation allowed the changes in barrier thickness observed to be described with rather good precision. But evidently, it relates to rat lung tissue, and the question must be left open whether it may also be applicable to other tissues, particularly if the depth of penetration required for fixation is greater than in these exceedingly thin specimens.

It is still difficult to decide on the correct osmotically optimal composition of fixatives. It is clear, however, that both the glutaraldehyde concentration and buffer osmolarity play a role, though not to the same extent. The formula of Barnard (equation 7.9) is a useful first approximation. Whether equation (7.11) is an important improvement remains to be shown; it works better than Barnard's simpler formula in extreme fixative–buffer combinations.

It is evident then that at the time of writing this issue is not yet settled altogether. The main point of this presentation has been to make the reader aware of this most important problem. The minimal moral to be derived:

keep the composition of the fixative, and of all subsequent solutions, strictly constant for a given set of experiments, lest your morphometric data reveal more about artefacts than about real experimental changes!

The end product of the specimen preparation procedures are the sections or the micrographs which are to be subjected to stereological analysis. This must be taken into consideration at all steps of the processing that follow on fixation. Particularly, the following points need attention and determined planning:

(a) *The structures should be preserved in their original shape and dimensions,* where we mean by "original" the state at which the reference (organ) volume was determined, but also, of course, the state that prevailed during life. To satisfy this condition the procedures of dehydration and embedding must be carefully chosen, monitored and standardized. For example, it was shown that cells may swell if dehydration is started with 50% ethanol whereas their size does not change with 70% ethanol (Weibel and Knight, 1964). A further important point is that the sections should be cut with a good knife to reduce compression, and finally one must be aware that some photographic papers may undergo considerable shrinkage during processing, thus distorting the micrographic image.

(b) *The micrographs must present a random sample*: appropriate sampling procedures must therefore be set up for all steps and performed with rigor. It is important that this be given greatest attention, because any fault in sampling cannot be remedied later on. General rules for various sampling schemes were given in Chapter 3, together with some practical suggestions.

(c) *Only one section per block is to be used* for stereological analysis, because the same structural component must be considered once only. This section should be technically perfect.

This leads to a final question with respect to specimen preparation: how good should a micrograph be? The quality of a descriptive morphological study is in part judged by the quality of the documents presented: only the best are good enough. One often says that for stereology even micrographs that are less than perfect will do. In a way this is correct, and even unavoidable. The micrographs are not selected and recorded for their esthetic quality, but rather through a strict random sampling procedure which does not permit the rejection of fields that "do not look nice". Also, smaller imperfections, such as knife marks or stain contaminations etc., often do not disturb the stereological evaluation, whereas they may render the picture unpublishable. However, one must demand that the material used for stereological analysis be as good as that destined for a descriptive study; in other words, one should be able to produce show-piece pictures from the sections used for stereological evaluation, otherwise the material is suspect of being of low quality.

7.4.   Instrumentation

Before finally giving the reader some advice on instruments that may facilitate his work two remarks are in order:

1) Stereology can be performed virtually without expensive equipment, at least to begin with (Elias and Botz, 1976). Test systems in the form of plastic overlays on micrographs or incorporated as graticules into microscope eye-pieces permit point counting methods to be applied (Section 4.2.7); the counts are easily recorded by pencil on a tally sheet, or perhaps with a simple mechanical counter device as used in hematology. These are of course not very efficient procedures, and one will definitely need to acquire adequate machinery if stereological methods become a regular part of the work pro-cedure. However, it is in principle advisable to start off in this kind of work by first doing it "by hand"; the choice of instruments is then made easier. The most expensive instrument is not necessarily the best equipment for a particular purpose.

2) The development of instruments for stereology proceeds at a fast pace. A detailed description of instruments available at the time of writing would with certainty be outdated at the time of reading. Therefore, this chapter will be limited to an incomplete description of some of the principles.

7.4.1.   SEMI-AUTOMATIC DATA ACQUISITION

7.4.1.1.   *Simple systems*

The stereological methods outlined in the previous chapters all work on the following principle: a test system of specified geometric properties and dimensions is made to interact with a specimen, the micrograph; this results in certain probabilistic events—point hits, intersections, profiles etc.—which need to be recorded in view of their subsequent use in calculating stereological parameters.

In a wide field of application image interpretation needs to be done by an experienced investigator. If the test systems are properly designed the data to be recorded are counts, more exactly differential counts of "hits" on the various components and features of the specimen. The recording of these primary data is greatly facilitated by using a keyboard which feeds the classified counts into a battery of counters. Figure 7.4 shows an older home-made model with ten electromechanical counters which can be cleared at the end of the operation, i.e. after the counts have been entered into a data sheet. Such counters are easy to make, but there are also several models available on the market, built mainly for use in hematology.

FIG. 7.4: Simple electrical counter unit.

The amount of data accumulated in this fashion is large; the next postulate for easing work is evidently efficient transfer of the data from the counters onto a data storage system, be it printing of a data sheet and/or transfer of data onto a carrier from which they can be retrieved automatically, such as punch cards, punch tape, magnetic tape or even computer discs. The technical solution is to use read-out counters, from which the primary data can be automatically transferred to the data carrier and to a teletype or printer for tabulation.

7.4.1.2. *Computer-assisted point counting*
The advent of mini- and microcomputers has recently made it possible to devise integral systems for stereological data processing which operate in four stages: (1) data accumulation by counting; (2) data storage and tabulation; (3) computation of stereological parameters; (4) statistics.

A system of this kind has been realized by Rätz *et al.* (1974); although the hardware solution adopted is outdated the general principle of the system still holds, and can actually be realized with much more ease using more advanced computers. For this reason the general principle of this system is presented here. It was designed to satisfy the following requirements:

(1) Point counting should be performed directly in the computer, the counts being fed into the computer from a peripheral keyboard with at least 10 counter keys.

(2) For purposes of continuous checking of the counting operation the counts from a group of classes (e.g. all point hits) should be summed up in one or two totalizers.

(3) The primary data (counts) must be printed in tabular form on operating a control key.

(4) It should be possible to perform statistical tests on the precision achieved (confidence interval) at any time point of the counting procedure.

(5) On operating a control key stereological parameters must be calculated with means and standard errors of the mean, and this should be printed out automatically.

(6) The user should be able to tailor his own specific program without knowledge of a special computer language and with minimum effort.

Figure 7.5 shows the general concept of the system which is made of three basic components: a *minicomputer* as a central unit which can be programmed from an alphanumeric keyboard; an *operating keyboard* which carries a number of counter keys, control keys for computer operation, and a display

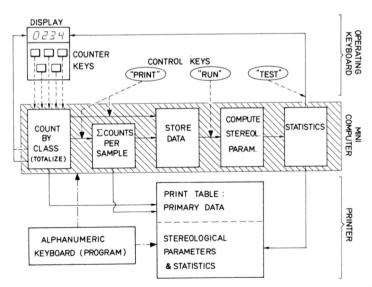

FIG. 7.5: Layout for dedicated minicomputer system for point counting stereology.

for some selected data; a *printer* for data output, which can evidently be coupled with a card or tape puncher (teletype) or with a magnetic tape recorder etc. In this system all operations, from counting to computation and formatting of print-out, are assigned to the minicomputer; they are controlled from a basic program which is stored in the computer memory (entered from some external carrier such as punch tape, etc.). The hardware solution for this system is, in fact, of secondary importance.

In designing the basic program, called "BAPOSTER", we had intended to allow for easy assembly of user programs with as great flexibility as possible (Rätz *et al.*, 1974). It was restricted to the case where all primary data entered from the operating keyboard were counts (point hits, intersections etc.). We required that some of these primary counts be summed and that such sums be used in subsequent calculations (auxiliary sums). It was found that most stereological parameters could be reduced to the general formula

$$A_i \cdot A_k/A_l \qquad (7.12)$$

where A may be either a constant, C_i, or a variable, U_i, such as P_i (point hits), I_i (intersections) etc. The program provided facilities for auxiliary sums or auxiliary parameters computed by formula (7.12) to be entered as variables in subsequent calculations.

Let us look at an example: calculating the surface density of rough endoplasmic reticulum membranes $S_V(\text{RER})$ in cytoplasm of liver cells (CY). The formula is

$$S_V(\text{RER}) = 2 \cdot I(\text{RER})/L(\text{CY}) \qquad (7.13)$$

Using a multipurpose test system (M 168) of unit line length d this becomes

$$S_V(\text{RER}) = 2 \cdot I(\text{RER})/[(\tfrac{1}{2}) \cdot d \cdot P(\text{CY})] \qquad (7.14)$$
$$= [4/d] \cdot I(\text{RER})/P(\text{CY})$$

The term in square brackets is a constant, I and P are variables, i.e. point counts.

Figure 7.6 shows a planning and programming sheet as it is used to instruct the minicomputer once BAPOSTER has been entered. The program identifies the input from the twelve counter keys of the operating keyboard as variables U∅1 to U12; one needs to identify what counts they represent (lines 1–12 in Fig. 7.6). The next two variables are totalizers which automatically sum up the counts entered into any of the primary counting variables (lines 13–14). Suppose we count point hits on cytoplasmic components in counters U∅1 to U∅6 and RER intersections in U∅8. Then we

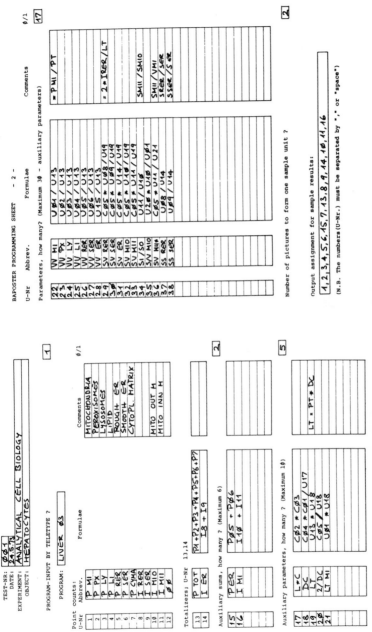

FIG. 7.6: Stereological programming sheet for dedicated minicomputer system.

form the cytoplasmic point hits by summing U∅1 to U∅6 in variable U13:

$$P(CY) = P(MI) + P(MB) + P(LY) + P(RER) + P(SER) + P(X)$$

$$U13 \quad = U∅1 \quad + U∅2 \quad + U∅3 \quad + U∅4 \quad + U∅5 \quad + U∅6$$

$$(7.15)$$

Entering the constant as C∅5 the parameter could be obtained from equation (7.14) by (line 29 in Fig. 7.6):

$$S_V(RER) = C∅5 * U∅8/U13 \tag{7.16}$$

and the volume density of mitochondria evidently by (line 22 in Fig. 7.6):

$$V_V(MI) = U∅1/U13 \tag{7.17}$$

It turns out that the provision of formulas for summing, subtracting, and multiplication-division satisfies most needs for computation of stereological parameters.

An important point in setting up this program was that the counts obtained on one field should be summed up over the several fields that constitute a representative sample before the stereological parameters are computed (see Section 4.3.1). For purposes of checking, all primary counts as well as the sample counts are printed in table form.

The computer was programmed to compute means and standard errors of the parameters. A statistical test program was also incorporated which calculates the 95% confidence intervals achieved for any of the parameters and displays it as a percentage of the mean on the keyboard display.

There is no point in pursuing the description of this program any further. Contemporary computer languages, e.g. BASIC, make it very easy to write such programs to suit the user's specific needs. For many purposes, medium sized desk-top computers are entirely adequate. In fact, even some of the programmable pocket computers can be used; the author has programmed a Hewlett-Packard HP65 to do the counting by special function keys, and to calculate volume and/or surface densities including standard errors by the ratio estimate method (see Sections 4.3.1 and 4.3.3).

The latest development (1977!) in computer-assisted point counting realized in our laboratory is to use a group of small table-top programmable calculators in conjunction with a compatible central computing facility (Fig. 7.7). The system we have chosen is to use the calculator Hewlett-Packard 9815 as a counter unit. This relatively inexpensive portable calculator has 15 special function keys which can be programmed as counting keys; its 2008 program steps are sufficient for the counting instructions, for controlling the storage of the counts on the built-in small tape cassette, for performing elementary statistical treatments of the data, and for printing out the results. For many purposes we have found this small unit to be entirely satisfactory. Since all data are stored on a magnetic tape cassette they can be transferred, through a suitable interface, to a larger computer, in our case a Hewlett-Packard 9830, with which more sophisticated statistical treatments of the data are possible. We have added two Diskette stations, a line printer, and a plotter to this computer which gives almost unlimited possibilities for data treatment and output-formatting. The attractive feature of this set-up is that as many stereological "work-benches" as are required can be made independent by providing them with a small inexpensive counter-computer (Fig. 7.7), and that the results from several such units can be evaluated centrally and, if required, in a combined fashion in the larger central unit. Most work-benches are simple point-counting units (Fig. 7.7) combining the HP 9815 with a film projector (Fig. 4.14) or with a light microscope (Fig. 4.13), but one of them is a computer-assisted tracing device, as discussed in the next section, whose data are likewise collected in a HP 9815. As long as dedicated stereological computers are not available such systems can be of great help.

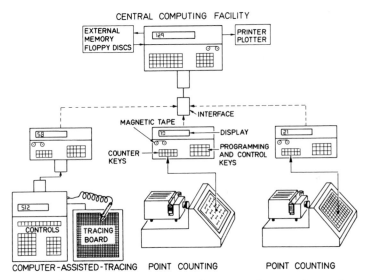

FIG. 7.7: Versatile data acquisition and computing system for stereology, using small calculators for point counting, as well as a computer-assisted tracing device.

7.4.1.3. Computer-assisted tracing devices

The advent of relatively inexpensive microprocessors with a large capacity has made it possible to exploit the micrographs with a much greater density of points than is feasible with visual-manual point counting. The computer-assisted tracing devices use a "digitizer board" as it is available with several computer systems; some work with ultrasound, others with magnetostriction, etc. The principle is to trace with an appropriate "pen" the contour of the object, in general a profile; the device then records the coordinates of the trace and computes by integration algorithms of various kinds parameters such as the profile area, its perimeter or boundary length, its largest and smallest diameter, its projected diameter in a given direction, some shape factors, etc. At the time of writing these devices are relatively new, but they can already record a large variety of parameters used in stereology. An instrument of this kind (Rohr, 1976; Haug, 1976) is shown in Fig. 7.7 integrated into a general stereological data acquisition system.

The advantage of these devices is that they measure the micrograph features with greater precision than the manual point counting procedures, and that one tracing operation yields several pieces of information. Their disadvantage is that the precision depends greatly on the precision of tracing, and in practice this may be a difficult task with small profiles or with features that have a complicated outline. Furthermore none of the profiles of a certain component present on a micrograph—even the smallest—may be missed, lest the information retrieved is not representative.

In comparing the relative merits of tracing versus point counting the following recommendations can be given:

(a) In using a tracing device the investigator must search for the features in the micrograph which are variable, recognize each of them, and then trace them; this works well with comparatively large features that occur in small numbers on each micrograph, features also that are easy to recognize.

(b) In point counting the investigator follows the points or lines of a predetermined test lattice, a lattice which is constant for all micrographs, and he is forced to decide on the location of the point etc.; this works better than tracing with components that are fragmented into many small parts and with features for which some profiles, e.g. small caps, may be difficult to recognize.

Clearly then, a good semi-automatic data acquisition system should allow both point counting and profile tracing; the instrument shown in Fig. 7.7 provides for both.

A word of caution is in order: these devices, in general, are *image analysers* in that they offer the investigator some comfort in interpreting and measuring the micrographs. *They are, however, not necessarily doing stereology.* The last step, the correct transformation of measurements obtained on two-dimensional images to information relating to the three-dimensional object, as well as the proper consideration of sampling conditions, remains the responsibility of the investigator.

7.4.2. AUTOMATIC IMAGE ANALYSERS

To adequately deal with automatic image analysers would mean writing a whole book on that topic alone. We will therefore again restrict ourselves to outlining some of the principles, recognizing that here again developments are very fast and recognizing also that the image analyser *per se* provides information on the two-dimensional image only and that stereology follows thereafter when this information is transformed into parameters describing the three-dimensional structure. Just as no biochemistry is done yet by possessing and using an automatic spectrophotometer so no stereology is done yet by having and operating an automatic image analyser.

All image analysers work on the same principle (Hougardy, 1976). The image is recorded by a video-type camera which converts the image into a parallel array of linear scans with the video-signal reflecting the light intensity of the specimen line by line. By sampling the analogue signal at regular intervals the linear scan is "chopped" into lines of points thus sub-dividing the image into a dense grid of discrete points (Fig. 7.8b). Each point of this "point picture" now carries two pieces of information: (1) its location in the specimen (x, y coordinates), and (2) its light intensity I related to the "gray level" of the specimen at this point.

There are two ways by which this electronic image can be processed further: (1) the entire image information (coordinates and gray level of each point) is transferred into the memory of a large computer and the information is processed by software (cf. Rink, 1976); or (2) further evaluation is done, at least in part, by some hardware incorporated into a dedicated image analysing computer system. We shall only outline some of the more fundamental procedures.

In image analysing computers dedicated to stereological work, such as Quantimet 720, Leitz TAS, Bausch and Lomb Omnicon, etc., the first analytic step is to discriminate the points into various classes. Basically, this is done, in the first instance, by sorting out the points on the basis of gray level values: the investigator prescribes a range of gray values within which all picture points belonging to one component should fall (Fig. 7.8c); the discriminator identifies all these points, and displays them on a screen (Fig. 7.8d), usually superimposed on the original image.

It is now, evidently, a straightforward matter to determine automatically some of the image information relevant for a stereological analysis: the relative number of detected points (P_P) is evidently an estimator for the relative volume of discriminated features; the number of intercepts of the discriminated image (N_L) or its boundary length (B_A) are related to the surface.

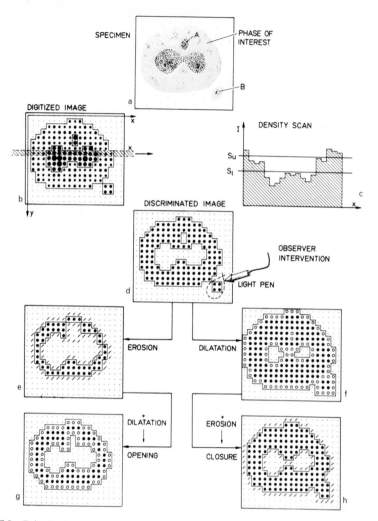

FIG. 7.8: Principal operations of automatic image analysers. (a) original half-tone image; (b) digitized image with light intensity of points indicated; (c) density scan of gray levels with upper (S_u) and lower (S_l) thresholds for point discrimination; (d) discriminated image (points in phase of interest enhanced, others suppressed); (e)-(f) principle of image transformation by erosion and dilatation.

Indeed, a large number of image descriptors, profile numbers, shape factors, projected diameters, etc. have been worked out.

The crux of the matter is that discrimination of image features depends, in the first place, on a discrimination of gray levels, and this introduces two difficulties: (a) the specimen must be evenly illuminated, which is not always easy with microscopes although there are means to correct for this deficiency so as to allow an analysis of pictures that are unevenly illuminated; (b) the gray level is not always an unambiguous criterion of a certain feature. This second difficulty poses many problems and is, in fact, the greatest obstacle to a wide-spread use of automatic image analysers, particularly with biological specimens.

One of the solutions offered by some of the instruments to overcome these difficulties is to allow for observer intervention. That means that the observer can check whether the automatically discriminated image faithfully represents the characteristic features of the specimen; if not, he is able to correct it. This can be done, for example, by a "light pen": through a tracing operation on the screen (Fig. 7.8d) he can add features to the detected image, erase some others, or separate erroneously connected features. The "automatic" image analyser thus becomes in a way semi-automatic and resembles the tracing devices described in the preceding section (Section 7.4.1.3). But this proves to be a very important addition because, with comparatively little effort on the part of the investigator, it allows the use of these powerful instruments even in cases where automatic image discrimination fails.

Advanced image analysers offer the possibility of modifying the detected image by certain transformations typified by what is called "erosion" and "dilatation" (Fig. 7.8e and f). Erosion means that the boundary of the feature is displaced inwards by a certain amount (one erodes away a layer of a certain thickness) and dilatation is the opposite, i.e. one adds a layer of a certain thickness. Ideally, this is done by moving a tangent circle along the boundary, either inside or outside the boundary; in practice the point lattice of the video image imposes some limitations so that the basic "structuring element" (Serra, 1972, 1978) is either a hexagon or an octagon, but any polygon, squares, rectangles or even line segments may also be appropriate for certain purposes. A very powerful process is to combine erosion and dilatation (Fig. 7.8g and h): by first eroding the image and then dilating the eroded image by the same amount one is "opening" the image; the reciprocal procedure leads to "closing". The operation "opening", for example, removes small image elements or cuts touching elements apart without changing the size of large elements. This operation can therefore be used for "cleaning" or "amending" the detected image. But these operations are much more powerful than that and can be exploited in many ways for a sophisticated approach to automatic image analysis. To dwell on this would clearly surpass the scope of this book. These operations of advanced texture analysis have been pioneered by J. Serra and G. Matheron at the Centre de Morphologie Mathématique in Fontainebleau, France. The reader is particularly referred to Serra's (1980) treatise on the theoretical basis of texture analysis.

What is in store for the future? It is likely that further developments in computer technology—making larger and faster computers available at lower cost—will open new possibilities for automatic image analysis. If an automatically digitized image can be stored in the (large!) memory of a computer it is certainly possible to write sophisticated programs which may do better than gray level discrimination in recognizing complex image constituents; and it might even become possible to write programs by which the computer can learn by itself to recognize certain features.

Much has already been achieved in this direction, particularly with respect to recognizing pathological cell types in smears; the reader is referred to a number of recent references for further information (Cheng, 1976; Rink, 1976; Bernroider, 1976, 1978; Moore, 1972; Bartels and Wied, 1977). The technological and programming sophistication required is very high, but it should be possible to overcome many of the problems—granted sufficient time and money is available! The trouble is that hopes for good automatic pattern recognition have been with us for a long time. When it is achieved, will it help us to do better stereology? In particular, will it do better than the human eye and brain, and at the same cost? It takes no more than a few minutes to teach an uninitiated person—I have tried it on a secretary and on a professor of mathematics—to recognize mitochondria, even grazing sections, with nearly 100% certainty! Can a pattern recognition computer do this? Maybe a quotation from G. C. Cheng's (1976) review entitled "What can Pattern Recognition do for Stereology?" summarizes the actual situation rather succinctly: "If your data is or can be controlled in quality and in quantity as in OCR (optical character reading), the structural and/or statistical approaches of Pattern Recognition can be of great assistance to Stereology". The crux of the matter—clearly recognized by Cheng—is that the "stereological data", i.e. the specimen image, cannot be controlled and that it contains very little redundancy. There is still a long way to go.

This is not to say that automatic image analysers do not have their place in stereology. Their powerful and sophisticated analytic systems, particularly when coupled to a computer and when provided with possibilities of observer control of the image, offer many possibilities which manual methods cannot provide. The cost and the requirements for operator skill are high, however. And it must again be remembered that these instruments provide an analysis of the two-dimensional image only; to do good stereology with them is the responsibility of the stereologist.

CHAPTER 8

Model cases from biological morphometry

The practical application of the general stereological methods presented in the preceding chapters in view of a morphometric analysis of various tissues requires a number of choices and adaptations to be made. These depend on the characteristic organization of the tissue and on the type of answers sought. This is evidently the place where the investigator will have to exercise his imagination and ingenuity. For the novice it might be useful to follow some guidelines which evolved from experiences of other investigators when they tried various approaches and finally succeeded in applying a particular scheme of study—even though, in good scientific tradition, the choice of a particular approach may ever so often have been based on intuition rather than on rigorous deduction.

In order to provide the reader with first hand information I have asked a number of experienced investigators to contribute a condensed outline of their approach. The space available for this was limited; accordingly the presentations had to be restricted to examples and could not cover the entire aspect of the studies so that the reader will have to refer to the original articles for more details.

8.1. Morphometry of the subcellular organization of liver cells

Contributor: Ewald R. Weibel, Bern

8.1.1. MODEL OF LIVER STRUCTURE

The major functional components of the liver are the hepatocytes; the liver parenchyma also contains relatively large numbers of smaller cells which are mostly associated with the sinusoids (Wisse, 1969). The parenchyma is organized into lobules, about 1 mm in size, by the arrangement of afferent and efferent blood vessels, of biliary canals and a small amount of connective tissue. Fig. 8.1.1 shows a model of the subdivision of the liver into a hierarchical sequence of components; the subcellular organelles are detailed with respect to hepatocytes only, but, evidently, the same organelles are expected to occur in all other cell types. A more complete version of this model has been given in Fig. 3.1.

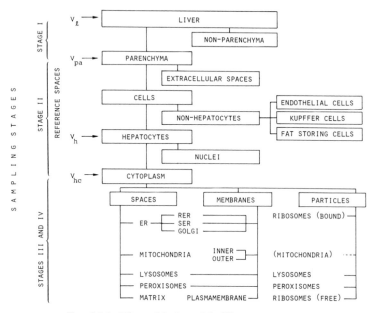

FIG. 8.1.1: Hierarchical model of liver structure.

8.1.2.   PURPOSE OF THE STUDY

The aim of the studies was to measure the content of liver cells in subcellular organelles, in order to establish the basis for a quantitative correlation of liver structure with some cell functions (particularly enzymes) determined by biochemical methods on subcellular fractions. In earlier studies (Weibel et al., 1969; Stäubli et al., 1969; Bolender and Weibel, 1973) attention was focussed on the hepatocytes' drug metabolizing enzymes associated with the endoplasmic reticulum membranes. Later the distribution of the various membrane classes to the different cell types of liver parenchyma was studied

(Blouin *et al.*, 1976). Bolender *et al.* (1978) have also attempted a correlation between a morphometric study of liver cell structure and a stereological and biochemical analysis of subcellular fractions.

8.1.3. GENERAL STRATEGY

Biochemical studies are performed on homogenate of whole liver; the results are expressed per gram liver tissue. Accordingly, the morphometric analysis was based on random sampling of liver tissue and the data were expressed per ml of whole liver; the specific gravity of liver tissue (1.067) was used to convert liver weight to liver volume. To obtain results comparable to biochemistry, most of the morphometric study (stages II–IV) was performed on small samples of exsanguinated liver fixed by immersion in 1% OsO_4, the remainder of the organ being homogenized and used for subcellular fractionation.

8.1.4. SAMPLING PROCEDURE

The range of structural dimensions varies widely: the diameter of lobules measures about 1 mm, that of hepatocytes 20 μm, whereas SER tubules are 60 nm wide. It was therefore necessary to adopt a four-stage sampling procedure (Fig. 8.1.1) using various levels of magnification and a sequence of reference spaces (see Section 3.1). The following is the procedure used by Bolender *et al.* (1978); it is a modification of the original scheme proposed by Weibel *et al.* (1969)

The evaluation of cumulative standard errors (see Fig. 3.22) has shown that the number of electron micrographs chosen for each sampling stage yields results with a coefficient of variation smaller than 10% of the mean for all components, except for Golgi membranes in stage IV (Bolender, 1978).

Stage I: Light microscopy, magnification × 200
        Reference volume: total liver tissue

Step sections of a rat liver were obtained by cryostat at 1600 μm intervals, fixed with 4% paraformaldehyde and stained for connective tissue according to van Gieson. Using a single test point on the screen of a Wild M 501 sampling stage microscope the stepping stage was stopped every 550 μm, and the location of the test point recorded. This allowed estimations of the volume densities of liver parenchyma (test points on parenchyma $P_{pa}{}^1$), central veins, portal triads, other connective tissue elements, blood vessels etc.

Calculated parameter: volume density of parenchyma in liver:

$$V_{Vpa,l} = \Sigma P_{pa}^{\ I} / \Sigma P_l \tag{8.1.1}$$

[NOTE: the roman numerals as superscripts identify the sampling stage at which the count was made. The counts are summed over a representative number of subsamples before being introduced into the formulas (see Section 3.7)].

Stage II: Electron microscopy, final magnification $\times 6200$
    Reference volume: liver parenchyma

For each specimen 48 electron micrographs were collected by systematic sampling and analysed with a multipurpose test system M 168 (Appendix 3) in the projector unit shown in Fig. 4.14 (test screen size 27×27 cm). This stage was used to count the points falling on hepatocytic nuclei (P_{hn}) and cytoplasm (P_{hc}), and to count nuclear profiles (N_{hn}); the points falling on parenchyma $P_{pa}^{\ II}$ served as reference.

Calculated parameters:
Volume density of hepatocyte cytoplasm and hepatocytes in parenchyma

$$V_{Vhc,pa} = \Sigma P_{hc}^{\ II} / \Sigma P_{pa}^{\ II} \tag{8.1.2}$$

$$V_{Vh,pa} = \Sigma (P_{hc}^{\ II} + P_{hn}^{\ II}) / \Sigma P_{pa}^{\ II} \tag{8.1.3}$$

Numerical density of hepatocyte nuclei by equation (4.107)

$$N_{Vhn,pa} = \frac{1}{1.38 \cdot \Sigma P_{pa}^{\ II} \cdot (k_2 \cdot d^2)^{\frac{3}{2}}} (\Sigma N_{hn})^{\frac{3}{2}} \cdot (\Sigma P_{hn})^{-\frac{1}{2}} \tag{8.1.4}$$

where 1.38 is β for spheres and $k_2 \cdot d^2$ the test system characteristics (Appendix 3).

Stage III: Electron microscopy, final magnification $\times 19,000$
    Reference volume: hepatocyte cytoplasm

This sampling stage was used to determine the plasma membrane surface of hepatocytes; it can also be used to estimate the plasma membrane of non-hepatocytes (Blouin *et al.*, 1977). 48 electron micrographs were collected from the same sections. A square double lattice test system of ratio 1:9 (C 64, Appendix 3) was used to count the coarse points falling on hepatocyte cytoplasm ($P_{hc}^{\ III}$), and the intersections of the plasma membrane with *all* horizontal and vertical test lines (I_{pm}).

Calculated parameter: surface density of plasma membrane "in" hepatocyte cytoplasm:

$$S_{V\,pm,hc} = 2 \cdot (\Sigma I_{pm})/(6d \cdot \Sigma P_{hc}^{III}) \qquad (8.1.5)$$

where $6d$ is the test line length per coarse test point (Appendix 3, C 64).

Stage IV: Electron microscopy, final magnification $\times 96{,}000$
Reference volume: hepatocyte cytoplasm (Fig. 8.1.2)

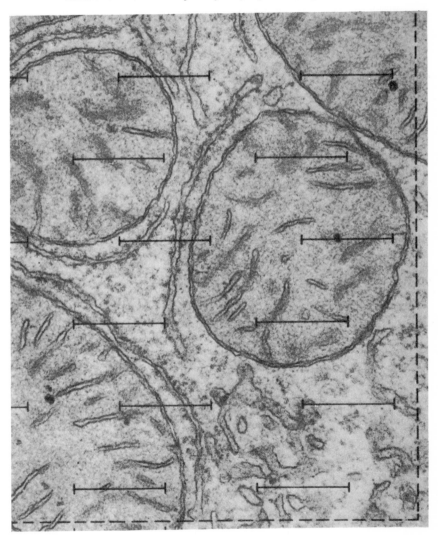

Fig. 8.1.2: Part of high power micrograph of hepatocyte ($\sim 90'000 \times$) as used in Stage IV. Multipurpose test system M168 superimposed.

A total of 72 micrographs were collected for each animal. Using test system M 168 (Appendix 3) test points falling on cytoplasm ($P_{hc}{}^{IV}$) and on cytoplasmic organelles, as well as intersections with cytoplasmic membranes were counted: rough ER (P_{rer}, I_{rer}), smooth ER (P_{ser}, I_{ser}), mitochondria (P_{mi}), mitochondrial outer membrane (I_{omi}) and inner membrane with cristae (I_{imi}), etc.

Calculated parameters: surface density of RER and ER membranes etc. in hepatocyte cytoplasm:

$$S_{Vrer,hc} = 2 \cdot [\Sigma I_{rer}]/[(d/2) \cdot \Sigma P_{hc}{}^{IV}] \qquad (8.1.6)$$

$$S_{Ver,hc} = 2 \cdot [\Sigma I_{rer} + I_{ser})]/[(d/2) \cdot \Sigma P_{hc}{}^{IV}] \qquad (8.1.7)$$

where $(d/2)$ is the test line length per test point. Similarly for mitochondrial membranes.

Volume density of mitochondria etc. in cytoplasm:

$$V_{Vmi,hc} = (\Sigma P_{mi})/(\Sigma P_{hc}{}^{IV}) \qquad (8.1.8)$$

8.1.5. FINAL CALCULATIONS

The volume and surface densities of the various components in the entire liver are obtained by combining the estimates of parenchymal volume density (equation 8.1.1) and hepatocytic cytoplasm in parenchyma (equation 8.1.2) with the density estimates in cytoplasm. For mitochondrial volume and RER surface one finds, for example:

$$V_{Vmi,l} = V_{Vpa,l} \cdot V_{Vhc,pa} \cdot V_{Vmi,hc} \qquad (8.1.9)$$

$$S_{Vrer,l} = V_{Vpa,l} \cdot V_{Vhc,pa} \cdot S_{Vrer,hc} \qquad (8.1.10)$$

If one has determined the total liver volume V_l (see Section 7.2) the absolute mitochondrial volume or RER membrane surface is found by

$$V_{mi} = V_l \cdot V_{Vmi,l} \qquad (8.1.11)$$

$$S_{rer} = V_l \cdot S_{Vrer,l} \qquad (8.1.12)$$

By dividing these absolute values by the animal's body weight one may calculate "specific" dimensions related to the unit body weight and thus normalize for variations in body mass (see Section 7.2).

From equations (8.1.3) and (8.1.4) one can estimate the volume of a "mean mononuclear hepatocyte" (Weibel et al., 1969) by

$$v(\hat{h}) = V_{Vh,pa}/N_{Vh,pa} \qquad (8.1.13)$$

The surface of RER membrane contained in a "mean mononuclear hepatocyte" is then found from equations (8.1.13, 6, 2 and 3) by

$$S_{er}(\hat{h}) = v(\hat{h}) \cdot S_{Ver,hc} \cdot V_{Vhc,pa}/V_{Vh,pa} \qquad (8.1.14)$$

8.1.6. ERROR CONSIDERATIONS

The most serious errors in estimating surface and volume density of endo-plasmic reticulum or of mitochondrial cristae, for example, are due to finite section thickness and to section compression. Whilst the estimation of surface of rough ER which occurs in broad sheets is comparatively little affected by section thickness, the surface and volume of smooth tubules will tend to be overestimated by some 30% (Weibel and Paumgartner, 1978). The correction procedures presented in Section 4.3.4 have to be taken into consideration when absolute values of these parameters are to be obtained. This is the case if the morphometric data obtained on whole tissue need to be compared with measurements done on homogenized cells (Bolender et al., 1978).

8.1.7. EXTENSIONS OF THE STUDY

This study on liver cells was recently extended to encompass all cell types of liver parenchyma (Blouin et al., 1977); a summary of the results is shown in Figs. 3.3 and 3.4. In a series of further studies an attempt was made to work out an integrated study of intact liver tissue and subcellular fractions in view of direct correlation with biochemistry (Bolender et al., 1978); this required the combination of stereology with cytochemistry and with freeze-fracture replicas (Losa et al., 1978). The dynamics of hepatocyte microbodies under experimental conditions was studied by a combination of stereology with autoradiography (Stäubli et al., 1977). The validity of needle biopsies of liver for morphometry was evaluated on dog livers (Hess et al., 1973). An alternative approach to sampling was used by Loud (1968) who studied the regional variations in hepatocyte structure within the liver lobule.

8.2. The exocrine pancreas: a stereological approach to membrane kinetics

*Contributor: Robert P. Bolender, Seattle, Washington**

8.2.1. PURPOSE OF THE STUDY

Individual membrane compartments have been used to define the steps of the synthesis, intracellular transport, and release of secretory proteins (Caro

* Department of Biological Structure, School of Medicine, University of Washington, Seattle, Washington 98195, U.S.A.

and Palade, 1964; Jamieson and Palade, 1967a, b; 1971; Hokin, 1968; Meldolesi, 1974). Although the route of the secretory proteins through the cell seems to be well established, surprisingly little is known about the membrane kinetics associated with the synthetic and secretory cycles. Membrane shuttles (Palade, 1959), flows (Morré et al., 1974) and recyclings (Meldolesi, 1974) are all thought to play a role. This chapter describes an experimental approach that attempts to study the kinetics of the secretory process stereologically. Morphometric data are used to characterize exocrine cell membranes within an analytical framework in order that changes in individual membrane compartments can be detected, as well as movements of membrane from one compartment to another (Bolender, 1974).

8.2.2. ANALYTICAL ULTRASTRUCTURAL MODEL FOR THE PANCREAS

The pancreas consists of morphologically distinct compartments that can be characterized by volumes and surface areas (Bolender, 1974). These are illustrated by the model in Fig. 8.2.1. When an appropriate reference has been chosen for a particular problem, the entire model or any part thereof can be used analytically to detect and interpret morphological changes. For

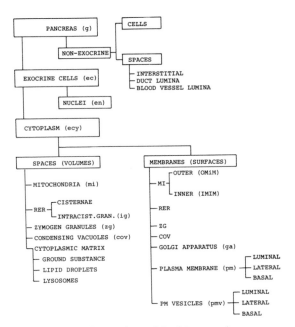

FIG. 8.2.1: Morphometric model of the exocrine pancreas.

example, in studying membrane kinetics in exocrine cells, the surface areas of individual and total membrane compartments are related to the volume of an *average exocrine cell*. This reference volume is particularly suitable for studying membrane changes since, for all practical purposes, the stereological estimates require corrections for only section compression and thickness.

The analytical interpretation of the membrane data, however, uses the individual and total surface areas of the membranes as the reference parameter. During an experiment, for example, the changes observed for an individual compartment might be the result of a movement of membrane from one compartment to another or the changes might reflect a more complex series of events that may also include changes in the rates of membrane synthesis and turnover. Consequently, the interpretation of individual changes becomes more comprehensive when done within the analytical framework of a total membrane compartment wherein balance sheets are kept for all membrane compartments.

8.2.3. SAMPLING PROCEDURE

Since the components estimated at both the tissue and cellular levels represent a broad range of sizes and frequencies, three magnifications were needed to establish adequate relationships between the size of the components being measured and the test systems.

One of the major decisions of any stereological study will be to choose the sampling levels that are to be tested statistically. In studying the pancreas, three were chosen. 1. The homogeneity of the *tissue* was determined by applying an analysis of variance to compare 5 different sampling regions. 2. For a given *animal*, each component at every sampling stage was evaluated by calculating cumulative means and standard errors. 3. The relationship between the individual animal values and the *animal population* was determined by calculating the means and standard errors for data derived from 5 different animals.

To accommodate this stepwise statistical analysis, samples were collected from 5 regions equally spaces along the length of the pancreas. From each animal, 12 micrographs were selected per zone per magnification stage. Consequently, the stereological estimates were based on 60 micrographs per animal and a total of 300 for the animal population. Details are given for the three magnification stages:

Stage I
Micrographs were made of entire grid squares (200 mesh) containing tissue and "counted" at 6200× with a multipurpose test system (Weibel, 1969), 27 × 27 cm, which contained 168 test points (P_T) spaced at 2.2 cm (Appendix

3, M 168). The use of the electron microscope at this level eliminated the necessity of making section thickness corrections when estimating nuclear diameters and facilitated the identification of the tissue compartments.

The volume density of a glandular component i ($V_{Vi,g}$), which is identified either as a cell or space compartment (see Fig. 8.2.1), was obtained by counting points in the test system associated with i (P_i) in the underlying micrograph, as well as those points associated with the entire gland (P_g). As an example, the volume density (V_V) of exocrine cell cytoplasm (ecy) in the gland (g) was estimated by

$$V_{Vecy,g} = P_{ecy}/P_g \qquad (8.2.1)$$

By counting the points falling on exocrine cell nuclei (en) the volume density of cytoplasm in exocrine cells is found by

$$V_{Vecy,ec} = P_{ecy}/(P_{ecy} + P_{en}) \qquad (8.2.2)$$

Stage II

Micrographs were collected from the upper left hand corner of a grid square or from alternative corners (moving in a clockwise direction) when the tissue was either absent or contained excessive artifacts. However, only *one* micrograph per grid square was taken. The multipurpose test system (see Stage I) was used to estimate volume and surface densities at $13,000 \times$; these compartments were identified in Fig. 8.2.1.

At Stage II the sampling was restricted to exocrine cell cytoplasm, with points falling on this space (P_{ecy}) serving as the reference. Volume densities were obtained accordingly.

$$V_{Vi,ecy} = P_i/P_{ecy} \qquad (8.2.3)$$

At this stage membrane surface densities for the cell surface, zymogen granules and condensing vacuoles were estimated by counting intersections (I_i) with the test lines, and, similarly, related to the exocrine cell cytoplasm reference:

$$S_{Vi,ecy} = 2I_i/(d/2) \cdot P_{ecy} = 4I_i/d \cdot P_{ecy} \qquad (8.2.4)$$

Stage III

Electron micrographs were again selected in the manner described for Stage II, however, a square double lattice test system with 36 coarse points and a test point ratio 1:9 (C 36), 27×27 cm, was used to collect point and intersection counts at $\times 96,000$. The total length of the linear probe (L_T) is flexible with this test system and can be adjusted to accommodate specific sampling requirements (see Section 4.2.3.1). For example, the plasma membrane vesicles required a longer unit probe length ($6d$) than the Golgi apparatus ($3d$), which in turn was longer than that for the rough-surfaced endoplasmic reticulum ($1d$). Since the general equations are similar to those used for Stage II (except for the variable L_T), only a few specific examples are given:

$$S_{Vpmv,ecy} = 2I_{pmv}/(6d) \cdot P_{ecy} \qquad (8.2.5)$$

$$S_{Vga,ecy} = 2I_{ga}/(3d) \cdot P_{ecy} \qquad (8.2.6)$$

$$S_{Vrer,ecy} = 2I_{rer}/(d) \cdot P_{ecy} \qquad (8.2.7)$$

8.2.4. RELATING PRIMARY ESTIMATES TO OTHER REFERENCE SPACES

Estimates of volume and surface density of subcellular components obtained in Stages II and III were related to exocrine cell cytoplasm. Using the estimate of exocrine cell cytoplasm volume density in the entire gland obtained in Stage I the surface density of RER in the gland could be obtained by:

$$S_{Vrer,g} = S_{Vrer,ecy} \cdot V_{Vecy,g} \qquad (8.2.8)$$

One can proceed likewise for the parameters obtained in Stage II.

Alternatively, the volume or surface of a component i contained in an average exocrine cell ($\overline{V}_{i,ec}$) can be estimated when the nuclear and cytoplasmic volumes of such a cell are determined. For this purpose nuclear profiles were measured at Stage I and used to estimate the mean nuclear diameter $\overline{D}$, according to the method of Giger and Riedwyl (1970) (see Section 5.2.6); the average nuclear volume ($\overline{V}_{n,ec}$) was calculated from $\overline{D}$.

$$\overline{V}_{n,ec} = \tfrac{4}{3}\pi \left(\frac{\overline{D}}{2}\right)^3 \qquad (8.2.9)$$

average cell volume ($\overline{V}_{ec}$) is derived from $\overline{V}_n$:

$$\overline{V}_{ec} = \overline{V}_n / V_{Vn,ec} \qquad (8.2.10)$$

The surface of RER in the average exocrine cell, for example, is obtained from equations (8.2.2, 7, and 10) by

$$\overline{S}_{rer,ec} = S_{Vrer,ecy} \cdot V_{Vecy,ec} \cdot \overline{V}_{ec} \qquad (8.2.11)$$

8.2.5. STATISTICAL EVALUATION OF DATA

Tissue: Figure 14 (% of pancreas volume) and Table II (analysis of variance) in Bolender (1974) were used to look for regional variations within the pancreas. Although the analysis of variance detected no regional variations in the volumes of the tissue components, the distributions of the individual regional data would suggest that a conclusion of glandular homogeneity may be applied only to the exocrine cells. Since the remaining tissue components have been sampled so sparsely (standard error $\gg 10\%$ of mean), the resolution of the data is too coarse to allow the differences suggested by the regional values to be detected statistically. Furthermore, regional variations for organelles within the exocrine cell cytoplasm might also exist, but this was not considered.

Animal Population: In Bolender (1974) data from 4 animals were used to estimate the animal population. However, one would like to know if this represents the minimum number of animals consistent with the statistical objectives of the study. Using the ER surface area data from 5 animals (numbered 1 through 5), all the possible combinations were formed taking 4, 3 and 2 animals at a time. These "selected data" were then expressed as a % of the "final mean" calculated with the total ER data from all 5 animals.

When only 2 or 3 animals are used, considerable variations were detected having a range as large as 30%. When the number of micrographs per animal was varied, for example, reducing it from 60 to 20, estimates for endoplasmic reticulum changed only slightly. The endoplasmic reticulum is the major membrane compartment in exocrine cells (60%), and seems to require at least four animals for an acceptable population estimate. Of course, a sample consisting of four animals might not be expected to apply equally to all morphological components, especially the rarer ones.

8.3.   Endocrine cells of adrenal cortex

*Contributor: Hanspeter Rohr, Basel**

8.3.1.   MODEL OF ADRENAL GLAND STRUCTURE (Rohr *et al.*, 1976)

The adrenal gland (*ag*) is composed of the medulla (*am*) and the cortex (*ac*) which consists of three zones: the zona glomerulosa (*gl*), the zona fasciculata (*fa*) and the zona reticularis (*re*). These zones were divided into morphlogically well defined compartments. Principally, each zone has two parts: (a) The extracellular space which comprises the blood vessels (capillaries) and the intercellular space together with some interstitial tissue, and (b) the adrenal cortical cells which were divided into the nuclear and the cytoplasmic compartments following the stereological model outlined in Fig. 8.3.1.

*Institut für Pathologie, Schönbeinstrasse 40, CH–8056, Basel, Switzerland.

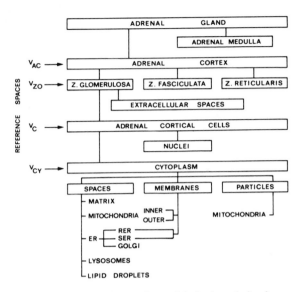

Fig. 8.3.1: Morphometric model of adrenal gland.

Five reference spaces are used in the present model (Fig. 8.3.1): The adrenal cortex (ac), each of the three adrenocortical zones (gl, fa, re), the zonal cells (c), the zonal cell cytoplasm (cy), and the volume of an average zonal cell.

8.3.2. PURPOSE OF THE STUDY

The stereological study of the adrenal cortex of rats allows a quantitative characterization of those cell compartments which are involved in steroid synthesis (mitochondria, smooth endoplasmic reticulum): this also permits a morphometric evaluation of cellular changes, which occur when the cortex is subjected to unspecific stresses or specific stimulations, or to inhibition of the various steroid synthesizing enzymes. A comparison of such data with stereological results from other organs would help to clearly distinguish between specific and unspecific cortical cell reaction patterns; they thus contribute to a better understanding of the physiology of the adrenal cortical cells under physiological or experimentally induced conditions. Finally, the extensive biochemical data available for the adrenal cortex can be complemented by associated morphometric data. The study will hence have to focus principally on the subcellular organelles of the various cell types which serve different functions.

8.3.3. GENERAL STRATEGY

The study was performed on samples of rat adrenal glands fixed by immersion in OsO_4. Zonal identification was done at level I and II according to the criteria proposed by Idelman (1970) and Rhodin (1971). Even in the unstimulated cortex zonation poses some problems for stereological analysis especially in the transitional zones, which cannot be definitely identified by their vascular structure. Therefore, these zones were included within the zona fasciculata.

The region close to the capsule, consisting of about three to six cell layers, was assigned to the zona glomerulosa, the area close to the adrenal medulla (six to ten layers) was evaluated for the zona reticularis and the middle region of the adrenal cortex for the zona fasciculata.

8.3.4. SAMPLING PROCEDURE

The cell compartments could not be determined at a single stage of magnification since they represented a broad range of sizes and frequencies. Therefore, sampling was done at four magnifications.

Level I (LM 30 ×)
For each of the five animals 10 paraffin sections were selected for light microscopic evaluation, giving a total of 50 sections. Only those sections which included the entire cross-section of the adrenal gland (cortex and medulla) were accepted. On each section 10 test areas were evaluated. Evaluation was done with the Wild sampling microscope (Wild AG, Heerbrugg, Switzerland) using a square double lattice test system (1:9, 121:1089); this test system would be labelled C 121 according to Section 4.2.2.2 and Appendix 3. The volume densities of the zones (*gl*) and (*re*) were determined by counting the fine test points. The volume densities of the zona fasciculata (*fa*), the adrenal medulla (*am*) and the areas outside the adrenal gland were evaluated by counting the heavy test points.

Level II (LM 1000 ×)
Semi-thin sections were analysed by light microscopy, using a total of 60 test areas for each zone for all five animals together. A multipurpose test system (see Fig. 4.10) was applied having a total number of 100 test points. This was the level at which the number of nuclear profiles per test area for the calculation of the numerical density by equation (4.106) was determined ($\beta = 1.45$, $K = 1.07$). At this level the number of points over the nuclei was counted. The shape factor β was graphically determined (see Fig. 2.30).

Level III (*EM magnification* 3400 ×)

For each animal 54 micrographs were recorded in each zone. Again the multi-purpose test system was applied ($P_T = 100$). The following compartments were counted: nuclei (n), extracellular space (ex), cytoplasm (cy) and fat droplets (f) as well as the number of the mitochondrial profiles within the test area. Furthermore, for each zone and animal the ratio of the axes of 100 mitochondrial profiles was measured in order to estimate the shape factor β which was found to be 1.45 to 1.50.

Level IV (*EM magnification* 8 000 ×)

For each animal 54 micrographs were recorded in each zone. A square double lattice system (1 : 9, 121 : 1089), 11.2 × 11.2 cm frame width, was used to estimate volumes and surfaces of organelles. The coarse test points were counted for mitochondria (m), ground cytoplasm (gcy), extra-cellular space (ex), nuclei (n) and fat droplets (f), and the fine test points for the other compartments (smooth endoplasmic reticulum, ser, Golgi apparatus, ga, lysosomes, ly).

At this stage the intersections of the 22 coarse test lines (both vertical and horizontal) with the membranes of the following structures were counted: smooth endoplasmic reticulum, I_{ser}, mitochondrial outer membrane, I_{mo}, mitochondrial inner membrane, including cristae, I_{mc}.

8.3.5.   EXAMPLES OF STEREOLOGICAL CALCULATIONS:

Level I:

Counted: Test points on cortical zones:

$\quad\quad\quad P_{fa}$ (coarse lattice) $\quad\quad P_{gl}$, P_{re} (dense lattice)

and on adrenal medulla

$\quad\quad\quad\quad\quad P_{am}$ (coarse lattice).

Calculated:

$$P_{ag} = P_{fa} + \tfrac{1}{9}(P_{gl} + P_{re}) + P_{am}$$

$$P_{ac} = P_{fa} + \tfrac{1}{9}(P_{gl} + P_{re})$$

volume densities of *gl*, *fa* and *re* in adrenal gland (*ag*) and adrenal cortex (*ac*) respectively:

$$V_{Vfa,ag} = \frac{P_{fa}}{P_{fa} + \tfrac{1}{9}(P_{gl} + P_{re}) + P_{am}} = \frac{P_{fa}}{P_{ag}} \tag{8.3.1}$$

and

$$V_{Vgl,ag} = \frac{P_{gl}}{[P_{fa} + \tfrac{1}{9}(P_{gl} + P_{re}) + P_{am}] \times 9} = \frac{P_{gl}}{9 \times P_{ag}}$$

K

$$V_{Vfa,ac} = \frac{P_{fa}}{P_{fa} + \frac{1}{9}(P_{gl} + P_{re})} = \frac{P_{fa}}{P_{ac}} \qquad (8.3.2)$$

and

$$V_{Vgl,ac} = \frac{P_{gl}}{[P_{fa} + \frac{1}{9}(P_{gl} + P_{re})] \times 9} = \frac{P_{gl}}{9 \times P_{ac}}$$

Level II:
Counted for each zone separately:
Test points on nuclei (P_n), cytoplasm (P_{cy}) and extracellular space (P_{ex}), and the number of nuclear profile (N_{An}), $P_T = 100$.

Calculated: respective volume densities in each adrenal cortical zone (e.g. *fa*)

$$P_{fa} = P_n + P_{cy} + P_{ex}$$

$$\text{e.g. } V_{Vcy,fa} = \frac{P_{cy}}{P_n + P_{cy} + P_{ex}} = \frac{P_{cy}}{P_{fa}} \qquad (8.3.3)$$

volume densities in adrenal cortical cell (e.g. *fac*):

$$P_n + P_{cy} = P_{fac}$$

$$V_{Vn,fac} = \frac{P_n}{P_n + P_{cy}} = \frac{P_n}{P_{fac}}$$

or

$$V_{Vcy,fac} = \frac{P_{cy}}{P_n + P_{cy}} = \frac{P_{cy}}{P_{fac}} \qquad (8.3.4)$$

Level III:
Counted for each zone (e.g. *fa*):
Test points on fat droplets (P_f) and cytoplasm, inclusive of fat droplets, (P_{cy}); alternatively, by counting test points on nuclei and extracellular space ($P_n + P_{ex}$) one finds

$$P_{cy} = P_T - (P_n + P_{ex})$$

Calculated: volume density of fat droplets in cytoplasm (e.g. *fa*):

$$V_{Vf,facy} = \frac{P_f}{P_{cy}} = \frac{P_f}{P_T - (P_n + P_{ex})}$$

Level IV:
Counted for each zone (e.g. *fa*):
Test points on cytoplasmic compartments:

Coarse lattice ($P_T = 121$);
Mitochondria (P_m), ground cytoplasm, inclusive fat droplets (P_{gcy})
Dense lattice ($P_T = 1089$):
Smooth endoplasmic reticulum (P_{ser})
Golgi apparatus (P_{ga}) and lysosomes (P_{ly})

Intersections of both horizontal and vertical coarse test lines with smooth endoplasmic reticulum, I_{ser}, mitochondrial outer membrane, I_{mo}, and mitochondrial inner membrane (inclusive cristae), I_{mc}.

Calculated:

$$P_{cy} = P_{mi} + P_{gcy} + \tfrac{1}{9}(P_{ser} + P_{ly})$$

volume densities of cytoplasmic compartments in cytoplasm (e.g. *fa*):

e.g. *mi*: $V_{Vmi,facy} = P_{mi}/P_{cy}$ (8.3.5)

or *ser*: $V_{Vser,facy} = P_{ser}/9 \cdot P_{cy}$

Surface density of, for example, *ser* in cytoplasm (e.g. *fa*):

$$S_{Vser,facy} = \frac{2 \times I_{ser}}{L_{T,facy}} = \frac{I_{ser}}{d \cdot P_{cy}}$$ (8.3.6)

Absolute values:
The mean volume of an "average zonal adrenal cortical cell", e.g. $\overline{V}_{fac}$, was calculated as follows:

$$\overline{V}_{fac} = \frac{V_{Vc,fa}}{N_{Vn,fa}}$$

The aggregate volume of an organelle (e.g. mitochondria, *mi*) per average cell was then obtained as (e.g. in *fac*):

$$\overline{V}_{mi,fac} = V_{Vmi,fac} \cdot \overline{V}_{fac}$$ (8.3.7)

and the average volume of a single organelle (e.g. *mi*) (e.g. in *fa*) as:

$$\overline{V}_{mi,fa} = \frac{V_{Vmi,facy}}{N_{Vmi,facy}} = \frac{V_{Vmi,fac}}{N_{Vmi,fac}}$$

and finally the mean number of an organelle per adreno-cortical cell (e.g. *fac*) as:

$$N_{mi,fac} = \frac{N_{Vmi,fa}}{N_{Vn,fa}}$$

Numerical densities were calculated by means of equation (4.106) developed by Weibel and Gomez (1962).

Finally for a better understanding of the structural and functional relationship the surface densities of the smooth endoplasmic reticulum (S_{Vser}) and of the mitochondrial membranes (S_{Vmo}, S_{Vmc}) were related to a unit volume of smooth endoplasmic reticulum (*ser*) and mitochondria (*mi*) respectively:

e.g. for outer mitochondrial membrane by equations (8.3.5 and 6)

$$S_{Vmo,mi} = \frac{S_{Vmo,facy}}{V_{Vmi,facy}} = \frac{I_{mo}}{d \cdot P_{cy}} \cdot \frac{P_{cy}}{P_{mi}} = \frac{I_{mo}}{d \cdot P_{mi}}$$

Example for cascade of calculations: e.g. zona fasciculata (*fa*), cytoplasm in zona fasciculata (*facy*), and mitochondria (*mi*).

Reference space	Phase of interest		
	fa	*facy*	*mi*
V_{fac}:	—	—	$\bar{V}_{mi,fac}$ (eq. 8.3.7)
facy:	—	—	$V_{Vmi,facy}$ (eq. 8.3.5)
fac:	—	$V_{Vcy,fac}$ (8.3.4)	$V_{Vmi,facy} \cdot V_{vcy,fac} = V_{Vmi,fac}$
fa:	—	$V_{Vcy,fa}$ (8.3.3)	$V_{Vmi,facy} \cdot V_{Vcy,fa} = V_{Vmi,fa}$
ac:	$V_{Vfa,ac}$ (eq. 8.3.2)	$V_{Vcy,fa} \cdot V_{Vfa,ac} = V_{Vcy,ac}$	$V_{Vmi,facy} \cdot V_{Vcy,fa} \cdot V_{Vfa,ac} = V_{Vmi,ac}$
ag:	$V_{Vfa,ag}$ (eq. 8.3.1)	$V_{Vcy,fa} \cdot V_{Vfa,ag} = V_{Vcy,ag}$	$V_{Vmi,fac} \cdot V_{Vcy,fa} \cdot V_{Vfa,ac} = V_{Vmi,ag}$

8.4.   Skeletal muscle fibers: stereology applied to anisotropic and periodic structures

Contributor: *Brenda Eisenberg, Chicago**

8.4.1.   MODEL OF MUSCLE STRUCTURE

Muscle tissue is composed primarily of long parallel muscle fibers that perform its main function of contraction. The tissue also contains the complements of nerves, blood vessels and connective tissue required to support this activity. The major constituents of the muscle fiber are shown as a hierarchical model in Fig. 8.4.1. Besides these major constituents the muscle fiber contains a variable amount of poorly structured sarcoplasm in which one may find glycogen granules and fat droplets.

* Brenda R. Eisenberg, Department of Pathology, Rush Medical College, Chicago, Illinois 60612, U.S.A.

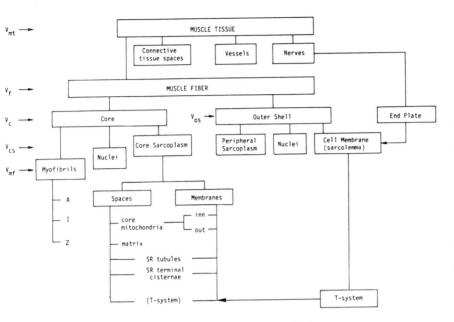

FIG. 8.4.1: A hierarchical model of muscle tissue.

Skeletal muscle fibers extend along the entire length of a muscle. They are multinucleated cells whose nuclei lie beneath the plasma membrane or sarcolemma. The major part of a single muscle fiber is occupied by the contractile filaments which are found in a periodically repeating striation pattern. The filaments are bundled together in highly oriented structures called myofibrils. Contraction uses energy and the mitochondria to supply ATP are dispersed amongst the myofibrils. Because of the myofibrillar shape and orientation the mitochondria are forced to lie transversely in the *I* band or longitudinally between the myofibrils (Fig. 8.4.2). In addition, packets of mitochondria lie peripherally beneath the sarcolemma. The myofibrils throughout a fiber contract simultaneously when the muscle fiber is stimulated because the electrical signals are rapidly carried inwards from the surface by a complex network of tubules called the T-system. A second network of tubules called the sarcoplasmic reticulum (*SR*) closely approaches the T-system at the triad junction. These two networks are arranged around the myofibrils (Fig. 8.4.2) such that the T-system lies mainly in the transverse plane and the *SR* lies mainly longitudinally to the fiber axis. We further sub-divided the *SR* into *A* band *SR*, *I* band *SR* and terminal cisternae (*tc*).

The muscle cell is hence a highly ordered structure, exhibiting at once parallel arrangement of strictly longitudinal myofibrils (anisotropy) and

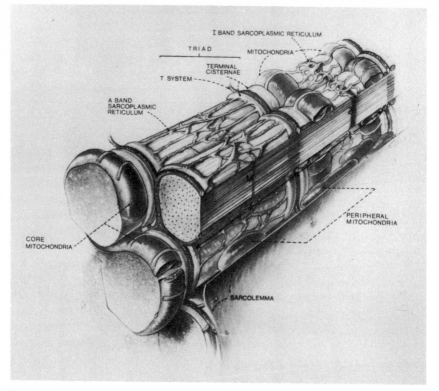

FIG. 8.4.2: A schematic drawing of part of a mammalian skeletal muscle fiber showing a few myofibrils close to the fiber surface (sarcolemma). The mitochondria and membrane systems lie between the parallel myofibrils and their surfaces exhibit a degree of aniso-tropy. Note also that mitochondria are distributed differently in the periphery and the fiber core. (Reproduced by permission from Eisenberg *et al.*, 1974.)

periodicity due to the alternating arrangement of actin and myosin filaments in the sarcomeres. This imposes some limitations on the sampling schemes that may be used for a stereological analysis.

8.4.2.   PURPOSE OF THE STUDY

Individual muscle fibers contract at different speeds and have different metabolic capabilities. These differences are reflected by the amounts and distribution of the organelles within any one fiber. On the basis of histo-chemical differences three fiber types have been defined which correspond to different functional units. The guinea pig has three muscles which are each composed predominantly of one type; namely the soleus muscle of slow-twitch-oxidative type, the white portion of the vastus lateralis of fast-twitch-glycolytic type and the red portion of the vastus lateralis of fast-twitch-oxidative-glycolytic type. In our early studies (Eisenberg et al., 1974; Eisenberg and Kuda, 1975, 1976) we therefore used these three muscles to find, through a stereological analysis, the quantitative structural information which could be correlated with the histochemical, biochemical and physiological properties of these three fiber types. Furthermore, statistical testing of these quantitative structural data allowed a critical analysis of the validity of the concept of three discrete fiber types as opposed to a continuous spectrum of the whole fiber population.

8.4.3. GENERAL STRATEGY

Because muscle fibers are anisotropic due to the full orientation of myofibrils, we did not feel justified in applying the simple methods of stereology and sampling valid for isotropic structures. Therefore, on the first muscle we studied (Eisenberg et al., 1974), we performed the entire stereological analysis in two orientations using both transverse and longitudinal sections. In addition, the effect of section thickness and the ability to recognize a small component is dependent on its orientation to the section plane. Furthermore, the highly ordered distribution of the organelles within the fiber and within striations of the contractile filaments had to be overcome. To all these technical problems one must add the requirement that sufficient data be collected from each fiber to perform the fiber population analysis.

The problems of stereological analysis in muscle are most challenging and we present below a method of compromise that enabled us to answer the specific question of fiber types. We therefore restrict the following discussion to the longitudinal sections of muscle and the examination of part of the SR, T-system and mitochondria in the fiber core and will not deal with the fiber periphery where local concentrations of organelles occur.

8.4.4. SAMPLING PROCEDURE

The components of interest were grouped into three size categories and three appropriate magnifications were chosen. The light microscope was used for objects of 2–40 μm and electron microscopy at two magnifications was used for objects 0.2–1 μm and for objects less than 0.2 μm. At each stage of magnification the equation (3.38) (Weibel and Bolender, 1973) was used to determine the total number of points necessary on each fiber to reduce the relative expected error to 10% (see Fig. 4.7). The grid spacing, d, was then determined by $d = an/P_T$, where a = area of one micrograph, n = number of micrographs and P_T = number of points. Practical limitations forced the choice of 0.5 cm

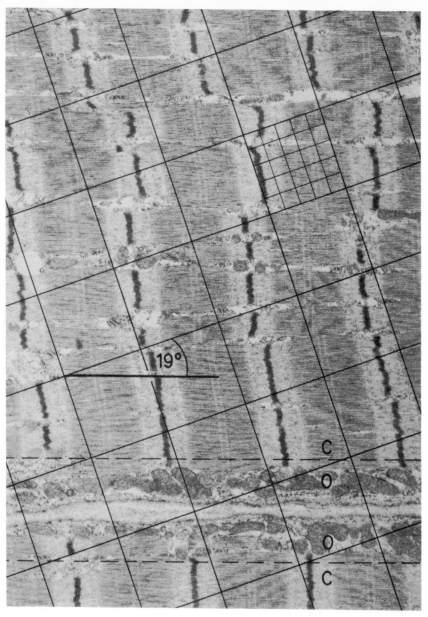

Fig. 8.4.3: A longitudinal section of the soleus muscle from a guinea pig showing part of two fibers. A dashed line is drawn 1 μm from the sarcolemma to separate the peripheral outer shell (o) which contains a higher concentration of mitochondria from the central core (c). The square test grid is oriented at 19° to the fiber axis to overcome systematic counting errors which anisotropy would otherwise cause in surface density estimates. The coarse grid spacing is 2 μm apart. The fine grid with spacing of 0.4 μm ($q = 5$) is drawn in only one coarse square.

on the micrograph as the smallest d and therefore very rare components could not be accurately determined fiber by fiber from any reasonable number of micrographs.

At all stages a square lattice test grid was used and oriented at 19° to the fiber axis (Fig. 8.4.3) to overcome systematic errors in the number of intersection counts on oriented membrane systems (Sitte, 1967; Eisenberg *et al.*, 1974). See also Sections 3.6 and 6.4.

Stage I

The three muscles were taken from each of five animals and fixed in glutaraldehyde and osmium. Five blocks were randomly selected from each muscle and embedded in Epon for sectioning at 1 μm thickness with a glass knife after being longitudinally oriented. The sections were stained with toluidine blue for evaluation with the light microscope whose eye piece was fitted with a 100 point square lattice grid. The apparent surface to volume ratio of the fiber seen as a smooth cylinder was measured at this stage ($S_{Vf\,\text{smooth},\,f}$). Data could not be collected fiber by fiber at this magnification.

Stage II

Ultra-thin sections (0.06 to 0.09 μm thick) were cut from the same blocks and mounted on 300 mesh copper grids. The micrographs were randomly chosen by systematic sampling rules (cf. Section 3.5). Alignment between the copper grid bars and the fibers can result in "random" selection of several areas of the same fiber. In such a case only the first area was photographed and the next fields lying within the same fiber were skipped. Fibers whose profile diameter was less than 13 μm were also rejected because these profiles contain too high a sample of the peripheral shell of cytoplasm being within 1 μm of the fiber surface. This peripheral shell contains local concentrations of mitochondria that can bias the data for composition of the core.

Micrographs were taken at $\times 4000$ magnification and printed at $\times 12,000$. More than six micrographs were taken to insure that six different fibers were included, each of which contained the core of the fiber. A 25 : 1 double lattice square test system was placed with one line at 19° to the fiber axis (Fig. 8.4.3) which allowed estimation of the surface area of the sarcolemma, S_f. The volume densities of the mitochondria and lipid droplets in the I and A bands were estimated by point counting. Note that the volume density measurements could be made with the grid at any angle.

Stage III

Micrographs were taken to give a print magnification of $\times 30,000$ for each of the six fibers identified at Stage II. Two suitable areas of fiber core could generally be found in the usual manner in the corners of the supporting grid.

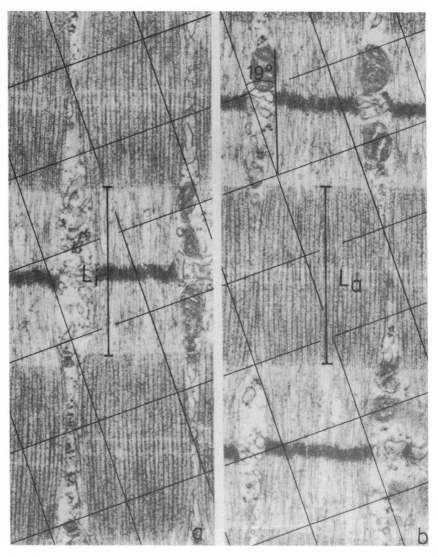

Fig. 8.4.4a and b: Longitudinal section through a few myofibrils of soleus muscle of the guinea pig. At higher magnifications representative samples of both the dark A band (L_a) and light I band (L_i) are not generally obtained because of the periodic repeat of the bands. In (a) the dark A band occupies about two-thirds of the area whilst in (b) it occupies only about one-third. The lengths of the bands, L_a and L_i, are used to correct this systematic sampling error, see text. A coarse grid with spacing of 0.8 μm is oriented at 19° to the fiber axis.

However, if the grid bars were parallel to the fiber axis this method of random selection can fail. We then took the micrograph near the grid bar and approximately in the fiber center. At this magnification representative sampling from both the A and I bands does not generally occur. Thus one micrograph may cover two A bands with only one I band between and thus give a biased overestimate of the A band or vice versa (Fig. 8.4.4). Therefore, the A and I filament lengths (L_a and L_i) were measured to determine the correct ratio of A to I band volumes. The same test lattice was used on Stage III as at Stage II. The coarse points (1/25th of the total points) were used for counting the actual I band (P_i) and the A band (P_a) volumes. All the points were used to count the membranous structures of the sarcoplasmic reticulum in the I-band and the T-system. The lines of this square test lattice grid were used to make intersection counts of these same membranous structures.

8.4.5.   EXAMPLE OF CALCULATIONS OF STEREOLOGICAL PARAMETERS

In the following the calculation procedure is illustrated using the example of volume and surface densities of the T-system and of the SR in the I band and the surface density of the sarcolemma.

Stage III
Counted: Coarse test points on I band (P_i), A band (P_a) which added together cover the entire fiber core, as well as fine test points which fall on the T-system (P_t) and on SR in the I band (P_{isr}); intersection counts of the membrane bound components were counted yielding I_t and I_{isr}. The length of the I and A filaments L_i and L_a were measured with a ruler.
   Volume densities in the core (c):

$$V_{Vt,c} = P_t / [q^2(P_i + P_a)]$$

$$V_{Visr,c} = \frac{P_{isr}}{q^2 P_i} \cdot \frac{L_i}{L_i + L_a}$$

where the ratio of $L_i/(L_i + L_a)$ corrects for biased sampling of the I band, and $q = 5$ is the fine to coarse grid line ratio.
   Surface densities in the core (c):

$$S_{Vt,c} = \frac{M}{q^2 d} \cdot \frac{I_t}{(P_i + P_a)}$$

$$S_{Visr,c} = \frac{1.06 M}{q^2 d} \cdot \frac{I_{isr}}{P_i} \cdot \frac{L_i}{L_i + L_a}$$

where M is magnification, d is the grid spacing between the fine lines. The factor $\times 1.06$ is empirically determined to correct for partial anisotropy of the SR tubules (Eisenberg et al., 1974; Eisenberg and Kuda, 1975, 1976).

Stage II
Counted: Test points falling on the fiber core (P_c), fiber peripheral outer shell (P_o), nucleus (P_n), and mitochondria in core and outer shell $(P_{cmi}$ and $P_{omi})$. Intersection of test lines with real, wrinkled sarcolemmal surface, I_f.
    Calculated: Volume densities of outer shell and mitochondria in the core, and surface density of sarcolemma.

$$V_{Vo,f} = P_o/(P_o + P_c)$$

$$V_{Vcmi,c} = P_{cmi}/P_c$$

$$S_{Vf,f} = \frac{1.1M}{q^2d} \cdot \frac{I_f}{P_f}$$

Stage I
    Counted: Test points falling on fiber (P_f) and intersection of fiber sur-surface with test lines $(I_{f\,\text{smooth}})$.
    Calculated: Apparent surface to volume ratio of the fiber:

$$S_{Vf\,\text{smooth},f} = \frac{1.1M}{q^2d} \cdot \frac{I_{f\,\text{smooth}}}{P_f}$$

Any difference between the real and apparent surface densities of the sarcolemma from Stages I and II can be ascribed to wrinkling of the sacrcolemma. The wrinkling factor $I_f/I_{f\,\text{smooth}}$ is known to vary with sarcomere length (see freeze-fracture study of Dulhunty and Franzini-Armstrong, 1975).

8.4.6. ERROR CONSIDERATIONS

Perhaps the most serious error in the analysis arises from an unmeasured alteration of muscle ultrastructure upon fixation. A study of this error on single semitendinosus fibers (Mobley and Eisenberg, 1975) showed that a striking net decline in the volume of the cell by approximately 45% occurred during the steps after osmium fixation. Of course it is impossible to monitor changes in individual components in the fiber under the same circumstances. The large fluctuations in volume and the extensive shrinkage observed in single fibers probably did not occur in the fibers of whole muscles or bundles studied here. However, this suspicion is based only on a comparison of the quality of the final embedded material and not on direct observation or testing of the tissue. It should be noted that if shrinkage in the fiber volume or its components occurred then the calculated parameters, though accurate for fixed material, might not accurately describe the living fibers studied in physiological experiments.

Before starting the analysis we recognized that there could be subjective differences in the identification of certain components, particularly longitudinal sarcoplasmic reticulum. Therefore, two of us independently counted several of the same micrographs without moving the grid. Our data differed by less than 10% for any component and this difference was random. The anisotropic nature of muscle made it necessary to modify the standard stereological formulae by multiplying by a numerical correction factor. A theoretical value of this factor is $\pi^2/8 = 1.23$ as given by Whitehouse (1974b) for fully oriented, right circular cylinders cut longitudinally. However, biological tissue usually results in partial orientation of varying degrees and therefore the factor needed will vary from 1.23 for full orientation down to 1 for isotropic surfaces. By an empirical study (Eisenberg *et al.*, 1974) we found appropriate factors to be 1.1 for the fiber surface and 1.06 for the *SR*.

Furthermore, the errors introduced by the thickness of the section, known as the Holmes effect, will be different for membranes in different orientations with respect to the section plane. We attempted to overcome these problems by empirical and theoretical corrections. If future research leads to a change in the correction factors, then our absolute values will have to be corrected by a constant factor. This correction is unlikely to invalidate comparative studies of fiber populations, since they deal with ratios between the same parameters in several fibers.

8.4.7. EXTENSIONS OF THE STUDY

Other skeletal muscles studied include frog (Mobley and Eisenberg, 1975) and human muscle (Hoppeler *et al.*, 1973; Jerusalem *et al.*, 1975; Cullen and Weightman, 1975). Some studies include evaluation of changes with exercise and age (Kiessling *et al.*, 1973; Hoppeler *et al.*, 1973) and in disease (Cullen and Fulthorpe, 1975).

We used the stereological data collected from 300 fibers in guinea pig muscle to perform fiber-typing in an objective manner. This was achieved by a statistical procedure called a discriminant analysis in which all the parameters are taken into account and weighted according to their usefulness; a single compound parameter is formed such that the final discrimination of the fiber types is maximized (Eisenberg and Kuda, 1976).

8.4.8. ALTERNATIVE METHODS

In Section 3.6 it has been proposed that the difficulties in sampling a structure which is at once anisotropic and periodic can be overcome by using oblique sections. On such samples Hoppeler *et al.* (1973) have studied the mitochondria in the case of human muscle fibers obtained by biopsy from normal subjects and athletes. Weibel (1972) has presented a method to estimate SR membrane area from oblique sections; this approach has been used by Kilarski (1973).

Cardiac muscle is another tissue which exhibits anisotropy and periodicity. Page et al. (1971) have chosen a different method of obtaining the average number of intersection counts (see Page, Section 8.5.5 and Mobley and Page, 1972). Alternatively, intersection counts can be obtained using an isotropic semicircular test system (see Fig. 4.11 and L 100 Appendix 3).

8.5.   Mammalian ventricular myocardial cells

Contributor: Ernest Page, Chicago*

8.5.1.   MODEL OF MYOCARDIAL CELL STRUCTURE

Mammalian myocardial cells are delimited by the sarcolemma, a plasma membrane with several functionally important areas of structural specialization (McNutt and Weinstein, 1973; Stewart and Page, 1978): the external sarcolemma or cell envelope, whose known secondary specializations include the areas of diadic couplings with underlying terminal cisternal membrane, vesicular invaginations, and areas of Z-band attachments; and the internal sarcolemma, which is made up of the T-system and the transverse cell boundaries ("intercalated disk"), including the three sub-specializations of the disk (the nexal or gap junctional membrane, the fasciae adhaerentes, and the desmosomes). These membranes enclose a volume containing sarcoplasm, myofibrils, mitochondria, nuclei, sarcoplasmic reticulum ("SR", including terminal cisterns and calcium-collecting tubular portions), and other structures. Myocardial cells from the atria and conduction system, as well as from embryonic hearts, differ in ultrastructural composition from post-natal and adult ventricular myocardial cells, and therefore require some modification of stereological approach (Mobley and Page, 1972).

8.5.2.   PURPOSE OF THE STUDIES

The aims of the studies were to measure the changes in ultra-structural composition of ventricular myocardial cells in response to the application or withdrawal of various stimuli to cell growth (Page et al., 1974b; Smith and Page, 1976, 1977; Page and Oparil, 1978), and to estimate the areas of membranous specializations of particular interest to cardiac cellular electrophysiology (Page et al., 1971; Mobley and Page, 1972; Page and Upshaw-Earley, 1977; Stewart and Page, 1978; Page, 1978).

* Department of Medicine, University of Chicago, Chicago, Illinois 60637, U.S.A.

8.5.3. GENERAL STRATEGY

Stereological measurements were made on electron micrographs of hearts perfused through the coronary vessels with a fixative approximately isosmolal with the cells. These measurements were correlated with parallel quantitative microchemical assays of myofibrillar mass and content of mitochondrial cristae (Page *et al.*, 1973, 1974a). Results were normalized to unit cell volume. For stereological measurement of extracellular and cellular volumes (Polimeni, 1974), tissues were fixed by immersion in isosmolal (312 mosmoles/kg water) solution of cacodylate-buffered saline containing about 200 mM glutaraldehyde, embedded in Epon as for electron microscopy and examined stereologically in prints of thick sections (0.75 μm) stained with toluidine blue.

8.5.4. SAMPLING PROCEDURE

To minimize sampling problems, sampling was limited to a restricted region of the ventricle from a small mammal (the subepicardial myocardium of rat left ventricle, exclusive of the apex) or from a small, structurally defined region of the heart of a larger mammal (the myocardial cells of right ventricular papillary muscles from the rabbit's right ventricle, exclusive of the endocardial and immediately subendocardial cell layers). When sampling was thus restricted, examination of three tissue blocks per ventricle showed no systematic variation between blocks with respect to the stereological results obtained.

Stereological analysis was carried out on prints at several stages of magnification. Thin, transparent sheets of plastic on which were imprinted square grids (0.319 or 1.0 cm/side) were superimposed on positive prints made from the negative film produced by the electron microscope camera. The sectioning plane was oriented systematically to obtain near perfect longitudinal sections, near perfect cross sections, or oblique sections, using a specially developed orientation procedure (Smith and Page, 1976). Magnification, number of points, grid spacings, and orientation for each stereological quantity are listed in Table 8.5.1. For procedures performed on longitudinal sections that require orientation of the test grid with respect to the orientation of the myofibrils and long axis of the cells, a line was drawn parallel to the mean myofibrillar direction (i.e., parallel to the long axis of the cell); a horizontal coordinate line of the grid was then oriented parallel to this line before starting to count points and intersections. A fine and a coarse grid (0.319 and 1.0 cm/side) were frequently applied successively in this way to the same prints.

Table 8.5.1.

Procedures for Stereological Analysis of Cardiac Muscle.

Measured Quantity	Magnification of Print $\times 10^{-3}$	Grid Spacing (cm)	Orientation of Section[a]	Number of Prints[b]	Calculation	Ref.
$V_{V ecs,t}$, $V_{V c,t}$	0.88–0.90	1.0	a	20–36	P_{ecs}/P_t, P_c/P_t	1
$V_{V mf,c}$, $V_{V mi,c}$	5–6	1.0	a	15–30	P_{mf}/P_c, P_{mi}/P_c	2
$V_{V tt,c}$	11–12	0.319 and 1.0	a	15–30	equation 106	3
$V_{V sr,mf}$	30–33	0.319	b	10–20	P_{sr}/P_{mf}	2
$V_{V mm,mi}$, $V_{V ics,mi}$, $V_{V cr+i,mi}$	130–200	0.319	c	10–12	equations 107–109	4
$S_{V es,c}$	5–6	0.319 and 1.0	c	20–30**	equation 104	3
$S_{V dk,c}$	5–6	0.319 and 1.0	c	20–30**	$I_{dk}^{\parallel} d_b M/(P_c d_a^2)$	3
$S_{V gj,c}$	5–6	0.319 and 1.0	c	20–30**	$\{(\pi/2)(I_{gj}^{\perp} - I_{gj}^{\parallel}) + 2I_{gj}\}\{d_b M/(P_c d_a^2)\}$	3
$S_{V tt,c}$	11–12	0.319 and 1.0	c	20–30	equation 105	3
$S_{V lsr,lsr}$	135–150	see footnote[c]	d	20	as for a right circular cylinder	5
$S_{V cr+i,mi}$	130–200	0.319	c	10–12	equation 110	
$\bar{D}_{mi}$	10.5–12	see footnote[c]	d	8–12	as for a right circular cylinder	4

Subscripts: *ecs*: extracellular space; *mf*: myofibrils; *mi*: mitochondria; *tt*: T-tubules; *sr*: sarcoplasmic reticulum; *ics*: intercristal space; *cr*: cristae; *es*: external sarcolemma; *gj*: gap junction; *lsr*: longitudinally oriented sarcoplasmic reticulum exclusive of terminal cisterns; *cr + i*: cristae + inner mitochondrial membrane; *c*: cell; *t*: tissue. [a] Orientations: *a*: any orientation usable, but conveniently done on longitudinal sections; *b*: slightly oblique cross section; *c*: near perfect longitudinal section; *d*: near perfect cross section. [b] Usually 3 blocks/ventricle, except for $V_{V ecs,t}$ and $V_{V c,t}$ where 5–11 blocks were examined. [c] Done with Zeiss Model TGZ3 Particle Size Analyzer. *References:* (1) Polimeni (1974); (2) Page *et al.* (1971); (3) Stewart and Page (1978); (4) Smith and Page (1976); (5) Page and Upshaw-Earley (1977).

8.5.5. EXAMPLES OF CALCULATIONS OF STEREOLOGICAL PARAMETERS

Volume densities of myofibrils, mitochondria, T-system, sarcoplasmic reticulum, and mitochondrial subcompartments were determined by point counting in rectangular photographic prints (25×20.5 cm) made from electron micrographs at four successive stages of magnification (Table 8.5.1). Surface densities of sarcolemmal, mitochondrial, and SR membranes were determined by combined point and intersection counts, the intersections of the test grid lines with one another being used as points. At successive stages of magnification, the results were referred, respectively, to unit tissue volume, unit myocardial cell volume, unit myofibrillar volume, and unit mitochondrial volume. Since the relationships of myofibrillar and mitochondrial volume to cell volume ($V_{Vmf,c}$ and $V_{Vmi,c}$) were rather precisely measurable by point counting, the relationship to cell and tissue volume of volumes and membrane areas referred at their initial measurement to myofibrillar or mitochondrial volume could be readily calculated. Two examples of such a conversion are $S_{Vsr,mf}$, the membrane area of SR, initially measured with respect to myofibrillar volume and $V_{Vmm,mi}$, The volume of the mitochondrial matrix, initially measured with respect to mitochondrial volume; thus

$$S_{Vsr,mf} \cdot V_{Vmf,c} = S_{Vsr,c}$$

and

$$V_{Vmm,mi} \cdot V_{Vmi,c} = V_{Vmm,c}$$

Similarly, conversion of volume and areal densities referred to myocardial cell volume ($V_{Vi,c}$ and $S_{Vi,c}$) to a reference unit of tissue volume is accomplished by multiplying these quantities by $V_{Vc,t}$.

The following paragraphs illustrate measurements which present special problems. These and other measurements are summarized in Table 8.5.1, which also gives the original references.

The membrane area of external sarcolemma per unit myocardial cell volume, $S_{Ves,c}$, was determined in near perfect longitudinal sections by orienting one axis of the test grid (1 cm/side) parallel to the mean direction of myofibrillar pull (i.e., parallel to the long axis of the cell). Then

$$S_{Ves,c} = \frac{(\pi/2)(I_{es}^{\perp} + I_{es}^{\parallel}) - I_{es}^{\parallel}}{P_c(d_a/M)^2}(d_b/M)$$

in which $I_{es}^{\perp}$ and $I_{es}^{\parallel}$ denote intersection counts of external sarcolemma with grid lines which are, respectively, perpendicular ($\perp$) and parallel ($\parallel$) to the myofibrils on the print; P_c is the number of points falling on all the myocardial cells in the print; M is the final magnification of the print; and d_a

and d_b are the distances between parallel grid lines of the grids used for point and intersection counting, respectively.

The following four additional quantities were obtained from the same prints as those used for measuring $S_{Ves,c}$: $V_{Vmf,c}$, $V_{Vmi,c}$, and the surface densities of intercalated disk ($S_{Vdk,c}$) and gap junction ($S_{Vgj,c}$). An equivalent cell length, $\bar{l}_c$, is given by $(S_{Vdk,c})^{-1}$; this quantity, which varies with sarcomere length, was converted to a standard value $\bar{l}_c(\text{std})$ for the (arbitrary) reference sarcomere length of 2 μm by the relation $\bar{l}_c(\text{std}) = \bar{l}_c \cdot 2/\lambda$, in which λ is the mean sarcomere length of the fixed myocardial cells measured in the same prints as those used for determining $S_{Vdk,c}$. A higher stage of magnification was next used to determine the area of T-tubular plasma membrane per unit cell volume, $S_{Vtt,c}$, from the relation

$$S_{Vtt,c} = \frac{(\pi/2)I_{tt}^{\|}}{P_c(d_a/M)^2} \cdot (d_b/M)$$

in which $I_{tt}^{\|}$ is the number of intersections by grid lines oriented parallel to the myofibrils with traces of T-tubular plasma membrane. The volume of T-tubular lumen per unit myocardial cell volume, $V_{Vtt,c}$, is given by

$$V_{Vtt,c} = (P_{tt}/P_c)(d_b/d_a)^2$$

in which P_{tt} is the number of points falling on the T-tubular lumen as measured with the fine grid ($d_b = 0.319$ cm), P_c being measured with a coarse grid ($d_a = 1.0$ cm). The surface to volume ratio of the T-tubules, $S_{Vtt,tt}$, can then be obtained by dividing $S_{Vtt,c}$ by $V_{Vtt,c}$.

The highest stage of magnification ($130–200 \times 10^3 \times$) was used for estimating the surface to volume ratio of the longitudinal SR ($S_{Vlsr,lsr}$) as well as the volume and surface densities of mitochondrial subcompartments and their delimiting membranes. $S_{Vlsr,lsr}$ was determined in near perfect cross sections cut as thinly as possible (< 60 nm). The outline of the inner membrane face for each of the roughly circular profiles was traced on tracing paper, and the distribution of the diameters was determined with the Zeiss Model TGZ3 Particle Size Analyzer. $S_{Vlsr,lsr}$ was then calculated from the mean diameter assuming cylindrical geometry.

For measurements on mitochondrial membranes and subcompartments, very thin sections of mitochondrial profiles were prepared from longitudinal sections. An area of part of the profile in which the traces of cristae were sharply delineated was outlined in red wax pencil, and the matrix was colored with yellow wax pencil. The area so outlined was subjected to point counts on the matrix, cristae + inner membrane, and space between inner and outer membrane ("intercristal space"), and to intersection counts on the cristae + inner membrane. The volume densities of matrix ($V_{Vmm,mi}$), of intercristal space ($V_{Vics,mi}$), and of cristae + inner mitochondrial membrane

$(V_{Vcr+i,mi})$, as well as the surface density of cristae + inner membrane $(S_{Vcr+i,mi})$, were approximated from the relations

$$V_{Vmm,mi} = (P_{tot} - P_{ics} - P_{cr+i})/P_{tot}$$

$$V_{Vics,mi} = P_{ics}/P_{tot}$$

$$V_{Vcr+i,mi} = P_{cr+i}/P_{tot}$$

$$S_{Vcr+i,mi} = (I_{cr+i}/P_{tot}) \cdot (M/d_b)$$

in which P_{tot} is the point count on the entire portion of the mitochondrion marked off by the red wax pencil and I_{cr+i} are the intersections of the cristae + inner mitochondrial membrane with both horizontal and vertical grid lines lying within this marked off area.

8.5.6. ERROR CONSIDERATIONS

Each of the determinations described here is subject to serious errors including overestimation, underestimation, sampling bias, and indeterminacies. Since a discussion of these errors is beyond the scope of this section, the reader is referred to the original articles.

8.6. Mucous membranes: stereology applied to layered structures

*Contributor: Hubert E. Schroeder, Zürich**

Among the various mucous membranes lining internal surfaces of the body, the oral mucous membrane is a variably structured integument which differs from site to site and frequently is subject to inflammatory and degenerative diseases and cancerous transformation. It consists of a stratified squamous epithelium, a variably dense lamina propria and, at most sites, a submucosa. The superficial layer of the lamina propria forms a papillary body, and its papillae vary in density, height and configuration (Klein-Szanto and Schroeder, 1977). The epithelium is variably thick, ranging from 0.08 to 0.5 mm, and to a variable degree interdigitated with the lamina propria by rows of the connective tissue papillae. All oral epithelia exhibit a differentiation pattern which, in part, is expressed in the form of structural gradients.

* Department of Oral Structural Biology, Dental Institute, University of Zürich, Plattenstrasse 11, CH–8028, Zürich, Switzerland.

8.6.1. PURPOSE OF STUDY

The aim of the stereological study of oral mucous membrane epithelium was (1) to characterize quantitatively the gradient of structural differentiation at different sites, (2) to compare the differentiation pattern of normal and pathologically altered epithelia, and (3) to establish a basis for correlating structure with chemistry and function of the various epithelial types.

8.6.2. MODEL OF ORAL EPITHELIUM

The principle structural and functional units of oral epithelia are keratinocytes. Because differentiation and structural gradients follow a direction from basal to surface layers, these cells differ from one another structurally at different levels or strata of the epithelium, basal cells being the least differentiated. As differentiation evolves simultaneously in cells leaving the stratum basale, areas of a rather homogeneous state of differentiation can be located reproducibly along an axis vertical to the epithelial surface. These areas constitute epithelial sheets which may or may not coincide with histologically defined strata. Irrespective of thickness, structural character and number of histologically defined strata, all oral epithelia, in spite of their anisotropic character, can be assumed to arise from a multilayer of polarized sheets each of them being characterized by an optimum of fine structural homogeneity (Schroeder and Münzel-Pedrazzoli, 1970a; Meyer and Schroeder, 1975). Therefore, the stereological model of oral epithelia is based on histologically defined or arbitrarily chosen strata (Fig. 8.6.1). Each of these strata was then subdivided into a hierarchical sequence of components (Fig. 8.6.2). This model provides relative (percentage) rather than absolute data and operates on a comparative basis. The reference compartments for calculating the various density parameters are unit volumes of epithelial stratum (V_{st}) and/or epithelial cytoplasm within a given stratum (V_{cy}) respectively.

8.6.3. SAMPLING PROCEDURES

Biopsies of normal or diseased human oral mucosa are harvested by a cornea trepan punch, cut vertically to the epithelial surface into several slices (blocks), and fixed by immersion in a mixture of glutar- and paraformaldehydes, as well as in OsO_4. Ultrathin (50 to 80nm) sections of the Epon-embedded tissue blocks, covering the entire epithelial thickness, are cut perpendicularly to the surface and mounted on carbon-coated R-150 copper grids with rectangular slits (VECO) in an oriented fashion which allows an uninterrupted cross-sectional scanning of the epithelium. Random sections

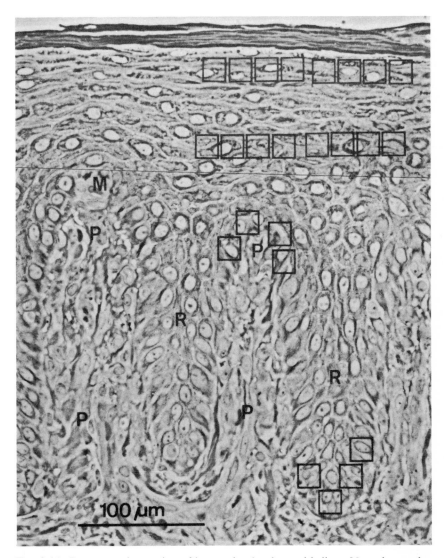

FIG. 8.6.1: Representative section of human hard palate epithelium. Note the regular interdigitation of connective tissue papillae (*P*) with epithelial ridges (*R*), clear cells and a corpuscle of Meissner (*M*). The rectangles indicate the stratified sampling at level *I* in the stratum basale, the upper stratum spinosum and the stratum granulosum of epithelial portions in ridges and over connective tissue papillae. (Reproduced by permission from Meyer, M. and Schroeder, H. E.: *Cell Tissue Res.* **158**, 177—203, 1975.)

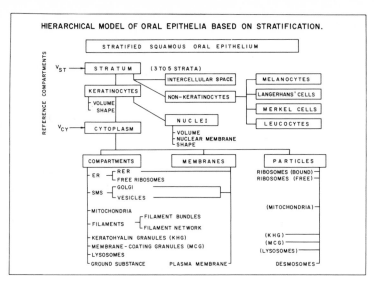

FIG. 8.6.2: Hierarchical model of oral epithelium.

derived from a minimum of 2 to a maximum of 4 tissue blocks per biopsy were subjected to electron microscopic sampling of micrographs. Because of the wide dimensional range of the epithelial components, a pattern of sampling at 2 different levels of magnification was chosen (Schroeder and Münzel-Pedrazzoli, 1970a; Meyer and Schroeder, 1975; Klein-Szanto et al., 1976b):

Level I

With the exception of a very narrow stratum basale (in which 3 times 2 micrographs were recorded at a primary magnification of about 3000 ×), 3 to 4 micrographs per stratum were taken at a primary magnification of about 1400 × (Fig. 8.6.3), the final magnification on the projection unit (see Fig. 4.14) being 11.4 times higher. While in the suprapapillary region the micrographs formed a consecutive field to field sample, the micrographs taken in basal and lower spinous regions followed the basement membrane or formed a square (Fig. 8.6.3). Sections were oriented within the electron microscope in order to produce an inclination of 20 to 30° of the epithelial surface to one side of the test field, thus avoiding any parallelism between cytoplasmic structures (such as plasma membranes of flattened cells) and the lines of the test system. Level I micrographs served for estimating the volume densities per unit volume of stratum of epithelial cell nuclei and cytoplasm, of the intercellular spaces and the non-keratinocytes. In addition, counts of intersections of horizontal test lines with profiles of the nuclear envelope and the

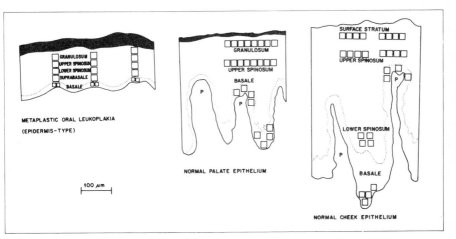

FIG. 8.6.3: Schematic illustration of various stratified oral epithelia and the respective sampling pattern used for collecting level *I* micrographs from 3, 4 and 5 different strata. *P* = connective tissue papilla; black surface zone = stratum corneum.

plasma membrane provided estimations of their surface density per stratum and the respective surface to volume ratios. The reference point set (P_T) was the total number of points per test field. Finally, the number of desmosomes was counted per test field, and subsequently calculated per unit area of $100 \, \mu m^2$.

Level II

A second sample of 9 electron micrographs was recorded within each of the areas and strata sampled at level I, the primary magnification ranging from $5–6000 \times$, the secondary magnification on the projection screen being about $70,000 \times$. In the stratum basale, this sample consisted of alternating fields taken from basal and distal cell portions. In all other strata, micrographs were recorded as a semi-horizontal field to field sample. Level II micrographs allowed the estimation of volume and, in part, surface densities per cytoplasmic unit volume of various cytoplasmic components such as rough endoplasmic reticulum, Golgi-apparatus and vesicles, free ribosomes and polyribosomes, mitochondria, bundled or network-like arranged filaments, keratohyalin and membrane-coating granules, lysosomal bodies, lipid droplets and the cytoplasmic ground substance (Fig. 8.6.2). The reference point set was the total number of points falling on epithelial cytoplasm. Occasionally, the level II samples served for estimating the number of bound ribosomes per unit surface area of rough endoplasmic reticulum (Landay and Schroeder, 1977). Depending on the density of interdigitating connective tissue papillae, on epithelial thickness and on the configuration of the epithelium-connective tissue interface, the sampling site, the mode of test field arrangement and the number of sampled strata varied (Meyer and Schroeder, 1975; Klein *et al.*, 1976b; Landay and Schroeder, 1977). Typical examples of how samples were taken in different oral epithelia are shown in Figs. 8.6.1 and 8.6.3. Because of differences in the distribution of mitotic activity in some epithelia, a stratified sample was taken each in regions of epithelial ridges as well as over connective tissue papillae (Meyer and Schroeder, 1975; Landay and Schroeder, 1977).

8.6.4. TEST SYSTEMS USED

Density estimations were achieved by using coherent test systems and a table projector unit in combination with an electronic data collector with automatic print-out facilities (Weibel, 1969). In earlier studies, either a multipurpose test system (M 168) or a double lattice test system (1:25) was used for level I, while level II micrographs were superimposed with the multipurpose test system M 168 (Schroeder and Münzel-Pedrazzoli, 1971a, b; Meyer and Schroeder, 1975). In order to avoid possible underestimation of rare cytoplasmic components and because of limited sample size, the analytic system used at present operates satisfactorily by applying a 1:9 (99/891) double lattice test system to both levels of magnification (Klein-Szanto *et al.*, 1976b; Hammer and Schroeder, 1977). In this test system, the fine lattice is used for rare components, such as intercellular spaces at level I and ergastoplasm, membrane coating granules, etc. at level II and the coarse lattice serves for estimating frequently occurring and rather homogeneously distributed components.

8.6.5. EXAMPLE OF CALCULATIONS OF STEREOLOGICAL PARAMETERS

For reasons of simplicity, the calculation procedure is demonstrated by using the example of the volume density of keratohyalin granules (KH):

Level II: Counted are light test points falling on KH (P_{kh}) and heavy points on total epithelial cytoplasm (P_{cy}^{II}). From these, the density in epithelial cytoplasm is calculated:

$$V_{Vkh,cy} = P_{kh}/P_{cy}^{II} \times 9$$

Level I: Counted are heavy test points falling on epithelial cytoplasm (P_{cy}^{I}) and on the total portion of the epithelial stratum (which is identical to P_T because the field is entirely within the stratum) as well as points on epithelial nuclei, intercellular spaces and non-keratinocytes. Calculated is:

$$V_{Vcy,st} = P_{cy}^{\;I}/P_T$$

From these data, parameters characterizing the relative volume density of keratohyalin granules can be calculated in relation to 2 different unit volume compartments which arbitrarily are set to $1000\,mm^3 = 1\,cm^3$.

(1) density in a unit volume of epithelial cytoplasm:

$$V_{Vkh,cy} = V_{Vkh,cy} \cdot 1000$$

(2) density in a unit volume of stratum:

$$V_{Vkh,st} = V_{Vkh,cy} \cdot V_{Vcy,st} \cdot 1000$$

The present system of analysing oral epithelia is served by an alternative off/on-line computer program allowing rapid and standardized parameter calculation by means of the Hewlett-Packard calculator HP-9830A, and providing automatic print-out of all parameters (averages per stratum over blocks, biopsies and biopsy groups plus respective standard deviations and standard errors) in table form (Hammer and Schroeder, 1977). This program, in addition, facilitates counting for example by automatic calculation of $P_{cy}^{\;I}$ from the difference between P_T and the sum of P_i other than P_{cy}, and $P_{cy}^{\;II}$ from the difference between P_T minus P_{ni} (i.e. every component other than epithelial cytoplasm). The final version of this program is available for those interested.

8.6.6. APPLICABILITY AND EXTENSION OF THE SYSTEM

The analytical system for stratified oral epithelia, in conjunction with automatic parameter calculation, can be applied to all epithelial integuments irrespective of whether they are stratified or not, i.e. disregarding the number of strata one would wish to analyse. Recently, the study on normal oral epithelia (Meyer and Schroeder, 1975; Landay and Schroeder, 1977; Bernimoulin and Schroeder, 1977) and the epidermis (Klein-Szanto, 1977) was extended to establish the alterations of the structural differentiation pattern in oral epithelia which are affected by autogenous connective tissue transplants or suffer from diseases such as leukoplakia and lichen planus (Klein-Szanto et al., 1976a, b; Bernimoulin and Schroeder, 1977, 1979). In addition, the stereological analysis of normal and/or pathologically altered oral epithelia can be supplemented and combined with a stereological study of the normal or inflamed and infiltrated subepithelial lamina propria (Schroeder and Münzel-Pedrazzoli, 1973; Schroeder et al., 1973, 1975; Petrzilka and Schroeder, 1976). Stereological analysis of the various cell types forming a chronic, immunopathologically regulated infiltrate provides information about the character and the destructive potential of the host response taking place (Schroeder and Graf-De Beer, 1976; Petrzilka and Schroeder, 1976).

8.7. Gut epithelium of insects: a variable single layered epithelium

Contributors: Hermann Hecker, Basel* and Peter H. Burri, Bern**

8.7.1. PURPOSE OF THE STUDY

Insects which take blood meals on vertebrate hosts are nutrition specialists. They have to digest a protein rich diet and the cells of their midgut epithelium therefore show some structural peculiarities.

Midgut cells of female mosquitoes, for example, perform very different functions during blood digestion: the formation of the peritrophic membrane, the synthesis and release of digestive enzymes, the resorption and transport of products of digestion (Gooding, 1972; Hecker et al., 1974).

The mosquito midgut consists of two distinct parts (Christophers, 1960) postulated to fulfill different functions: a narrow anterior (A-part) and a larger posterior part ("stomach" = P-part, Fig. 8.7.1). We investigated the structural composition of the epithelial cells of the two parts by morphometric techniques (Hecker et al., 1974) in order to define special functions for each of them. We wanted also to investigate the structural modifications of the midgut cells during blood digestion.

8.7.2. METHODOLOGICAL PROBLEMS

(a) The smallness of the midgut (diameter: 0.2–1 mm) and its subdivision into two parts makes it impossible to work with randomly oriented sections. A special sampling procedure had to be applied.

(b) The midgut epithelial cells exhibit a marked structural and functional polarity. Basis and apex of the cells differ distinctly in their ultrastructure (Figs 8.7.2 and 8.7.3); it is therefore essential that all parts of the cells have an equal chance to be sampled.

(c) The epithelium of the midgut can greatly vary in height (Figs 8.7.2 and 8.7.3) depending on the midgut part or/and the physiological stage examined. This has to be considered in determining the sample area, which should be of equal size for proper statistical analysis.

* Schweizerisches Tropeninstitut, Socinstrasse 57, CH–4051, Basel, Switzerland; ** Anatomisches Institut, Bühlstrasse 26, CH–3000, Bern 9, Switzerland.

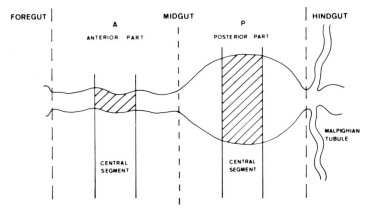

FIG. 8.7.1: *Aedes aegypti*, female, schematic drawing of midgut. Position of investigated central segments of *A* (anterior part) and *P* (posterior part = "stomach"). Reproduced from Hecker *et al.*, *Cell Tissue Res.* **152**, 31–49 (1974) with the permission of Springer Verlag, Heidelberg.

(d) In order to assess the functional capacity of a whole organ or of a part of it, it is desirable to know, besides its relative composition, its absolute volume. In the case of the insect midgut it was not possible to measure this latter parameters. An alternative solution was found in the measurement of nuclear sizes and the calculations of nuclear volumes, which were then used as reference parameters.

8.7.3. SAMPLING PROCEDURE AND MORPHOMETRIC EVALUATION (Hecker *et al.*, 1974; Hecker, 1978)

(a) The mosquitoes of each investigation were derived from the same egg batch in order to obtain a homogeneous population of insects.

(b) Midguts of insects in various physiological conditions (i.e. stages before, during and after a blood meal) were prepared for electron microscopy and embedded *in toto* in an oriented fashion.

(c) Only central portions of the A- and P-parts respectively were used, thus avoiding possible transitional zones (Fig. 8.7.1).

(d) Six midguts per stage were cut perpendicularly to their long axes within the A- and P-part. This sectioning procedure does not comply to the basic condition that random sections should be used in applying stereology. It takes into account, however, the smallness of the organ, its separation into two distinct parts and the polarity of the midgut epithelium.

(e) The stereologic analysis was performed in 3 steps.

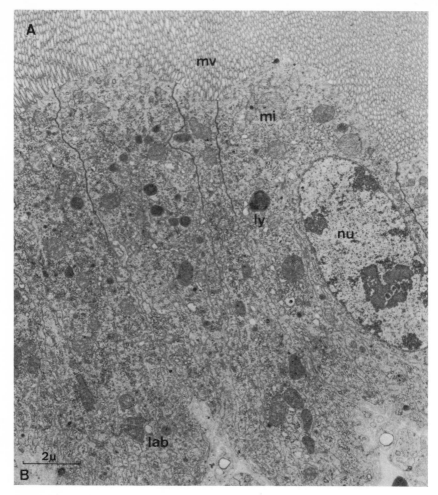

Figs 8.7.2 and 3: *A. aegypti*, female midgut epithelium 4 hours after blood-intake (2: high columnar epithelium of *A*-part, 3: flat epithelium of *P*-part, distended by the blood meal). *A* = cell apex, *B* = cell basis, *lab* = basal labyrinth, *ly* = lyosome, *mi* = mitochondria, *mv* = microvilli, *nu* = nucleus.

Step 1: *Determination of mean nuclear volume* $\bar{v}_{nu}$

In order to obtain an absolute reference parameter we determined the mean nuclear volume ($\bar{v}_{nu}$) in each midgut part. Camera lucida drawings from five of the largest nuclear profiles of each part per insect were made from semithin sections at a magnification of 2200 ×. Using 6 midguts a total of 30 nuclei per part and stage were measured. Assuming the nuclei to be prolate rotation ellipsoids we measured their long and short axes and calculated the

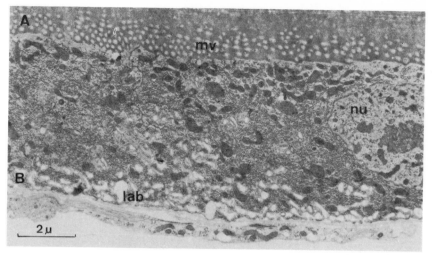

FIG. 8.7.3 (see FIG. 8.7.2 for legend).

mean nuclear volume (Hecker *et al.*, 1972). Under the assumption of a single population of nuclei the largest profiles yield the best estimate of the true diameter. The small number of nuclear profiles available in a transection of the midgut did not allow a size distribution analysis to be performed.

$\bar{v}_{nu}$ is influenced by the preparation techniques for electron microscopy. Shrinkage occurs and therefore uncorrected absolute values are biased. As we were however mainly interested in the comparison between groups, we disregarded such influences. The same applied to the Holmes-effect.

Step 2: *Estimation of the nuclear cytoplasmic ratio,* V_{nu}/V_{cy}
The nuclear cytoplasmic ratio V_{nu}/V_{cy} was estimated for each zone and for each insect on one low power electron micrograph depicting the epithelium from base to apex. The final magnification on the table projector was 5200 ×. Estimation was done by point counting on cytoplasmic and nuclear areas, using a square test lattice with a period $d = 15$ mm, corresponding to 2.9 μm on the specimen.

Step 3: *Estimation of volume and surface densities of organelles and of numerical densities of ribosomes.*
Volume and surface densities of organelles of midgut epithelial cells were estimated by point and intersection counting at a final magnification of 84,000 ×. We used a square test lattice with a period $d = 4$ cm, corresponding to 0.48 μm on the specimen. Numerical densities of membrane-bound and free ribosomes were estimated according to Simar (1973) (see Section 8.12).

In order to investigate representative and corresponding areas in both parts (and also to be able to compare different stages), a systematic random sampling was applied, which had to fulfill two conditions:

(1) In the case of a high columnar epithelium (Fig. 8.7.2) the sampling had to be done field by field from base to apex, traversing at least twice the whole thickness of the epithelium. This yielded a minimal cytoplasmic area covered by 500 test points.

(2) In the case of a flat epithelium (Fig. 8.7.3), we attempted to sample a cytoplasmic area equivalent to that used for a high columnar epithelium; the sample area hence had to cover a surface of at least 500 test points independently of how many times the height of the epithelium was traversed by the sampling procedure.

8.7.4. CALCULATIONS

The morphometric analysis according to step 3 allowed the calculation of the volume, surface and numerical densities of various organelles of midgut cells, e.g. mitochondria, lysosomes, rough endoplasmic reticulum (*rer*), ribosomes.

Owing to the fact that we measured the mean absolute nuclear volumes and the nuclear cytoplasmic ratios, we were able to calculate the mean cytoplasmic and cell volumes of the average epithelial cell of each part and stage:

$$\bar{v}_{cy} = \frac{\bar{v}_{nu}}{V_{nu}/V_{cy}}$$

This enabled us to derive mean organelle volumes and surfaces and mean numbers of ribosomes per epithelial cell. Consideration of absolute as well as relative results proved to be very valuable, since comparison of A- and P-parts and/or different physiological stages revealed differences not only in relative parameters but also in absolute cell mass.

8.7.5. FINDINGS AND EXTENSIONS OF THE STUDY

This study has shown that there are distinct quantitative structural differences between the cells of the A- and P-parts which reflect functional aspects. For example it could be shown that synthesis of digestive enzymes is likely to prevail in the midgut P-part of the female mosquito. This is indicated by the quantitative predominance of the *rer*. Furthermore it could be shown that during blood digestion the increase in *rer* surface area and the increase in the ratio of membrane-bound to free ribosomes corresponded to the measured augmentation of protease activity (Gooding, 1973).

Hopefully, using these methods, we shall be able to find typical ultra-structural criteria for the midgut cells of various blood digesting insects. This could lead to the establishment of functional models of the midguts of hematophagous insects. Ultimately such knowledge contributes to the understanding of the transmission mechanisms of pathogenic micro-organisms by blood sucking insects.

8.8. Bone tissue

Contributor: Robert K. Schenk, Bern*

8.8.1. STRUCTURE OF CANCELLOUS BONE

Adult cancellous bone comprises two major compartments: bone tissue and bone marrow. The bony elements form a three-dimensional scaffold of either plates or more or less cylindrical bars. Both plates and bars appear in the sectional plane as continuous or discontinuous profiles referred to as "trabeculae", regardless of their actual three-dimensional shape. They sub-divide the marrow cavity into intertrabecular spaces that contain either red or white bone marrow. Both bone tissue and bone marrow form two inter-laced continua; the subdivision of the marrow space by bone is always incomplete.

The interface between bone and bone marrow merits special attention in our projects. All the bony elements are completely enwrapped by a special layer of marrow tissue called "endosteal envelope" (Frost, 1963). The composition of this envelope varies in accordance to local bone remodeling activities, i.e. areas where bone tissue is resorbed or where new bone is laid down. Since the whole interface between bone and bone marrow participates in the extremely well regulated exchange of calcium ions between bone and extra-cellular fluid, the quantitative evaluation of this surface as a whole, as well as of its fractions engaged in bone formation and bone resorption becomes one of the most important problems of this study.

8.8.2. PURPOSE OF THE STUDY

The aim of this study was to analyse structure and turnover of human cancellous bone, its physiological variation and its changes with age. Various

* Anatomisches Institut der Universität Bern, Abt. für Systematische Anatomie, Bühlstrasse 26, CH–3000, Bern 9, Switzerland.

structural parameters have been elaborated as a base for a possible differentiation between physiological bone loss with age, and pathological bone atrophy leading to mechanical insufficiency as in osteoporosis (Merz and Schenk, 1970a; Olah, 1974). Since loss of bone can be due to increased bone resorption, decreased bone formation or unidirectional but disproportional changes in both activities, an attempt has been made to judge the momentary state of the respective bone remodeling activities.

Intensity of bone resorption is evaluated by measuring the surface where bone resorption actually occurs or has taken place previously (Schenk et al., 1969). The notion that bone is resorbed by multinucleated giant cells— the osteoclasts—is now commonly accepted. These cells are easy to identify in undecalcified, properly stained microtome sections. Their bone resorbing activity is restricted to the direct contact area between their cytoplasm and the underlying bone substance, the osteoclast–bone interface (Fig. 8.8.1). The localized action of every single osteoclast leaves behind a shallow groove in the bone surface: a Howship's lacuna. These lacunae are also taken as a parameter for the intensity of bone resorption. But one has to bear in mind that they persist until they are buried by newly formed bone and therefore depend on bone formation as well as on the intensity of bone resorption.

Intensity of bone formation has to be considered in two different respects, since it proceeds in 2 steps: (1) deposition of unmineralized bone matrix (osteoid) by bone forming osteoblasts and (2) subsequent transformation of osteoid into calcified bone by the mineralization process. The morpho-

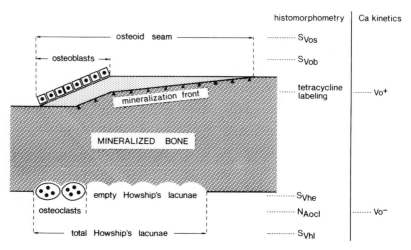

FIG. 8.8.1: Diagrammatic representation of bone turnover parameters (Courtesy of Dr. A. J. Olah).

metric evaluation of bone formation therefore includes parameters which reflect the mineralization process and its abnormalities separately (Merz and Schenk, 1970b). This approach has turned out to be effective in the detection of mineralization disturbances as they occur in certain metabolic bone diseases and under the influence of some therapeutic agents (Olah, 1974; Schenk et al., 1973).

8.8.3.   GENERAL STRATEGY

Cancellous bone in the axial skeleton participates intensively in the skeletal turnover. The anterior part of the iliac crest has proven to be the most appropriate site for cancellous bone biopsies and in most investigations cylindrical drill biopsies are taken ca. 3 cm behind the spina iliaca anterior superior. The cylinders are either taken in a vertical direction, thus comprising only a small layer of cortical bone, or transversally through both cortices and the middle cancellous part of the iliac bone about 1 cm below the iliac crest. The diameter of the cylinders varies from 5 to 10 mm, their length from 10 to 20 mm. Since cortical bone and sub-cortical layers have to be excluded from quantitation, the field of section that comprises true cancellous bone is somewhat smaller.

Standard values have been determined in a collection of iliac crest bone from random autopsy cases. In these cases, wedge-shaped pieces of iliac crest bone have been taken from the same location as the biopsies. In the cross section of the autopsy specimens, a measuring field is delineated that corresponds in size and location to the actual biopsies.

8.8.4.   SAMPLING PROCEDURE

Undecalcified, 5 micron thin sections are cut on a special microtome after embedding in methyl-methacrylate (Schenk, 1965). This method and the unique mechanical consistency of undecalcified bone eliminate any distortion due to shrinkage and sectioning. The comparatively small section thickness avoids any corrections for section thickness for the parameters under consideration. Most measurements are carried out on sections stained with a modified Goldner's trichrome method (Schenk et al., 1969). This allows a reliable differentiation of the various bone cells and in most cases a satisfactory distinction between mineralized and unmineralized matrix. In critical specimens, where mineralized bone and osteoid are not clearly demarcated after the Goldner stain, a modified von Kossa-reaction (Krutsay, 1963) is used for supplementary estimation of the mineralized and unmineralized tissue fraction (Merz and Schenk, 1970b). This method, however, is not suitable for differentiation of bone and bone marrow cells.

L

Cancellous bone has an anisotropic structure (Whitehouse, 1974a). Its plates and bars are oriented preferentially according to the prevailing stress and strain pattern in the proper part of the skeleton. Anisotropy, however, excludes intersection counts by means of test systems with straight and parallel sampling lines. Merz (1967) proposed a solution of this problem by introducing a test grid with points and a set of semicircular sampling lines that represent any direction in the test field with equal probability (see Fig. 4.11). For practical use, a square test field with 10 semicircular lines and 100 points is used on the screen of a projection microscope; in integrating eye-pieces, a grid with 6 lines and 36 points is preferable (L 36 and L 100, Appendix 3). All measurements are carried out at 250 × magnification. A total test area of 16 mm^2 is coherently covered with 100 sample fields in 3 non-consecutive sections of one specimen. This corresponds to a total number of 30,000 test points applied and yields, on empirical grounds, an error of ($\pm$) 5% for the estimate of the smallest volume fraction, the osteoid, that amounts to approximately 0.5% under physiological conditions. In a complete analysis of cancellous bone structure and turnover 9 or 10 parameters are recorded. Two values are obtained by point counting, 5 or 6 by intersection counts. In addition, the number of osteoclasts and the total number of test fields examined are registered.

8.8.5. VALUES OBTAINED BY POINT COUNTS

The values recorded by point counting are (for abbreviations of histological features see Table 8.8.1):

P_{min} —points on mineralized bone matrix
P_{os} —points on unmineralized matrix (osteoid)

The remaining points overlay bone marrow. They are counted separately only if the test field is incompletely covered by the section.

8.8.6. VALUES OBTAINED BY INTERSECTION COUNTS

In view of the evaluation of bone turnover 5 or 6 surface fractions or interfaces are measured by intersection counts. They are explained in Fig. 8.8.1 and defined as follows:

I_{ne} —intersection with neutral bone surface (fully mineralized, no traces of previous resorption)
I_{he} —intersection with Howship's lacunae without osteoclasts
I_{ocl-b} —intersection with osteoclast-bone interface (or Howship's lacunae with osteoclasts)
I_{ob-os} —intersection with osteoblast-osteoid interface

$I_{os\text{-}m}$ —intersection with osteoid-marrow interface
$I_{os\text{-}b}$ —intersection with osteoid-bone interface (in conditions with very irregular bone structure only).

Table 8.8.1.
Abbreviations for histological features of bone tissue.

b	bone tissue (mineralized and unmineralized matrix incl. osteocytes)
cb	cortical bone
he	empty Howship's lacunae
hl	Howship's lacunae, with or without osteoclasts
ho	Howship's lacunae with osteoclasts
m	bone marrow
min	mineralized bone
ne	neutral bone surface
ob	osteoblast
oc	osteocyte
ocl	osteoclast, osteoclast–bone interface
os	osteoid
s	osteoid seam thickness
tb	trabecular bone
$trab$	trabeculae

8.8.7. EVALUATION OF CANCELLOUS BONE STRUCTURE

Volume density V_V of cancellous bone is the most common parameter and calculated as

$$V_{Vb} = \left[\left(\sum_{}^{m} P_b\right)/m \cdot P_T\right] \cdot 100\% \qquad (P_b = P_{min} + P_{os})$$

where m is the number of microscope fields evaluated.

The mineralized and unmineralized fraction of the bone matrix can be measured separately:

$$V_{Vmin} = \left[\left(\sum_{}^{m} P_{min}\right)/m \cdot P_T\right] \cdot 100\%$$

$$V_{Vos} = \left[\left(\sum_{}^{m} P_{os}\right)/m \cdot P_T\right] \cdot 1000 \ (\text{mm}^3/\text{cm}^3)$$

The volume density of osteoid is very small in healthy bone and therefore given in mm^3/cm^3. Osteoid can replace a considerable amount of the bone substance in case of mineralization disturbancies. For diagnosis of such osteomalacic conditions, it is also calculated as a fraction of the actual bone volume:

$$V_{Vos,b} = \left[\left(\sum_{}^{m} P_{os} \right) / \sum_{}^{m} P_b \right] \cdot 100\%$$

Volume density yields the same information as other densitometric determinations, the volume density of mineralized bone being comparable to density measurements as by X-ray absorption and similar methods. But unlike other densitometric methods, histomorphometry offers additional parameters for quantitative assessment of cancellous bone structure.

The *surface density* of the total interface between bone and bone marrow S_{Vb} reflects the size of the endosteal envelope. It is calculated from intersection counts as

$$S_{Vb} = 2 \cdot \left[\left(\sum_{}^{m} I_b \right) / m \cdot L_T \right] \text{mm}^2/\text{mm}^3$$

($I_b = I_{ne} + I_{hl} + I_{ocl-b} + I_{ob-s} + I_{os-m}$ = all intersections with bone surface.)

It is known from the formula $S_V = 2 \cdot I/L_T$ that the surface density depends directly on the number of intersections with test lines of length L_T. Each crossing between a test line and a trabecular profile results in 2 intersections. (L_T/I_b) thus reflects the average distance between the trabeculae or, in other words, the width of the intertrabecular spaces. The trabecular surface becomes even more interesting if it is not related to total volume of cancellous bone, but to the volume of the trabecular substance alone:

$$S_{Vb,b} = 2 \cdot \left[\left(\sum_{}^{m} I_b \right) \cdot P_T / \left(\sum_{}^{m} P_b \right) \cdot L_T \right] \text{mm}^2/\text{mm}^3$$

The resulting surface-to-volume ratio $S_{Vb,b}$ depends directly on the spatial distribution of a given bone volume within a unit total volume and therefore on the number and thickness of the trabeculae. Cancellous bone with a given V_V, but with an array of numerous, but thinner trabeculae has a higher surface-to-volume ratio (or "specific surface") than a specimen with fewer, but coarse trabeculae (Fig. 8.8.2). Obviously, this reverse proportionality impedes communication with non-morphologists and may even be misleading, and an index for the mean trabecular diameter that is based on a volume-to-surface ratio seems to be more appropriate. To a certain degree, the mean diameter (or an index of the mean diameter) of a structural component can be derived from the volume-to-surface ratio, if the structure in question is geometrically simple and uniform (see Section 6.2). In true trabecular bone we can theoretically calculate two diameter-dependent values. In the case of a continuous framework of cylindrical trabeculae, their

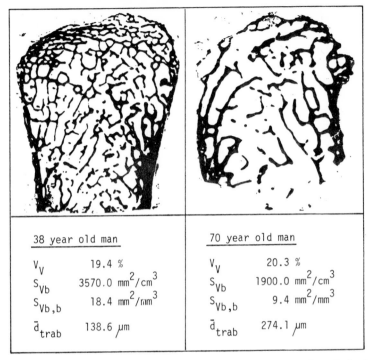

38 year old man		70 year old man	
V_V	19.4 %	V_V	20.3 %
S_{Vb}	3570.0 mm^2/cm^3	S_{Vb}	1900.0 mm^2/cm^3
$S_{Vb,b}$	18.4 mm^2/mm^3	$S_{Vb,b}$	9.4 mm^2/mm^3
$\bar{d}_{trab}$	138.6 μm	$\bar{d}_{trab}$	274.1 μm

FIG. 8.8.2: Comparison of cancellous bone structure and various values derived from surface to volume ratios in two cases with identical volume density. Cross sections through iliac crest. Magnification 3 ×.

mean diameter, d, is calculated from the volume-to-surface ratio as

$$\frac{V}{S} = \frac{L \cdot \pi r^2}{L \cdot 2\pi r} = \frac{r}{2} = \frac{d}{4} \quad \rightarrow \quad d = \frac{V}{S} \cdot 4$$

If we suppose that the trabeculae have a square cross section, the side length a is

$$\frac{V}{S} = \frac{a^2 \cdot L}{4 \cdot a \cdot L} = \frac{a}{4} \quad \rightarrow \quad a = \frac{V}{S} \cdot 4$$

The application of this principle to cancellous bone poses one problem: There is a considerable variation in the geometrical shape of the elements, since both cylindrical or prismatic bars but also plates may be present. To a certain degree, a distinction between plates and bars is possible in a section because plates always appear as continuous profiles, whereas bars are recognized by their discontinuity (Elias, 1972). But this difference is so far not reliable enough for an incontestable decision.

At the moment, the only way out seems to be to restrict this attempt to the two-dimensional sectional plane and to calculate an index for the thickness of the trabeculae referred to as the mean diameter of trabecular profiles. It is calculated as an area-to-perimeter ratio from the areal density A_{Ab} and the boundary length density B_{Ab} of the trabecular profiles as

$$\bar{d}_{\text{trab}} = 2 \cdot A_{Ab}/B_{Ab}$$

From $A_A = P_b/P_T$ and $B_A = (\pi/2) \cdot I_b/L_T$ results

$$\bar{d}_{\text{trab}} = (4/\pi) \cdot \left[\left(\sum_{}^{m} P_b \right) \cdot L_T \middle/ \left(\sum_{}^{m} I_b \right) \cdot P_T \right]$$

This value indeed has turned out to be very useful for the morphometrical evaluation of structural changes in cancellous bone with age and in bone disease, especially in the diagnosis of osteoporosis (Fig. 8.8.2).

8.8.8. EVALUATION OF CANCELLOUS BONE TURNOVER

The histological features that allow the recognition of bone resorption and bone formation sites are outlined on p. 302 and in Table 8.8.1. Further details about the microscopic identification of cells and bone surface properties and of the reliability of the selected parameters are given in Merz and Schenk (1970a,b) and Schenk et al., 1969. The following remarks are related mainly to some methodological aspects.

Basically, we determine the extent of the various surface and interface characteristics (i) by means of differentiated intersection counts (cf. p. 304). The results are calculated both in % of the total trabecular surface

$$S_{Si} = \left[\left(\sum_{}^{m} I_i \right) \middle/ \sum_{}^{m} I_b \right] \cdot 100\%$$

or as surface density of surface fraction (i) in mm^2/cm^3:

$$S_{Vi} = 2 \cdot \left[\left(\sum_{}^{m} I_i \right) \middle/ m \cdot L_T \right] \cdot 1000$$

Expressed as a percentage, they reflect perceptually the balance between formation and resorption and allow, to a certain extent, a comparison between past and actual intensity of these processes (e.g. osteoclast–bone interface versus empty Howship's lacunae). On the other side, surface densities are indispensable for the calculation of values that are based on combinations of surface and volume densities, and for correlation tests with other morphometrical and clinical data.

The *quantitation of bone formation* is based on several parameters, mainly because it proceeds in two steps: osteoid deposition and osteoid mineralization (cf. p. 302 and Fig. 8.8.1). Physiologically both processes are coupled but mineralization is often affected independently in all kinds of osteomalacic conditions. The different surface fractions and interfaces that have to be measured are listed on p. 304. The osteoblast–osteoid interface reflects the area where matrix deposition takes place. When the osteoblasts have disappeared the osteoid production stops and the osteoid seams become gradually thinner until they are fully converted into mineralized bone (Fig. 8.8.1). The osteoid–bone interface corresponds to the "calcification front" of other authors (Bordier and Tun-Chot, 1972). In healthy bone, it has almost the same extent as the surface of the osteoid seam facing the bone marrow. In cases of very irregular bone structure, as in Paget's disease, it differs and has to be measured separately. Finally, the mean width of the osteoid seams is calculated. This leads to the following parameters:

—surface density of osteoid seams S_{Vos} (mm^2/cm^3)
 (total osteoid surface facing the marrow space)
—surface density of osteoblast–osteoid interface $S_{Vob\text{-}os}$ (mm^2/cm^3)
—surface density of osteoid–bone interface $S_{Vos\text{-}b}$ (mm^2/cm^3)
 (in case of irregular bone structure only)
—mean width of osteoid seams $\bar{w} = V_{Vos}/S_{Vos}$
 or (irregular structure): $\bar{w} = 2 \cdot V_{Vos}/(S_{Vos} + S_{Vos\text{-}b})$

If one assumes that all cells identified as osteoblasts are producing new bone matrix at a fairly constant rate, $S_{Vob\text{-}os}$ will be the only parameter that gives an estimate of the actual bone formation. All the other parameters depend on the rate of matrix production as well as on the rate of osteoid mineralization, and the slightest delay of the mineralization process results in an increase of S_{Vos}, V_{Vos} and $\bar{w}$. But an increase of any of these osteoid parameters must always be related to the number of osteoblasts present. This can be done by correlation tests. In normal bone, the most significant correlation exists between the extent of the osteoblast–osteoid interface $S_{Vob\text{-}os}$, and the osteoid volume density V_{Vos} (Fig. 8.8.3). It has been shown that any mineralization delay leads to values that are clearly outside the confidence range.

Surface areas undergoing *bone resorption* are detected from the presence of multinucleated osteoclasts. The osteoclast–bone interface, however, is very small and amounts normally to only 0.5 % of the total trabecular surface. Precise measurements of such a small surface fraction require an adequately high number of intersections to be counted. This can be accomplished by increasing the total length of test lines L_T, by increasing the number of test fields and/or by using a higher microscopic magnification, but always at the expense of working time. For that reason other parameters have been introduced (Schenk et al., 1969; Olah, 1974). Thereby all osteoclasts present in the test fields are counted. Their number is either related to a unit area of the

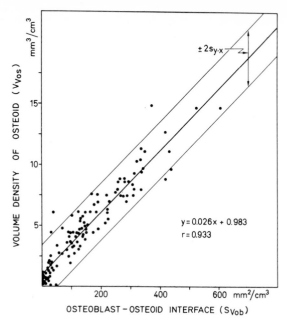

FIG. 8.8.3: Correlation between osteoblast–osteoid interface and osteoid volume density in 114 random autopsy cases.

test field or to a unit boundary length of the trabecular profiles found in the same area:

N_{Aocl} = numerical density of osteoclast profiles per test area of 1 cm^2:

$$N_{Aocl} = \left[\left(\sum_{}^{m} N_{ocl} \right) / m \cdot A_T \right] \cdot 100$$

Since this value is independent of the amount of trabecular bone surface present in the test field, an additional value, the ^{0}steoclast index $O\,I$, has been introduced and defined as number of osteoclast profiles per cm boundary length of trabecular profiles.

$$OI = \frac{N_{ocl}}{B_b} = \frac{N_{Aocl}}{B_{Ab}} = \frac{2}{\pi} \cdot \frac{N_{ocl} \cdot L_T}{I_b \cdot A_T} \cdot 10$$

Note that both parameters are strictly two-dimensional. A transformation into a three-dimensional reference system would necessitate an estimation of the mean caliper diameter of the osteoclasts, a rather cumbersome task in view of the rarity and the considerable variation in size and shape of these giant cells. The presence of osteoclasts cannot be taken as a sign of active bone resorption in all circumstances. The notion that osteoclasts may be turned on and off by parathyroid hormone and calcitonin as well as by other compounds is substantiated by recent electronmicroscopic findings (Holtrop *et al.*, 1974). But still the number of osteoclasts is thought to reflect the presence of stimuli for bone resorption in chronic metabolic bone disease, where the described resorption parameters, especially the indices based on osteoclast numbers, have proven to be a very useful tool for diagnostic purposes (Schenk *et al.*, 1973; Olah, 1974).

8.9.   Nervous tissue

Contributor: Herbert Haug, Lübeck*

8.9.1.   GENERAL ORGANIZATION OF NERVOUS SYSTEM

The nervous system consists of two parts, the central nervous system (CNS) and the peripheral nervous system (PNS). The basic structures of both parts show many similarities. However, the CNS of a higher animal shows a rather complex organization. Each of its major parts (cerebrum, cerebellum, and brain stem) is basically composed of gray and white matter. The gray matter contains the nerve cell bodies and the neuropil, i.e. the complex dendritic ramifications of neurons as well as parts of the neurites and all synapses. The white matter, on the other hand, is made up of bundles of myelin-ensheathed neurites which course from one part of the nervous system to another. The only cellular elements of the white matter are the glial cells which are also found regularly in gray matter; as ependymal cells they line the ventricular wall, whereas they assume various specialized forms in the PNS. The PNS is wide-spread throughout the body and also shows two parts, the ganglia with nerve-cells and the peripheral nerves with bundles of free and myelinated axons (Fig. 8.9.1).

Stereological methods are particularly useful when studying the make-up of nervous tissue, especially concerning the various grays which occur as either layers, e.g. in the cerebral and cerebellar cortex, or as nuclei of varying size and shape embedded in white matter, as for example in the brain stem. Each area of gray matter serves very specific functions and may, accordingly show a specific structural organization and composition. The cerebral cortex is, for example, composed of several layers where neuropil and cells show characteristic types and densities. This will evidently impose certain

* Anatomisches Institut, Medizinische Hochschule, Ratzeburger Allee 160, D-24, Lübeck, West Germany.

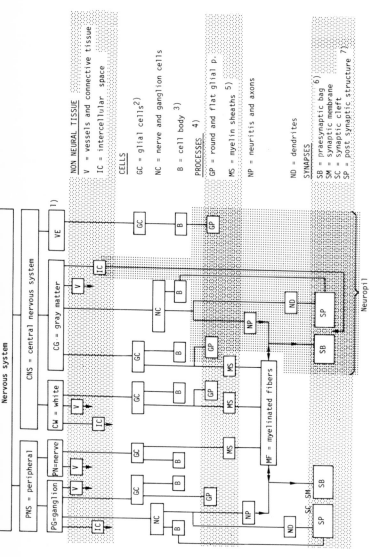

FIG. 8.9.1: Hierarchical model of the nervous tissue. Notes: (1) VE = ventricle and ependyma. (2) GC = glial cells of different types in CNS including ependymal cells, as well as satellite and Schwann cells in PNS. (3) The cell compartments of NC and GC are similar to those of other tissue, therefore see section 8.1. (4) The compartments of processes are tubuli, filaments, mitochondria and rarely other cellular components. (5) Including Schwann cell cytoplasma in myelin-free nerve fibers. (6) The presynaptic compartments consist of tubuli, mitochondria, synaptic vesicles of different types and some other special structures like ribbons. (7) The post-synaptic compartments are the same as in cells and processes, sometimes a subsynaptic cisterna can be found.

The labels within the figure include:

NON NEURAL TISSUE
V = vessels and connective tissue
IC = intercellular space

CELLS
GC = glial cells [2]
NC = nerve and ganglion cells
B = cell body [3]

PROCESSES [4]
GP = round and flat glial p.
MS = myelin sheaths [5]
NP = neuritis and axons
ND = dendrites

SYNAPSES
SB = praesynaptic bag [6]
SM = synaptic membrane
SC = synaptic cleft
SP = post synaptic structure [7]

Nervous system
CNS = central nervous system
PNS = peripheral
Pg=ganglion   PN=nerve
CW = white   CG = gray matter
MF = myelinated fibers

Neurop11

sampling conditions and require appropriate adaptations of the general stereological methods.

8.9.2. MEASUREMENT OF OVERALL SIZE AND COMPOSITION OF THE BRAIN

The basic information on the size of the brain and its parts is its volume which is best derived from the weight of the total brain and its parts after dissection. The specific weight of brain tissue is 1.04.

8.9.2.1. *The estimation of the macroscopic structure of CNS and folded bodies*

The complicated macroscopic configuration of the brain and its components leads to many difficulties in estimating the absolute and relative volumes of the different gray structures, and in estimating the surface of folded cortices in order to calculate the mean thickness of the cortex (Elias *et al.*, 1967; Schlenska, 1969; Haug, 1970, 1972).

The brain is a closed body; measurements regarding its total composition and surface construction must be performed on serial sections across the entire organ (total brain sections) by means of test lattices which are larger than the area of a total brain section. Furthermore it should be noted that the folded brain surface shows certain orientations of the gyri and sulci which may introduce some errors due to anisotropy. However, this anisotropy decreases with increasing cortical convolution as it occurs with evolution and enlargement of brain and body size. In order to eliminate such errors the following procedure should be adopted. First, the brain should be divided into two hemispheres and each hemisphere must then be cut in different planes (for example in a frontal and horizontal direction). Second, each slice must be measured with test lines oriented in two (perpendicular to one another) or in three (with angles of $0°$, $60°$ and $120°$) directions.

Measurements on composition and surface of the brain can be made by superimposing a clear plastic film directly on the slices, or indirectly on photographs of the slices; note that the magnification of the photographs must be known for estimating the surface area (Haug, 1970; Elias and Schwartz, 1971).

The estimation of the volumes is independent of orientation. Relative volumes are obtained directly by point counting with all points falling on brain tissue (gray and white matter) as reference point numbers; alternatively, points falling on the ventricle may also be included in the reference points. Let $P(CG)$, $P(CW)$ and $P(VE)$ designate points falling on gray, white or ventricle respectively. The relative volume of gray matter is then

$$V_V(CG) = \frac{P(CG)}{P(CG) + P(CW) + P(VE)}$$

It is evidently easy to obtain the relative volume of any subpart of gray by an analogous formula; knowing total brain volume one can then calculate its absolute volume.

The surface of the folded structure of the brain can be measured by the formula (Hennig, 1958; Elias and Schwartz, 1971)

$$S = 2 \cdot t \cdot d \cdot I$$

where t is the thickness of the slice or the distance between the section planes on which measurements are made, d is the distance of the measuring lines in the test lattice and I is the number of intersections between test lines and brain surface. The average thickness of the slices should be calculated by dividing the total length of the brain in the direction perpendicular to the cutting plane by the number of slices. Small differences in the thickness of the individual slices can be ignored. It should be noted that measurements made on all slices must be pooled to calculate either V or S.

If the intersection counts for surfaces in the sulci and on the free surface are recorded separately, one may calculate a convolution index by dividing all intersections by those of the free surface. Note that it is also possible to estimate the length of gyri and convolutions (Hennig, 1963a, b; Elias and Schwartz, 1971).

8.9.3. COMPOSITION OF GRAY MATTER

The application of stereological methods to sections of gray matter may yield information on their volumetric composition, on the number of nerve cells (perhaps differentiated by cell type) or of glial cells, on the length and surface density of dendrites, neurites and glia processes in the neuropil, on the number and surface density of membrane specializations in the synapses, etc. This is obtained by either light or electron microscopy.

8.9.3.1. Sampling problems

Nearly all gray structures, whether gray nuclei or cortical grays, are subdivided into subnuclei, areas and layers all having different cell compositions and cellular details. The aim of many investigations is to obtain information concerning the detailed structural composition in the subnuclei, areas and layers as well as on the total gray. This requires some careful considerations on sampling procedures.

(1) The borders of the subnuclei must be marked for light microscopy in all those grays which are not composed of parallel-layered substructures. This marking can be done with india ink under a stereo-microscope. Within such marked areas one can measure by using random sampling procedures.

(2) In parallel layered grays such as the cortex cerebri, cortex cerebelli, or structures seen in the corpus geniculatum laterale and tectum opticum a different procedure should be used. As shown in Fig. 8.9.2 the results on cell numbers will be different if one considers individual random fields, or horizontal or vertical strips of fields. If the counts of cells in fields arranged in a vertical row are pooled, the cell numbers show a normal distribution. Also the values of fields having the same distance from the surface have a normal distribution. On the other hand, pooled counts in horizontal strips at different distances from the surface reveal differences in cell density in the various layers provided that the breadth of the strip is smaller than the thickness of the smallest layer. Figure 8.9.2 shows a third distribution which indicates that values measured by random sampling of individual fields have an atypical distribution (Haug et al., 1971). Vertical strips, hence, yield an overall estimate of cell number in this particular gray, whereas horizontal strips reveal its layered nature (cf. also Section 3.6).

It is also possible to obtain such results by using narrow rectangular test lattices.

This kind of measurement is useful only if the counting fields are regularly spaced from each other and from the surface. The location of each field and its coordinates must be recorded in tables or stored by a computer system. A computer program in BASIC which transposes the matrix of values for calculation in rows and layers is available (Haug, 1978). This form of sampl-

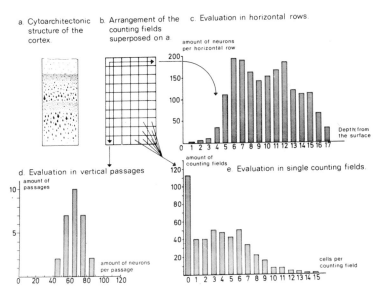

FIG. 8.9.2: Sampling in layers on an example of nerve cell counting.

ing and sequential analysis can be used in all layered tissues other than nervous tissue for cell counting by light microscopy, as well as for cytological analysis by electron microscopy.

Other evaluation procedures for subdivided nuclei and layered tissues need to be mentioned. Bok (1960) has developed a graphic procedure in which the distance from the surface and the size of each single cell will be recorded on a graph. But this graphic representation does not give direct quantitative results. Hollander *et al.* (1976) have developed a system in which the size (area) and the coordinates of the single cells are optomanually analysed by a special sampling stage and stored on tape. A special computer program displays on a screen the distribution of either all cells, or of those of a certain size only. It also can calculate the volume density values for the total structure or for a restricted area.

Inhomogeneity of the cellular structures and choice of magnification. A major sampling problem arises due to minute inhomogeneities which can be found in all CNS tissues. Nerve cells are arranged mainly in small clusters; also the processes of cells with different diameters are inhomogeneously distributed throughout the neuropil. This general problem of sampling in inhomogeneous structures has been discussed in Section 3.6. For the present purpose we need to know that the field observed must be large enough to yield a representative sample, i.e. it must cover more than one cluster.

In electron microscopical evaluation of the neuropil particular problems arise due to the fact that the low magnification necessary for adequate sampling leads to a loss of resolution which may introduce problems in discrimination of the structural details. A compromise must be made and I myself work with a final magnification of 30,000 to 40,000 which is not optimal for sampling, but allows good distinction of structural detail. The problem of sampling thus introduced can be solved by making a composite collage of high power micrographs into a large poster and using oversized and possibly rectangular test lattices for stereological evaluation, or by simply pooling the data derived from a number of micrographs.

8.9.3.2. Nerve cell counts

Nerve cells are best counted by light microscopy. Two reasons make it necessary to use thick sections (20 to 25 μm): (1) the borders of the different subnuclei or layers of a gray can only be reliably observed by the high cell density that occurs in such thick sections, and (2) the counting error diminishes in thick sections (Haug, 1967a; Konigsmark, 1970; Aherne, 1975).

The numerical density N_V of nerve and glial cells can be obtained by counting the number of nuclei N_{AT} occurring on the unit area of a slice of thickness t if the mean nuclear diameter $\overline{D}$ is known (cf. section 4.4). In order to reduce the counting error (Holmes-effect) we use the formula of Floderus (1944) and Haug (1967a):

$$N_V = N_{AT} \cdot \frac{1}{t + \overline{D} - 2h}$$

where h is the height of the structural cap that cannot be recognized on a grazing section (see Section 2.6.3).

It should be noted that in counting in thick sections the entire depth of the section must be evaluated by continuously changing the focal plane. In contrast, if we wish to evaluate the volume density by point counting on the same sections, we are not allowed to move the micrometer screw in order to restrict this analysis to a thin "optical section". This is accomplished by using an oil immersion objective with a fully opened condensor diaphragm (Haug, 1955).

The diameter of nerve cell nuclei is about $7-20 \mu m$, that of nucleoli $2-3 \mu m$; it would hence be advantageous to count nucleoli rather than nuclei since this would reduce the error due to section thickness. Nerve cells usually have only one nucleolus. If we count only those nerve cells which have their nucleolus in the section, the counting error decreases with the diameter of the nucleolus (Haug, 1967a). In early development stages, however, some nerve cells may have two or more nucleoli. Zilles et al. (1976) have developed a mathematical procedure which allows not only the elimination of the counting error but also an estimation of the percentage of one, two or three nucleolated cells. Glial cells do not have easily detectable nucleoli and must be enumerated by their nuclei.

8.9.3.3.   Investigations on the neuropil

In human gray matter the neuropil makes up 90 to 95% of the volume. It consists of densely interwoven cell processes and their interconnections (synapses). Because of the smallness of these processes electron microscopy is usually needed for a determination of their length, surface and volume density by the stereological principles outlined in the previous chapters.

The electron microscopic analysis of the different cell processes in the neuropil is complicated by the problem that the inclination of the cell membranes to the section plane renders it often difficult to recognize the boundary of a nerve cell process because they are often poorly provided with internal cell structures (Fig. 8.9.3). Therefore the position of the borders between two processes can sometimes only be indirectly defined (Haug, 1978). To this end· many investigators draw all cytomembranes on the print. We, and others, use, for this purpose, a thin transparent film as overlay, which allows not only the enhancement of blurred borders of the cell processes, but also the marking of other details in order to prepare for an efficient stereological evaluation (Foh *et al.*, 1973) (Fig. 8.9.3).

Such partially marked or amended images must be measured by visual techniques. The use of a fully automated system would require a complete drawing of all cytological patterns. This is a very time-consuming operation that cannot be compensated for by the high measuring speed of the scanning procedure.

An Example: Evaluation of the neuropil
In order to obtain as much information as possible on the processes within the neuropil the following sizes should be evaluated by stereological pro-

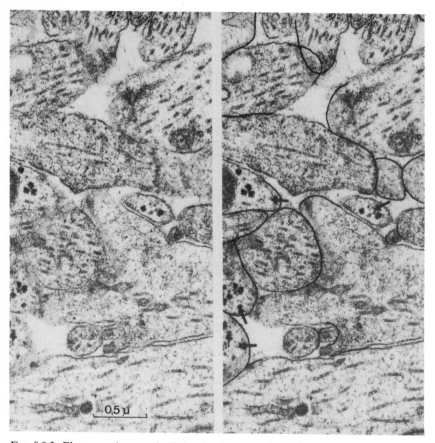

0,5 µ

FIG. 8.9.3: Electron micrograph of the visual cortex of a 12 hour old kitten. Figure on the right amended to show clear outline of membranes. Arrows indicate glial cell processes.

Table 8.9.1.

Estimation of the structure parameters in the neuropil of central grays: + values are meaningful, (+) values are relative, ? are values of questionable interpretation, − values make no sense.

Processes	V_V	S_V	L_V	N_A	S/V
Nerve cells together	+	+	+	?	+
Dendrites (NP)	+	+	+	?	+
Axons (ND)	+	+	+	?	+
Glial cells together (GP)	+	+	−	?	+
round	+	+	+	?	+
flat	+	+	−	?	+
Intercellular spaces (IC)					
embryonal	+	+[a]	−	?	?
adult	?	+[a]	−	?	?
Synapses	+	+	?	(+)	+

[a] Surface of surrounding cell membranes.

cedures (see Table 8.9.1):

(1) number per area of short structures (e.g. synapses) (Section 4.2);
(2) length density of the round processes (Section 2.4);
(3) surface density of all processes (Sections 2.3 and 4.3);
(4) volume fraction of all processes (Sections 2.2 and 4.3);
(5) diameter distribution of the round processes;
(6) the mean distance between cell membranes.

The estimation of most of these parameters follows stereological procedures, as indicated by the references to previous chapters. The measurements are done on electron micrographs at a magnification around 30,000 ×. The following comments outline some of the problems encountered.

(a) The volume fraction of the intercellular space in adult central gray structures can be indirectly calculated by multiplying the surface (sum of all processes) with one-half of the external distance of processes (value 6). The volume fraction of the intercellular space cannot be estimated by point counting, due to the narrow space between the cell processes (10 to 20 nm) and to the different inclination of the cell membranes to the section plane. However, the larger intercellular space in early developmental stages can be measured by point counting. Some difficulties arise in the intermediate and later stages of the development in which narrow as well as wide separations of the cell processes can be found. In this case, one should evaluate by distinguishing between surfaces with narrow and wide separation and calculate the volume fraction of intercellular space from a mixed calculation, although it may not be very accurate.

(b) With the values (2), (3) and (5) one can obtain information on the circularity of the cell processes by comparing their stereological surface values with those values calculated from their proportional length and referring to the appropriate circumference of their diameter distribution. However, one must bear in mind that average values are not permissible in such calculations. By using this procedure we have found that the nerve-cell processes of the cortical neuropil in the adult cat are almost cylindrical, whereas those of the glial cells are non-cylindrical (Haug *et al.*, 1976). Therefore it is suggested that a distinction be made between the flat and round glial processes during the estimation.

(c) The evaluation of the different stereological parameters of synapses is somewhat difficult due to their various shapes. Their shape can be almost cylindrical in the passage synapses, spherical like a flattened pear in common synaptic endings, or irregular within the complex glomerular synapses. Therefore actual stereological values can only be estimated for V_V and S_V whereas the latter can be divided into common cytomembranes $S_V(SCM)$ and thickened synaptic membranes $S_V(STM)$. The actual ratios between both types of membranes as well as the surface-to-volume ratio of both membranes can be calculated. The ratio of both parts of membranes $S_V(STM)/S_V(SCM)$ amounts to about 5%. All other data have only a limited value as the size distribution, numbers (per area), etc. This is due to the fact that it is sometimes difficult to differentiate between the various types of synapses in a section.

8.9.3.4. *The topological properties of the dendritic tree*

The knowledge of the kind and degree of dendritic arborization gives information about the functional organization of the neuronal network (Bok, 1936). The dendritic tree can only be examined after Golgi impregnation. It is fortunate that this procedure impregnates only a small fraction of neurons but completes impregnation of all parts of their dendritic trees. The small fraction of impregnated neurons avoids an overlapping of the different dendritic trees. The evaluation must be done in very thick sections because the dendritic tree usually has extensions in all directions of the space which can spread in large neurons up to 500 μm.

All estimation procedures of the dendritic tree are based on the measurement of spatial coordinates. The position of the special sampling stage and the location of the micrometer screw are recorded while tracing the dendrite.

The Dutch school (Schadé and Meeter, 1963; Smit and Colon, 1969; Uylings *et al.*, 1975), Berry *et al.* (1972) and Wann *et al.* (1973) work with an optomanual system where values of the local parameter of the dendritic tree and its branching patterns are transferred to a computer. The results describe the length of the dendrites (of first order, second order, etc.) their distribution in space, branching and angles of branching.

Berry *et al.* (1971) as well as Wyss (1971) have developed fully-automated devices for the estimation of dendritic trees. But Berry *et al.* (1972) have since switched to the above mentioned optomanual procedures. All investigations on the dendritic tree have common difficulties. First the measurement is lengthy, and only relatively few neurons can be investigated. Second the impregnation procedure is capricious and it is uncertain whether all types of neurons that should be measured have been impregnated.

Some quantitative investigations estimate only the course of the dendritic tree in order to enumerate the so-called spine synapses, which can clearly be observed in every Golgi preparation (Winkelmann *et al.*, 1974).

8.9.4. STEREOLOGICAL EVALUATION OF BUNDLED NERVE-FIBER SYSTEMS

Examinations of the bundled nerve-fiber systems in the CNS and PNS have, in most cases, the aim of estimating the total number of fibers in the bundles, the fractions of the various fiber types, as well as the fiber size distribution (Bishop and Smith, 1964; Haug, 1967b; Samorajski and Friede, 1968).

The evaluation can best be made on sections cut perpendicular to the main fiber direction. The size of the area in which the investigated fiber system is crossing through the plane allows one to calculate the total amount of fibers. However, the fibers of nearly all systems are twisted and therefore each fiber shows a cross-section ranging from a circular to an elliptical shape. In this case, the smallest diameter of the fiber corresponds to the real fiber diameter.

Some fiber systems of PNS which have large diameters (up to $20\,\mu\text{m}$) can be measured by high power light microscopy; those which have smaller diameters, particularly in the CNS ($0.3\,\mu\text{m}$), can only be evaluated on electron micrographs enlarged 2000 to 5000 times.

The measuring of fiber diameters can be done with the help of the following procedures: (a) with stencils of circles (see Fig. 5.6); (b) with the Zeiss particle size analyser TGZ 3 (Haug, 1967b); (c) with different kinds of optomanual systems; (d) with scanning instruments.

In small nerves all fibers occurring in the entire cross-section can be measured, larger systems require adequate sampling procedures. It should be noted that many nerves (PNS) and fiber systems (CNS) are composed of different subsystems which consist of various fiber sizes. In these cases the cross-sectional areas of the subsystems should be evaluated separately (Donovan, 1967; Treff *et al.*, 1971).

Dunn *et al.* (1975) have measured the area size distribution of fibers by estimating the local coordinates as well as the diameters of the single fibers with a computerized system. This enables one to plot the results for the different fiber size classes separately.

The influence of the fiber orientation on the results has been examined by Haug and Rast (1972). It is now possible to calculate the degree of small orientation preferences within an almost randomly distributed fiber system with the help of various stereological evaluations by simply determining two kinds of densities. One consists of the density of length and diameter, and the other utilizes the point-sampling procedure to determine volume density. The first value depends on the preferential orientation while the second is independent of orientation (Haug *et al.*, 1976).

8.10.   The lung and its gas exchange apparatus: an example of the use of stereology in studies of structure–function correlation

Contributor: Ewald R. Weibel, Bern

8.10.1.   FUNCTIONAL MODEL OF LUNG STRUCTURE

The lung's main function is to allow for efficient exchange of gases (O_2 and CO_2) between the air and the blood. For this purpose the air is led through the bronchi into a large volume made up of many small chambers, the alveoli; similarly blood flows through arteries into a dense capillary network which is contained in the alveolar walls and is then collected and conducted back to the heart by veins. In the alveoli air and blood become very closely approximated over a large surface, separated from each other only by a very thin barrier of tissue (Fig. 8.10.1) composed of three layers: the alveolar epithelium, the capillary endothelium, and an interstitial space of varying composition. The combination of large surface and thin barrier ensures efficient gas exchange between air and blood.

The fact that a free surface exists between a delicate tissue structure and air poses some major problems. The forces due to surface tension need to be minimized by smoothing the surface texture. This is achieved (1) by a certain plasticity of the tissue barrier which allows deformation by the effect of local surface forces, and (2) by the presence of a fluid extracellular lining layer which is unevenly distributed to fill pits and crevices (Gil and Weibel, 1969), as shown in Fig. 8.10.2.

From this general description of the lung's structure it follows that a variety of different problems of structure–function correlation could be approached. We shall here concentrate on three of them:

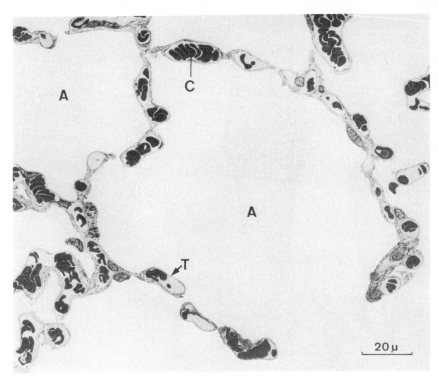

FIG. 8.10.1: Thin section of alveoli (A) from dog lung. The major part of the septa is occupied by capillaries (C). Blood and air are separated by very thin barrier of tissue (T).

(1) The *tissue structure* of lung parenchyma (Fig. 8.10.1) establishes a *gas exchange apparatus* whose morphometric properties—surface area, barrier thickness, blood volume in contact with air—must in some way determine the (maximal) amount of O_2 which can be transferred from air to blood in the unit time. Through a morphometric analysis of lung structure a quantitative structure–function correlation can be approached (Weibel, 1973b; Weibel and Gil, 1977; Gehr *et al.*, 1978). By experimentally varying functional conditions we might learn about the factors influencing the structural size of the pulmonary gas exchange apparatus (Burri and Weibel, 1977).

(2) The *plastic deformability* of lung tissue allows for folding of some of the alveolar septa as a result of the action of surface forces in the air-filled lung (Fig. 8.10.2); this entails a transient loss of gas exchange surface, together with a modification of other parameters, which in turn limits the availability of alveolar capillaries for gas exchange under actual physiological conditions (Weibel *et al.*, 1973).

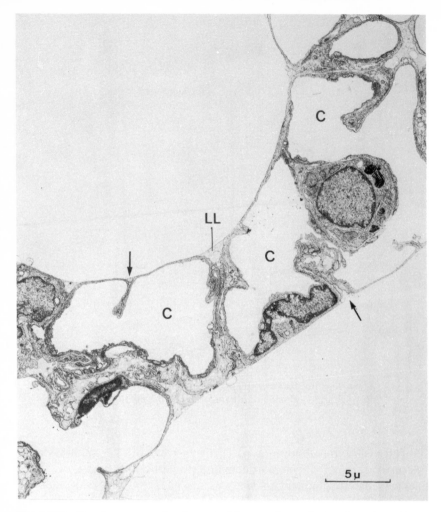

FIG. 8.10.2: Alveolar septum of rat lung fixed by vascular perfusion through capillaries (*C*). The alveolar surface is smoothed by plastic deformation and folding of the barrier (arrows), as well as by pools of extracellular lining layer (*LL*).

(3) Among the *mechanical forces* acting in the lung surface forces are most important. They are related to the mean curvature of the alveolar surface (see equation 6.18) and must be counteracted by air pressure or tissue tension to maintain a stable system. The measurement of mean air space curvature at variable levels of inflation can provide information on the interaction of lung structure and airway mechanics (Gil and Weibel, 1972; Weibel and Gil, 1977).

The preparations required to perform these studies are different. For (1) we require fully expanded alveoli and capillaries filled with blood; this is achieved by fixing the lung through instillation of a fixative into the airways under controlled conditions (Gehr *et al.*, 1978). For (2) and (3), however, a free air–tissue surface must be maintained throughout fixation; this is achieved by perfusing the fixative through the blood vessels, while controlling air inflation of airways (Gil and Weibel, 1969; Weibel *et al.*, 1973; Gil *et al.*, 1978).

8.10.2. DIFFUSION CAPACITY: A MODEL FOR STRUCTURE–FUNCTION CORRELATION OF GAS EXCHANGE

8.10.2.1. *The model*

In physiology, the fundamental measure of the lung's capacity to transfer O_2 from air to blood is the O_2 diffusion capacity, D_{LO2}, defined as the ratio of O_2 flow per min, $\dot{V}_{O2}$ to the partial pressure gradient, ΔP_{O2}, between air and capillary blood:

$$D_{LO2} = \dot{V}_{O2}/\Delta P_{O2} \qquad (8.10.1)$$

This is evidently based on Ohm's law; D_L is hence the reciprocal of the lung's resistance to O_2 diffusion, or its conductance. It should therefore be possible to find an alternative formulation for D_L based on the dimensions and material properties of the gas exchange apparatus, as shown in detail elsewhere (Weibel, 1970a, 1973b). The gas exchange apparatus is conceived as made of three resistances in series: tissue (t), blood plasma (p), and erythrocyte (e). We find (Weibel, 1970a):

$$D_t = K_t \cdot (S_a + S_c)/2\tau_{ht} \qquad K_t = 3.3 \cdot 10^{-8} \qquad (8.10.2)$$

$$D_p = K_p \cdot S_c/\tau_{hp} \qquad 3.2 \cdot 10^{-8} < K_p < 4.3 \cdot 10^{-8} \qquad (8.10.3)$$

$$D_e = \theta \cdot V_c \qquad 0.9 < \theta < 2.5 \qquad (8.10.4)$$

$$1/D_L = 1/D_t + 1/D_p + 1/D_e \qquad (8.10.5)$$

The factors K_t and K_p are permeability coefficients for O_2 of tissue and plasma respectively; θ is the rate of O_2 binding to whole blood (see Weibel, 1970a). The morphometric parameters involved are:

S_a total alveolar surface area
S_c total capillary surface area
V_c total capillary volume
τ_{ht} harmonic mean thickness of tissue barrier
τ_{hp} harmonic mean thickness of plasma barrier.

Note that S_a may be either the alveolar epithelial tissue surface [case (1)], or the free surface of the alveolar lining layer [case (2)]. The barrier thickness must be measured under the same conditions as it may be different for (1) or (2) respectively (Weibel et al., 1973).

8.10.2.2. The methods

The specimens used for this analysis are fixed by instillation of a fixative into the airways under standardized conditions (see Weibel, 1970a). After fixation the lung volume V_L is determined by the method of Scherle (1970) (see Section 7.2).

Sampling of tissue blocks requires particular attention because of the possibility of regional variations (Glazier et al., 1967; Gehr and Weibel, 1974). For larger lungs we usually proceed to stratified random sampling (see Section 3.5) respecting the various lung regions (Gehr et al., 1978). Within the strata an A-weighted random sample is obtained by the method shown in Fig. 3.11.

The measurement of the morphometric parameters requires a two-level sampling procedure:

Level 1: Light microscopy, Magnification × 100.
    Reference volume: total lung tissue.

The volume density of lung parenchyma, V_{Vpa}, (gas exchange region) is estimated by counting the test points (P_{np}) falling on non-parenchymatous components such as bronchi, arteries, or veins, with their connective tissue sleeves. The volume density of parenchyma is then obtained by

$$V_{Vpa,L} = 1 - V_{Vnp} \qquad (8.10.6)$$
$$= 1 - [(\Sigma P_{np})/(\Sigma P_L)]$$

where P_L are the total number of points falling on lung tissue.

Because non-parenchymatous components are unevenly distributed this must be done on large step sections which are cut through the entire lung. It is convenient to use a stepping stage on the microscope (Weibel, 1970b) and to use one or very few test points per field only, so as to spread the test points over a very large sectional area of the lung. [With very large lungs, the volume of the coarsest components may have to be estimated by naked eye on whole lung slices (Weibel, 1963), using the procedure here described for the smaller components.]

Level II: Electron microscopy. Final magnification × 10–12,000. Reference volume: lung parenchyma (see Fig. 6.3).

(a) Using a multipurpose test system M 168 (Appendix 3) we count test points falling on air (P_a), tissue (P_t) and capillary blood (P_c), as well as intersections with the alveolar (I_a) and capillary (I_c) surface of the barrier. The sum of the test points counted represent those falling onto lung parenchyma. Calculated parameters: Volume and surface densities in lung parenchyma:

$$V_{Vc,pa} = [\Sigma P_c]/[\Sigma(P_a + P_t + P_c)] \qquad (8.10.7)$$

$$S_{Va,pa} = 2 \cdot [\Sigma I_a]/[(d/2) \cdot \Sigma(P_a + P_t + P_c)] \qquad (8.10.8)$$

$$S_{Vc,pa} = 2 \cdot [\Sigma I_c]/[(d/2) \cdot \Sigma(P_a + P_t + P_c)] \qquad (8.10.9)$$

The arithmetic mean barrier thickness can be calculated from these data according to equation (6.2). If point hits on erythrocytes and plasma are separately recorded the capillary hematocrit (volume density of erythrocytes in the blood) can be obtained.

N.B. (1) The point and intersection counts are done on 30–70 micrographs derived from 6–12 sections depending on lung size; the counts are summed over at least the micrographs derived from one section to ensure representative sampling (see Section 3.7).

(2) The choice of magnification at level II is crucial since the estimation of S_V is influenced by resolution (see Section 4.3.7).

(b) The harmonic means of barrier thickness are determined on the same electron micrographs. A square line lattice (A 100, Appendix 3) is superimposed (Fig. 8.10.3), and the intercept length l_t in tissue or l_p in plasma (from the capillary surface to the nearest red cell) are measured in both directions by means of a special ruler based on a logarithmic scale (see Fig. 6.5). The harmonic mean barrier thickness is calculated by the formula of Weibel and Knight (1964) described in Section 6.1:

$$l_h^{-1} = (1/n) \sum_{}^{n} l^{-1} \qquad (8.10.10)$$

$$\tau_h = (\tfrac{2}{3}) \cdot l_h \qquad (8.10.11)$$

The reciprocal values of l are summed over all micrographs used for this analysis.

8.10.2.3. *Calculation of D_L*

The values of S_a, S_c, and V_c required in the model for D_L (equations 8.10.2–5) refer to the whole lung. They are calculated from equations (8.10.6) to

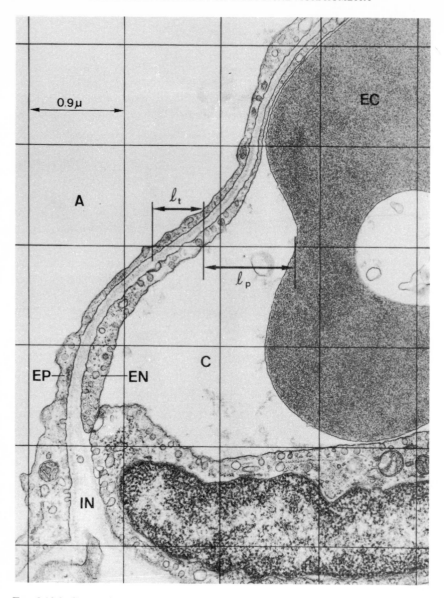

FIG. 8.10.3: Square lattice of test lines used to measure intercept lengths in tissue (l_t) and plasma (l_p). Note erythrocyte (*EC*) in capillary (*C*) and that barrier separating blood from alveolar air (*A*) is made of the three layers epithelium (*EP*), interstitium (*IN*) and endothelium (*EN*). (Reproduced by permission from Weibel, 1970/71.)

(8.10.9) together with the lung volume V_L, according to

$$S_a = V_L \cdot V_{Vpa,L} \cdot S_{Va,pa} \qquad (8.10.12)$$

$$V_c = V_L \cdot V_{Vpa,L} \cdot V_{Vc,pa} \qquad (8.10.13)$$

The harmonic mean thickness τ_{ht} and τ_{hp} are obtained by equation (8.10.11). If the parameter estimates (densities, etc.) are separately obtained for the various strata a weighted average should be calculated, accounting for the varying stratum volume.

The estimation of D_L by this procedure has allowed a number of conclusions on the relation between the morphometry of the lung and its gas exchange capacity. In comparative studies it was shown that D_L increases in proportion with body weight, but that it was influenced by different levels of O_2 requirements of the organism related to the behavioural patterns of physical activity of certain species (Weibel, 1973b). An experimentally induced increase in O_2 need resulted in a proportional enlargement of the gas exchange apparatus (Geelhaar and Weibel, 1971; Hugonnaud et al., 1976). Likewise, a reduction or increase in ambient O_2 resulted in an increase or reduction, respectively, of D_L (Burri and Weibel, 1971).

A comparison with physiological determinations of D_L has shown that morphometric D_L is larger than is found in physiology, particularly if physiological estimates are done, as usual, under conditions of basal metabolism. Even maximal physiological D_L, measured under conditions of heavy exercise, is smaller by about a factor of 2. This is partially explained by the effect of plastic deformation of the air–blood barrier in the air-filled lung under the influence of surface tension; this was found by performing the same type of analysis on lung tissue fixed by vascular perfusion (Weibel et al., 1973).

8.10.2.4. Estimation of D_L by light microscopy

Of the morphometric parameters required for calculation of D_L only the alveolar surface area can be obtained by light microscopy with some degree of reliability; for all other parameters the resolution afforded by the electron microscope is required. An approximate method for D_L based on a light microscopic estimation of S_a and of a factor of "capillary loading" of the alveolar septa has been worked out (Weibel, 1973a). $S_{Va,pa}$ is obtained from a line intersection count at about $200\times$ magnification. It must be noted that S_a thus obtained will be about 30% smaller than that determined on electron micrographs (Keller et al., 1976); this is due to the resolution effect on estimates of S_V (see Section 4.3.7).

8.10.3.   ALVEOLAR CURVATURE AND SURFACE FORCES

8.10.3.1.   *The model*

The maintenance of an air–tissue–blood interface depends on an equilibrium between the retractive (or elastic) forces of lung tissue and the distending force, i.e. the pressure gradient between the air and the pleural cavity ("negative pressure"). An important part of the retractive force is due to the surface pressure P_s generated at the air–tissue interface as a function of surface tension γ of the surface lining and of the curvature of the surface (Weibel and Gil, 1977). By the formula of J. W. Gibbs it is found that, at equilibrium,

$$P_s = \gamma \cdot \overline{K} \qquad (8.10.14)$$

where $\overline{K}$ is the average mean curvature of the surface which can be estimated by a stereological method (see Section 6.3).

8.10.3.2.   *The methods*

Lungs intended for this analysis must be fixed in a state of controlled air-inflation by perfusing the fixative through the blood vessels (Gil and Weibel, 1969; Weibel and Gil, 1977; Gil *et al.*, 1978).

The estimation of $\overline{K}$ was done by light microscopy (magnification $\times 400$) using a Wild M501 sampling stage microscope and $1\,\mu\mathrm{m}$ thick plastic sections (Gil and Weibel, 1972). The following measurements were obtained:

I_a : intersections of free alveolar surface, using a test system M 168 (Appendix 3).

$\left.\begin{array}{l} T_+ \\ T_- \end{array}\right\}$: counts of positive and negative tangents in the area of the test system (see Fig. 6.8).

The average mean curvature was calculated by equation (6.19). With d as the length of the unit test line of the M 168 test system equation (6.19) becomes equation (6.21):

$$\overline{K} = [\pi/\sqrt{3} \cdot d] \cdot [\Sigma(T_+ - T_-)]/[\Sigma I_a] \qquad (8.10.15)$$

8.10.3.3.   *Results*

This analysis revealed (Gil and Weibel, 1972) that the average mean surface curvature decreased inversely with the cube root of the air space volume, i.e. the higher the lung inflation the smaller the curvature. Since the distending pressure increased with increasing inflation it was concluded that the surface tension coefficient γ also had to increase in order to maintain equilibrium; this was in agreement with several predictions on the basis of physiological experiments (cf. Weibel and Gil, 1977).

In recent studies (Gil *et al.*, 1978) it was found that in the range of normal breathing the free alveolar surface did not change isometrically with air volume; this could be explained by alterations in alveolar geometry. At low lung volumes the surface became "crumpled" and very high local curvatures were sustained which suggested that surface tension was very low (Bachofen *et al.*, 1978).

8.10.4. CONCLUSIONS

The general conclusions of this series of experiments are that the conditions under which tissue is prepared for a morphometric analysis are of greatest importance, and that the methods of preparation must be carefully chosen. Functionally meaningful results can only be expected if the tissue is prepared under physiologically well controlled conditions. Stereological methods are very powerful tools, but they cannot distinguish artifacts from real properties of a structural system!

8.11. Isolated peritoneal macrophages: component-biased sampling

*Contributor: Terry M. Mayhew, Sheffield**

8.11.1. INTRODUCTION AND MODEL OF MACROPHAGE STRUCTURE

Most stereological extrapolations are performed on cells and tissues which are readily identified by virtue of their morphological appearance, diagnostic location or histochemical (staining) properties. When other criteria are deficient, identification of cells on the basis of their morphological characteristics assumes greater importance. This is the case for many isolated cells studied by electron microscopy (e.g., different types of cells in centrifuge pellets and mixtures of dissociated cells in tissue culture).

The experience of many workers has demonstrated that unfavourable planes of sectioning make a certain proportion of cell profiles difficult to classify unambiguously. Lymphocytes, for instance, have been mistakenly identified as monocytes or eosinophils for this very reason. To circumvent this problem, it is sometimes expedient to select specific section planes which transect a particular subcellular component, i.e. one which facilitates unequivocal identification. I propose to designate such a conscious selection "component-biased sampling".

* University of Sheffield, Department of Human Biology and Anatomy, Sheffield S10 2TN, England.

Component-biasing has been applied to studies of lymphocytes, plasma cells, hepatocytes, mast cells (component: the nucleus) and neurons (component: the nucleolus). The procedure and its consequences will be illustrated here by reference to nuclear-biased sampling of peritoneal macrophages although it can be applied to many isolated cells.

The functional compartments of the peritoneal macrophage are made up of organelles found in other types of cells. A morphometric model which suited our personal needs is defined in Fig. 8.11.1. Two basic reference volumes can be envisaged: (1) the volume of the macrophage, (2) the volume of its cytoplasm.

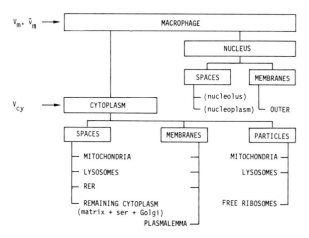

FIG. 8.11.1: Morphometric model of peritoneal macrophage. (Items of secondary interest to original studies given in parenthesis.)

8.11.2. PURPOSE OF THE STUDY

A wide range of substances with a variety of physico-chemical properties induce alterations of macrophage structure and functional activity. In our original studies (Mayhew and Williams, 1974 a,b; Williams and Mayhew, 1973) we wished to define quantitatively those changes which occurred after stimulation by triolein and Freund's adjuvants. To this end, we considered changes in subcellular organelles (their compartmental densities and size distributions) and in the surface areas of plasmalemmal and nuclear membranes. These parameters were of particular interest because earlier, purely descriptive studies reported in the literature were inconsistent in their interpretations of the structural alterations involved.

8.11.3. GENERAL STRATEGY

Peritoneal cells were harvested from rats and mice by lavage and fixed by immersion in glutaraldehyde or formaldehyde followed, where necessary, by post-osmication. Cytometric analyses were based on systematic random sampling of intact cells and of macrophage profiles sectioned across nuclear planes. The data obtained were expressed in terms of the "average cell" in the population.

8.11.4. SAMPLING PROCEDURES

The range of structural dimensions (cells about 7–10 μm in diameter; free ribosomes about 18 nm in diameter) and the requirements of the investigation were such that analyses could be performed comfortably at only three levels of magnification.

Stage I
Freshly withdrawn peritoneal fluid cells were spotted (rather than smeared) on to glass slides, fixed by neutral formaldehyde and stained by H and E. These preparations were scanned systematically using a filar micrometer eyepiece on an optical microscope at 600 × magnification. This was done to obtain estimates of cell diameters from which an estimate of mean macrophage cell volume, $\bar{v}_m$, was derived. Cells were assumed to be spheres as a convenient approximation and this may have introduced a certain degree of error into $\bar{v}_m$ due to departures from sphericity.

Similar measurements could be made on macrophage profiles appearing on toluidine blue-stained, Araldite sections cut at 1 μm using glass knives. For this purpose, a random sample of three blocks was taken from a pool of glutaraldehyde-fixed material representing cells from twelve animals. In this case it was necessary to reconstruct cell size distributions from observed profile size distributions using one of the appropriate stereological methods (see Chapter 5).

Stage II
From the pool of fixed, Araldite-embedded blocks a sample of three blocks was selected at random and sections were cut at 1 μm. Fifty photomicrographs were recorded from toluidine blue-stained sections at a primary magnification of ∼ 500 ×. These were printed at a final magnification of ∼ 2500 × and the micrographs evaluated to determine the number of (lysosomal) granule profiles in a selection of nucleate cell profiles. From these counts, and from point counting estimates of cytoplasmic profile areas made at stage III,

the areal density of granule profiles per unit area of cytoplasm was calculated. These densities were compared with similar estimations made at stage III.

Stage III
Ultrathin sections (about 70 nm thick) cut from the same blocks were mounted on 200-mesh, Formvar-coated copper grids and stained with lead citrate. Thirty to forty electron micrographs at primary magnifications of $\sim 7500 \times$ and $10,000 \times$ were taken by systematic sampling of these sections using the corners of the grid squares as approximate points of reference. Adjacent sections on the same grid were not monitored in order to avoid serial sectioning errors. All micrographs were printed at a final magnification of $\sim 30,000 \times$.

Only complete macrophage profiles containing a nuclear compartment and lying in the proximity of grid corners were recorded (Fig. 8.11.2) and

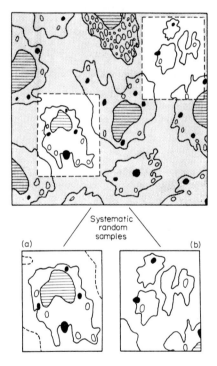

FIG. 8.11.2: Definition of nuclear-biased sampling at stage III. The top portion of the diagram represents a field of view under the electron microscope. For the present purposes, we consider only nuclear-biased profiles (a) which always contain a nucleus in sharp profile. In these samples, obtrusive macrophage fragments (broken lines) are ignored. For details see text and Mayhew and Williams (1971). (Reproduced with permission from *J. Microsc.* **94**, 195–204, 1971.)

any macrophage fragments obtruding into the micrographic fields were ignored. This gave a "nuclear-biased" sample.

A coherent, quadratic test lattice of spacing 1.0 cm was superposed on each cell profile and the points (line intersections) falling on specified subcellular compartments were counted to allow the estimation of nuclear and cytoplasmic profile areas and compartmental volume densities. We determined the volume densities in the cell of nucleus and cytoplasm, and the volume densities in the cytoplasm alone of rough endoplasmic reticulum, lysosomes, mitochondria and remaining cytoplasm. The latter comprised ground substance, free ribosomes, the Golgi apparatus and smooth endoplasmic reticulum.

At this stage also a discontinuous line test lattice (of the type M 168; Appendix 3) was employed for estimating cell and nuclear surface–volume ratios. The length of line used, d, was equivalent to 1.08 μm.

Also at this magnification we performed counts of the numbers of free ribosomes and of mitochondrial and lysosomal profiles to estimate organelle numerical densities in the cytoplasm. Because of the size, distribution and high areal densities of ribosomes, areas of cytoplasm containing these organelles were, in large measure, preselected: a number of circles (7.5 mm diameter) was impressed on to PVC sheeting using commercial instant lettering and these were superimposed onto micrographs in areas of cytoplasm free of the other main compartments. Free ribosomes occurring in these circles were then counted. Finally, we measured the sizes of lysosomal profiles to reconstruct lysosomal size distributions.

8.11.5. EXAMPLES OF CALCULATIONS OF STEREOLOGICAL PARAMETERS

In the following, the computational procedures are illustrated, by reference to the volume density of the nucleus and the surface–volume ratio of the cell. These particular parameters have been chosen because of the potential errors that nuclear-biasing may introduce into their estimation. Nuclear-biasing is known to over-estimate nuclear volume density (Konwinski and Kozłowski, 1972; Mayhew and Williams, 1971) and, at least in theory, may distort estimates of cell surface–volume ratio. Fortunately, nuclear-biased estimates of these parameters can be corrected if necessary. In the case of macrophages, the cell surface–volume ratio appears not to require correction.

Level III
Counted: with quadratic lattice, test points on nucleus (P_n) and on whole macrophage (P_{m1}); with discontinuous lattice, intersections of plasmalemma with test lines (I_p) and points on whole cell (P_{m2}).

M

Calculated: observed nuclear volume density in cell

$$V_{VOn,m} = P_n/P_{m1}$$

and cell surface–volume ratio

$$S_{Vm,m} = 4 \cdot I_p/P_{m2} \cdot d$$

From the observed nuclear volume density, we calculated the true volume density using the equation

$$V_{VTn,m} = \left[\frac{V_{VOn,m} + \sqrt{(33(V_{VOn,m})^2 + 48V_{VOn,m})}}{4V_{VOn,m} + 6} \right]^3$$

since the nucleus of the macrophage tends to be eccentrically disposed (see Mayhew and Cruz, 1973).

Knowing mean cell volume, $\bar{v}_m$, from level I the absolute nuclear volume and plasmalemma surface area could be calculated:

$$v_n = V_{VTn,m} \cdot \bar{v}_m$$

$$S_p = S_{Vm,m} \cdot \bar{v}_m.$$

Alternatively, an estimate of the membrane amplification factor (F_a) of the plasmalemma could be obtained by comparing the surface–volume ratio of the cell with that of a sphere of equivalent volume, $S_{Vs,s}$:

$$F_a = S_{Vm,m}/S_{Vs,s}.$$

This gave an indication of cell "roundness". A useful practical alternative to this can be found in Williams and Mayhew (1973).

8.11.6. ERROR CONSIDERATIONS

The most obvious source of error in component-biased sampling is the over-estimation of component volume density, in this case nuclear volume density. Methods for correcting this are approximations since they assume that the component and its containing volume are both spheres. However, this inherent error due to shape simplification is often not appreciable (Mayhew and Cruz, 1973). The actual correction formula applied also depends on the spatial relationship between the two compartments (Konwinski and Kozłowski, 1972; Mayhew and Cruz, 1973). In addition, the reader will appreciate that as the nuclear volume density needs correcting, the volume density of the cytoplasm should be adjusted accordingly.

Fortunately, nuclear-biasing of peritoneal macrophages does not affect the volume densities of intracytoplasmic organelles provided they are related to the cytoplasm as reference containing volume (Mayhew and Williams, 1971) and the same may be true for lymphocytes (Konwinski, personal communication) and other cells.

Nuclear-biasing clearly restricts measurement of cell surface–volume ratio to transnuclear section planes. While this still gives reliable estimates for macrophages and other isolated cells (Fritsch, 1971; Mayhew and Williams, 1971) it can in theory give values which require correction.

8.11.7. EXTENSION OF THESE STUDIES

The original studies have given scope for deriving new stereological correction formulae which specifically deal with problems arising from the adoption of component-biased sampling regimes. In the neurological field, for instance, neuron profiles exhibiting prominent nucleoli are often chosen for neurometric analysis and this leads to over-estimation of nuclear as well as nucleolar volume density. We have developed principles for correcting observed volume densities in such situations (Mayhew and Cruz-Orive, 1973, 1975). More recently, Cruz-Orive (1976a,b) has extended these studies to deal with geometric shapes other than the sphere and has developed corrections for surface-volume ratios and volume densities.

8.11.8. ACKNOWLEDGEMENTS

This work was made possible by personal support to the author from S.R.C. (London) and The Wellcome Trust to whom I am most grateful.

8.12.   Plasma cell differentiation in lymphoid tissue

Contributors: *Léon J. Simar, Liège* and*
Ewald R. Weibel, Berne

8.12.1. PURPOSE OF THE STUDY

The plasma cells play an essential role in the immunological reactions since they synthesize humoral antibodies. They are characterized by densely packed endoplasmic reticulum cisternae and a large Golgi zone. In the lymph nodes these cells are especially concentrated in the inner part, i.e. in

*Laboratoire d'Anatomie Pathologique, Institut de Pathologie, Université de Liège au Sart-Tilman, 4000 Liège, Belgique.

the medullary cords. They originate from centrocytes and centroblasts, lymphoid cells which show an abundance of free ribosomes and which are mainly localized in the cortical germinal centers of the lymph node outer part. The injection of an antigen to induce an immunological reaction is followed first by the appearance in the germinal centers of centrocytes which transform into centroblasts; these cells later become immunoblasts which migrate towards the medullary cords traversing the paracortical area (Simar, 1975).

The cells which have accumulated in medullary cords through this 4 day process of migration appear as fully differentiated plasma cells. The aim of the study was to define the changes in size and sub-cellular composition of the transforming immuno-competent cells in order to determine if the process of their differentiation from centrocytes to plasma cells is continuous or whether it proceeds through different steps related to the different localizations of the cells in the lymph nodes. The immunological reaction was induced by a single injection into the mouse of antigenic sheep red blood cells.

8.12.2. MODEL OF IMMUNO-COMPETENT CELL

Figure 8.12.1 shows a model of the subdivision of the immuno-competent cells into compartments. The term "endoplasmic reticulum" concerns all membrane-bounded endoplasmic cisternae, excluding the perinuclear cisterna. In counting ribosomes, we did not distinguish between free and bound particles.

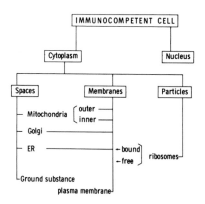

FIG. 8.12.1 : Model of immuno-competent cells.

8.12.3. GENERAL STRATEGY

The antibody titer was determined in the serum from 12 to 120 h after the antigen injection to check the adequate reaction. At 12, 24, 72, 96 and 120 h after this administration, 10 animals were sacrificed and their lymph nodes were fixed in 4% glutaraldehyde and 1% osmium tetroxide by a vascular perfusion method (Forssmann et al., 1967). The morphometric study was performed on samples of these lymph nodes.

8.12.4. SAMPLING PROCEDURE

The general strategy of sampling was to obtain a random sample of immuno-competent cells at each time point. Since the aim of the study was not a total characterization of the lymph nodes, but merely an analysis of the immuno-competent cells themselves, the sampling procedure could be directed to the selection of individual cell profiles.

Four animals showing adequate reaction to antigen administration were picked at random from each group. Six lymph nodes of each animal were removed and diced into a large number of small cubes. Semi-thin sections examined with the light microscope permitted the selection of 4 blocks containing the lymph node regions rich in immuno-competent cells. Of three of them, ultra-thin sections were cut for stereological analysis.

Because of the wide range of dimensions of the investigated components, three levels of magnification had to be used. The films were examined in a table projector unit yielding a nine fold secondary magnification (see Fig. 4.14).

Stage I

At a final magnification of × 2800, the nuclear diameters were measured. The scheme of random sampling employed called for a maximum of 6 cells to be recorded from each of the squares of the supporting grid, the cells being selected sequentially, along the edge of the square. The number of micrographs recorded for each group of four animals varied between 540 and 1080.

A test system consisting of 15 concentric circles (see Fig. 5.6c) was centered on nuclear profiles and the diameter of the circle of best fit by area was recorded.

Stage II

A second set of 35 micrographs per animal was recorded at a magnification of 4900. This level was used to estimate the nucleo-cytoplasmic ratio and the surface density of Golgi membranes. In each square of the grid, only one cell, namely that closest to the square edge, was photographed, irrespective of whether the profile contained a nucleus or not. This method ensured that the sample of the lymph node studied was evenly distributed in the lymph node region under investigation, irrespective of the immuno-competent cell concentration.

A multipurpose test system M 168 was used to count points falling on nucleus and cytoplasm, and intersections with Golgi membranes.

Stage III

For higher resolution of cytoplasmic structures, such as endoplasmic reticulum, mitochondria and ribosomes, 35 electron micrographs per animal were recorded at a final magnification of 28,000. A portion of cytoplasm of one cell per grid square, selected by the same procedure as mentioned at stage II, was recorded.

To analyse cytoplasmic composition (stage III), a double lattice test grid, with lattice ratio 9:1 was superimposed on the micrographs (Fig. 8.12.2). The volume density of mitochondria and cytoplasm was estimated with the coarse lattice whereas the fine total points were used to estimate the volume density of endoplasmic reticulum. Both the vertical and the horizontal lines of the dense lattice were used to count the number of intersections with endoplasmic reticulum membranes in order to determine their surface density. In view of the very large number of ribosomes present on each micrograph, these particles were counted in the upper left hand subsquare of each large square of the double lattice test system as indicated by white arrows in Fig. 8.12.2. All subsquares containing cytoplasm were used for counting. The test area used for counting was given by the number of coarse points falling on cytoplasm multiplied by the area of the unit subsquare.

8.12.5. EXAMPLE OF CALCULATIONS

The methods used to calculate the volume and surface density of the cytoplasmic organelles are identical to those described in Section 8.1 for liver cells. Only the determination of the numerical density of ribosomes is therefore illustrated.

Level III

Counted: number of ribosomes (N_{rib}) in a certain number of subsquares which have a total area A_1; test points on total cytoplasm (P_{cy}^{III}).

Measured: the mean diameter $\overline{D}$ of a ribosome (radius R); the mean diameter $\overline{d}$ of the smallest profile still recognizable as a ribosome (radius r).

Calculated:

—profile area A_1 on which the number of ribosomes was counted:

$$A_1 = P_{cy}^{III} \cdot d^2 \quad (d = \text{the subsquare side})$$

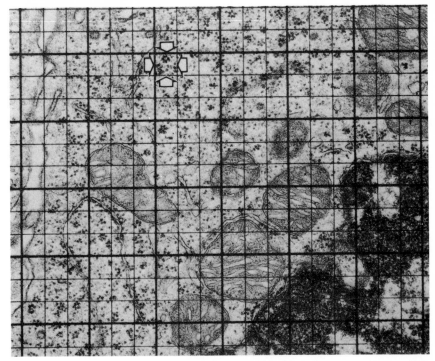

FIG. 8.12.2: Electron micrograph (stage III) of immunocyte showing abundance of free ribosomes in cytoplasmic ground substance and some profiles of RER. Arrows indicate sub-square used for counting ribosomes. Coarse grid spacing 0.4 μm.

—h, the depth by which a ribosome has to penetrate into the section before a profile becomes recognizable, is related to ribosome radius R and radius of smallest observable profile by:

$$h = R - \sqrt{(R^2 - r^2)}$$

—numerical density of ribosomes in cytoplasm:

$$N_{Vrib,cy} = \frac{N_{rib}}{P_{cy}{}^{III} \cdot d^2} \cdot \frac{1}{\overline{D} + t - 2h}$$

where t = section thickness (see 4.3.4.5).

Level II

Counted: number of test points falling on cytoplasm ($P_{cy}{}^{II}$) and on nucleus ($P_n{}^{II}$).

Calculated: volume density of nucleus in cell:

$$V_{Vn,c} = \frac{P_n^{II}}{P_n^{II} + P_{cy}^{II}}$$

Level I

Measured: nuclear profile diameter.

Calculated: the size distribution of nuclei and the mean nuclear diameter $\overline{D}_n$ applying the transformation of Wicksell (1925) adapting the computer program of Baudhuin and Berthet (1967) or the method described by Giger and Riedwyl (1970) (see Section 5.2). Approximate calculation of nuclear volume:

$$\overline{V}_n = \frac{\pi}{6}\overline{D}_n^3$$

From these data, one can calculate the cell volume:

$$\overline{V}_c = \overline{V}_n \cdot \frac{P_{cy}^{II} + P_n^{II}}{P_n^{II}}$$

as well as the volume of cytoplasm per cell:

$$\overline{V}_{cy} = \overline{V}_n \cdot \frac{P_{cy}^{II}}{P_n^{II}}$$

The absolute number of ribosomes per cell then becomes:

$$N_{rib,c} = N_{Vrib,cy} \cdot \overline{V}_{cy}$$

8.12.6.   ERROR CONSIDERATIONS

A source of error evidently lies in the procedure of regional sampling to harvest cells of different developmental stages; it is fortunate that auto-radiographic studies have shown systematic migration of cells from the cortex to the medulla (Simar, 1975).

The effect of section thickness has only been considered for estimating the number of ribosomes which are smaller than section thickness; evidently if absolute results on *ER* surface are to be obtained this effect must be corrected (cf. Section 4.3.4). For comparative studies, this effect, however, is negligible.

One of the major errors lies in the calculation of the mean nuclear volume from the mean diameter; this is evidently strongly affected by a possible distribution of diameters. Using the Wicksell transformation, one could obtain directly the mean volume.

8.13.   Stereological study of trypanosoma, a small protozoon

Contributors: Peter H. Burri, Bern*
and Hermann Hecker, Basel**

8.13.1.   PURPOSE OF THE STUDY

Trypanosoma brucei is the causative agent of the Nagana-disease in cattle and resembles the sleeping-sickness parasite of man. It is transmitted to the vertebrate host (Mulligan, 1970) by tsetse flies. *T. brucei* has a complex life cycle during which it goes through several stages in the blood of its host as well as in different organs of the tsetse flies (Steiger, 1973; Vickerman, 1965). In the blood of infected animals *T. brucei* shows a distinct pleomorphism: the trypanosomes occur as slender, intermediate and stumpy forms which differ in their morphology (Fig. 8.13.1) and in their metabolism (Vickerman, 1965). In the transmitting insect they undergo further structural changes. All these alterations concern mainly size and form of the trypanosomes, their chondriome and respiratory activity. In order to correlate structure and function, we analysed quantitatively different stages of *T. brucei's* life cycle. The methods of investigation applied to the pleomorphic bloodforms are presented in this chaper (Böhringer and Hecker, 1974; Hecker *et al.*, 1972, 1973).

8.13.2.   PARTICULAR   PROBLEMS   IN   INVESTIGATION   OF TRYPANOSOMAL BLOODFORMS

The difficulty in analysing *T. brucei* bloodforms is twofold:

(a) One is faced with the general problem of dealing with single polymorphic cells, where the reference volume needed for the calculation of absolute values from the relative parameters, cannot be directly measured.

(b) Working with trypanosomes it is difficult to get pure populations of slender forms and almost impossible to obtain pure stumpy forms. Furthermore it is not always possible to identify the pleomorphic type of a randomly sectioned trypanosome in an electron micrograph (Fig. 8.13.2). Since the aim of the study was to differentiate the slender and stumpy forms of *T. brucei* on the basis of quantitative morphological criteria, and to correlate morphological findings with physiological differences between the two forms, some methodological tricks had to be introduced for this analysis.

*Anatomisches Institut der Universität Bern, Bühlstrasse 26, CH–3000, Bern 9, Switzerland.
**Schweiz. Tropeninstitut, Socinstrasse 57, CH–4051, Basel, Switzerland.

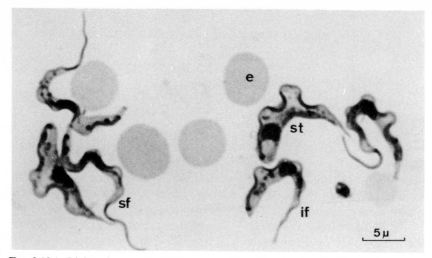

Fig. 8.13.1: Light micrograph of thin smear of rat blood infected with trypanosomes. Pleomorphic bloodforms of *Trypanosoma brucei*: *sf* = slender form, *if* = intermediate form, *st* = stumpy form, *e* = erythrocyte.

8.13.3. PREPARATION OF TRYPANOSOMES

Blood samples were drawn from heavily infected white rats around the time point of the first peak of parasitaemia. Seven series were used and processed for light and electron microscopy.

8.13.4. LIGHT MICROSCOPY

From each series five thin smears (Fig. 8.13.1) were prepared; in each slide up to 300 trypanosomes were classified as slender, intermediate and stumpy forms and their mean percentages calculated. Because the three forms cannot be properly identified in random transections on electron micrographs the ultrastructural morphometric results were expressed as functions of the percentage of slender forms ($f(SF)$; Fig. 8.13.3) in each preparation.

In order to obtain absolute data for an average slender and stumpy trypanosome, we determined on the smears the mean nuclear volume of each form and used it as a reference parameter. This was done by measuring the nuclear projections on camera lucida drawings of 50 nuclei (n) per pleomorphic form in the Giemsa stained thin smears. Assuming the nuclei to be prolate ellipsoids with the long and short half axes a_i and b_i, the mean nuclear volume ($\bar{v}_{nu}$) could be calculated for each form according to the formula:

$$\bar{v}_{nu} = \frac{4\pi}{3}\left(\frac{\sum_{i=1}^{n} a_i}{n}\right) \cdot \left(\frac{\sum_{i=1}^{n} b_i}{n}\right)^2$$

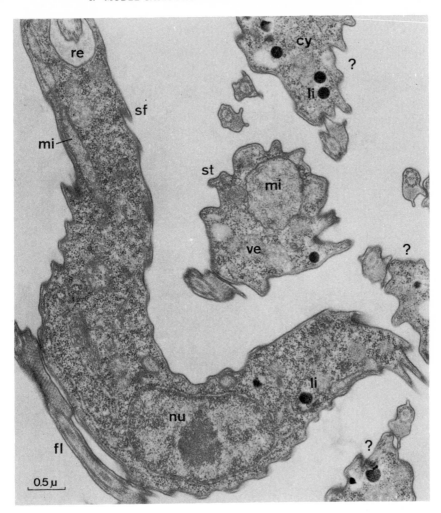

FIG. 8.13.2: Randomly sectioned pleomorphic trypanosomes. *sf* = slender form, *st* = stumpy form, ? = undetermined pleomorphic types, *nu* = nucleus, *cy* = cytoplasm, *fl* = flagellum, *re* = reservoir (flagellar pocket), *mi* = mitochondrium, *ve* = vesicles (glycosomes), *li* = lipid inclusions.

8.13.5. ELECTRON MICROSCOPY

For each series the trypanosomes were first separated from the blood cells by centrifugation at 400 g, and then by the isolation method of Lanham and Godfrey (1970, quoted after Hecker *et al.*, 1972) on a DEAE-cellulose column.

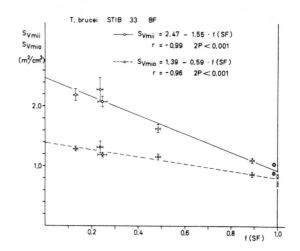

FIG. 8.13.3: Calculated regression lines for S_{Vmii} and S_{Vmio} (for explanation see text). (Reproduced from Hecker *et al.*, *Experientia* 29,901 (1973) with the permission of Birkhäuser Verlag, Basel.)

The isolated trypanosomes were pelleted by centrifugation at 1700 g and fixed in cold cacodylate buffered glutaraldehyde. After a brief wash in buffer they were embedded in warm agar (55°C, 2% Difco Bacto). Before the agar solidified the trypanosomes were concentrated at the bottom of the agar drop by centrifugation. The more or less densely packed trypanosomes appeared randomly distributed and oriented in the blocks. Small cubes of agar were then cut out and further processed for electron microscopy using standardized procedures. From each series 5 random blocks were sectioned, and the technically best section was selected for recording the micrographs on roll film. These were contact-printed on film and evaluated on a table projector (see Fig. 4.14).

8.13.6. MORPHOMETRIC EVALUATION

A two step procedure was used for the morphometric analysis.

Stage I
In a first step at a final magnification of 33,000 ×, analysing 100–150 micrographs per series, we determined the nuclear cytoplasmic volume ratio (V_{nu}/V_{cy}) and the volume density of the flagellum in the cytoplasm $(V_{Vfl,cy})$. On a square coherent double lattice test system (see Section 4.2.2.2) with a lattice point ratio of 1:9, V_{nu}/V_{cy} was measured with the coarse and $V_{Vfl,cy}$

with the fine point system. Each series was subdivided into 4 samples, a sample representing an area covered by 500 coarse points ($=200\,\mu m^2$) falling onto trypanosomal profiles; thus, the evaluated area per sample was constant.

Stage II
In the second step at a higher magnification ($83,000\times$) and using a simple square lattice test system the volume densities of mitochondria ($V_{Vmi,cy}$), vesicles ($V_{Vve,cy}$), and lipid inclusions ($V_{Vli,cy}$) were estimated by point counting. The surface densities of mitochondrial inner ($S_{Vmii,cy}$) and outer ($S_{Vmio,cy}$) membranes were obtained by counting the intersections with the test lines in two directions. In each of the 6 samples per series a cytoplasmic area of $60\,\mu m^2$ was analysed.

8.13.7. CALCULATIONS

From the *first step* the V_{nu}/V_{cy} ratio was calculated from the number of coarse points falling onto nuclei (P_{nu}) and cytoplasm (P_{cy}):

$$V_{nu}/V_{cy} = P_{nu}'/P_{cy}'$$

where

$$P_{nu}' = 9 \cdot P_{nu} \text{ and } P_{cy}' = 9 \cdot P_{cy} + P_{fl}'$$

P_{fl}' corresponding to the number of fine points falling onto the flagellum. Using V_{nu}/V_{cy} and $\bar{v}_{nu}$ and mean cytoplasmic volume ($\bar{v}_{cy}$) and the mean cell volume ($\bar{v}_c$) of each form was calculated by:

$$\bar{v}_{cy} = \bar{v}_{nu}/(V_{nu}/V_{cy})$$

and

$$\bar{v}_c = \bar{v}_{nu} + \bar{v}_{cy}$$

From the *second step* the volume densities of the various organelles (V_{vo}) in the cytoplasm were estimated by forming the simple ratio:

$$V_{vo,cy} = P_o^*/P_{cy}^*$$

P_o^* and P_{cy}^* being the number of points in the second magnification step falling onto an organelle system and onto cytoplasm respectively.

Surface densities of mitochondrial membranes S_{Vmi} were obtained according to the general formula (see equation 4.16):

$$S_{Vmi,cy} = I_{mi}/d \cdot P_{cy}^*$$

where d is the distance between lattice points. S_{Vmi} could be differentiated into surface densities of inner (S_{Vmii}) and outer mitochondrial membranes (S_{Vmio}).

Linear regressions were calculated for the above parameters with respect to the percentage of slender forms in the preparations ($f(SF)$), enabling the parameter values for 100% slender and 100% stumpy form populations to be extrapolated (Fig. 8.13.3).

8.13.8. ERROR CONSIDERATIONS

In the morphometric analysis of *T. brucei* stages we have disregarded the effect of section thickness, as we were especially interested in a direct comparison of the 2 forms. The calculated absolute values have hence to be taken with caution, since they are influenced by the Holmes-effect. They also depend on a reliable determination of $\bar{v}_{nu}$, which in our case may be biased by the measurement on smears, cells being possibly flattened and nuclei deformed in smear preparations. Despite these reservations the morphometric analysis allowed us to assess important differences between the trypanosomal stages.

8.13.9. EXTENSION OF THE STUDY

This example of morphometric analysis of the blood-forms of *T. brucei* has shown, that it is possible to encompass the difficulties inherent in the study of single cells and of mixed populations of pleomorphic forms. It can be assumed that the methods presented here can be applied to various types of unicellular organisms or populations of isolated cells.

In the case of *T. brucei* the method was successfully applied also to the stages in tsetse flies, whereby only minor changes were necessary in the sampling procedure (Böhringer and Hecker, 1975).

APPENDIX 1

Alphabetic list of symbols used in this text and their definition

Numbers in brackets refer to equation numbers where the symbols are introduced or defined. See also Table 1.1.

Note: the same letter may occasionally have to be used for different meanings, but context, actual definition, and subscripts avoid ambiguities.

A: area of profiles

A_A: areal density (2.20)

A_T: test area

a: area of individual profile

B: boundary length of profiles

B_A: boundary length density on area (2.29)

b: boundary length (perimeter) of individual profile

C: curvature of plane figure

C_A: curvature density of plane figures on area

D: diameter (of spherical particles)

d: diameter of profile

d: spacing of test points

$E\{x\}$:   expected or average value of x

$F(x), f(x)$: function of x, specifically frequency distributions

G: total or Gaussian curvature (2.66)

G: grain-associated point number (autoradiography)

H: caliper diameter (tangent diameter) of solid

h: caliper diameter of plane figure

h: height of lost cap section (2.92)

$\bar{H}$:   mean caliper diameter (2.14)

I: intersections between test line and surface or boundary

I_L: intersection density on test line (2.30)

349

J:	length of linear feature in space
J_V:	length density of linear features in space (2.62)
i, j:	general indicators of phase or size class etc.
K:	mean curvature of surface (integrated) (2.65)
$\bar{K}$:	average (or mean) mean curvature (6.19)
K_V:	density of integral mean curvature per unit volume (2.69)
K_S:	density of integral mean curvature per unit surface
k:	constants or coefficients
K_t:	correction coefficient for section thickness (4.64–65)
K_c:	correction coefficient for section compression (4.89)
L:	length of line or intercept (straight line)
L_L:	length density of intercepts on test line (2.23)
L_T:	length of test line
l:	length of intercept or chord
$\bar{l}$:	mean linear intercept (5.9–11)
M:	magnification (4.40)
M_k:	k-th moment (5.35)
m:	number of micrographs
N:	number of discrete objects or profiles
n:	number of profiles
n:	number of samples in statistics
N_A:	numerical density of profiles on test area (2.78)
N_L:	numerical density of intercepts on test line
N_N:	relative number (numerical frequency)
N_V:	numerical density of particles in volume (2.78)
P:	test points
P_T:	test point set
P_P:	test point density (2.26)
$\Pr\{x\}$:	probability of event x (2.4–5)
p:	element of probability measure (2.6)
Q:	transection of spatial lineal features with plane
Q_A:	transection density on unit area (2.63)
q:	lattice ratio for coherent test systems (4.10)
R:	radius of sphere
r:	radius of circle
R_V:	radiation label density (autoradiography) (6.36)
S:	surface area of spatial object
S_V:	surface density in unit volume (2.28)
S_S:	relative surface (6.44)
S.D.:	standard deviation (4.57)
S.E.:	standard error of the mean (3.24; 4.58)
T:	subscript designating test system

T:	tangent count (sweeping line on area)
T_+ (T_-):	positive (negative) tangents (2.67)
T_A:	tangents per unit area (2.68)
t:	section or slice thickness
V:	volume of components or structure
V_V:	volume density (2.19)
V_T:	test volume
v:	volume of individual object
W:	body weight
θ:	orientation angle between vector and z-axis (co-altitude)
ϕ:	orientation angle between x-axis and projection of vector onto x, y plane (latitude)
τ:	sheet (foil) thickness (6.1)
τ_h:	harmonic mean sheet thickness (6.5)
α, γ:	constants or coefficients
σ^2:	variance (3.19)
δ, λ, ρ:	shape factors
β:	shape dependent coefficient (2.84)
λ_a:	mean free distance of components a
Δ:	diameter class interval
κ:	local curvature

Glossary of terms

anisotropic: exhibiting different properties in different directions of space.
arithmetic mean: sum of the sample values divided by the number of samples (= average).
average: sum of the sample values divided by the number of samples (= arithmetic mean).
biased estimate: the estimate deviates from the true value by a certain amount and in a certain direction.
boundary: outline of a profile (= perimeter).
chord: straight line segment between two surface points (see intercept).
coherent test system: test system in which the elements occur in a precise quantitative relationship.
component: distinguishable part of a structure.
consistent estimate: a method of estimation is called consistent if the estimate becomes equal to the population value when the sample consists of the entire population.
containing space: structure space to which the parameters describing the internal components are related.
convex solid: a solid for which any straight line connecting any of two points of the solid lies totally within the solid.
curvature: deviation of curved surface (line) from plane (straight line) measured by reciprocal of radius of the largest tangent circle.
density: the quantity per unit volume, unit area or unit length.
feature: characteristic element of structure.
harmonic mean: sum of the reciprocals of the sample values divided by sample number.
intercept: segment or chord of test line contained within object and extending between two points on the object's surface.
intersection: intersection between test line and object surface or profile boundary.
isotropic: having the same properties in all directions of space.
morphometry: quantitative morphology; the measurement of structures by any method, including stereology.
normal: line perpendicular to a surface (or line), indicates orientation of surface.
object: discrete element of structure.
orientation: direction in space (of normal to surface, e.g.) measured by polar coordinates (angles θ, ϕ to coordinate system).
particle: discrete element of structure.

perimeter: outline of a profile (= boundary).

phase: aggregate of all parts of structure which are identical in nature.

probe: any geometric entity (plane, line, point, slice, etc.) used to penetrate into interior of structure.

profile: sharply outlined plane figure resulting from sectioning (or projecting) an object with a plane (onto a plane).

random sample: sample obtained by random process (by means of random numbers), i.e. without regard to the properties of structure.

random section: section of structure obtained by random sampling, whereby all regions of the structure and all orientations must have an equal chance to be cut.

reference system: quantity to which feature or component characteristics are related (containing space, test system, etc.).

section: plane intersecting a structure; sometimes thin slice of thickness *t*.

set: a collection of things belonging together which are called the elements of the set (a set of points, a set of lines, etc.).

slice: "section" of finite thickness *t*.

specimen: structure or sample of the structure available for study.

standard deviation: square root of variance.

standard error of mean: standard deviation of sample means, estimated by standard deviation divided by square root of sample number.

stereology: set of mathematical methods relating three-dimensional parameters of structure to measurements obtainable on sections.

stratified random sample: sample obtained by random process within sub-parts (strata) of the structure.

structure: something made of interdependent parts in a definite pattern of organization.

systematic sample: sample obtained according to a predetermined process, e.g. at pre-determined equal distances.

test system: set of geometric elements (points, lines, areas) used to obtain stereological estimates.

unbiased estimate: the average value of all estimates taken over all possible samples is equal to the true value.

uniform random sample: all points in space are equally likely to be included in the sample.

variance: the average of the square of the deviations from the mean.

APPENDIX 3

Synopsis of coherent test systems

The following collection of test systems have been found to be particularly useful in applying point counting methods of stereology as they were presented in Chapter 4. They are all coherent and are labeled by codes of the following type:

C 64

where "C" defines the test pattern (Table A.3.1) and "64" the number of (main) test points P_T. The coefficients assigned to the patterns in Table A.3.1 are based on the following relations (see Chapter 4.2 for details):

$$A_T = \mathbf{P}_T \cdot k_2 \cdot d^2$$
$$L_T = P_T \cdot k_1 \cdot d$$
$$P_T' = P_T \cdot q^2$$

The boundary of the test area frame is drawn in two different line types. The short-lined frame edge is the "forbidden line" which extends beyond the frame (to infinity) in the direction of the arrows: profiles intercepting this line at any point are rejected, those intercepting the other frame edge are accepted (see Fig. 4.12).

Note: these test systems are printed in such a way that they can be photographically reproduced on high contrast film (e.g. Kodalith) to serve as overlays etc.

Table A.3.1
Characteristics of coherent test systems.

Code	Test pattern	q	q^2	k_1*	k_2
A	Simple square lattice	1	1	2	1
B	Double square lattice	2	4	2	1
C	Double square lattice	3	9	2	1
D	Double square lattice	4	16	2	1
L	Curvilinear lattice (Merz)	1	1	$\pi/2$	1
M	Multipurpose hexagonal	1	1	1/2	$\sqrt{\tfrac{3}{2}}$

* k_1 refers to the use of all lines as test line; if in square lattices only one direction of lines is used $k_1 = 1$.

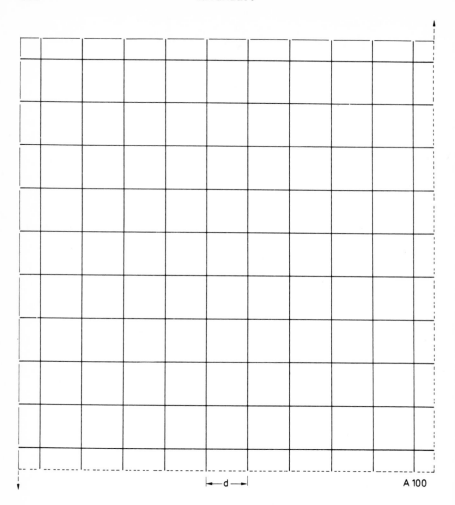

Simple square lattice test system A 100

Calibration d = distance of test lines

Frame width = $10d$

A 100

Test point number P_T $= 100$

Test line length $L_T = P_T \cdot 2 \cdot d = 200 \cdot d$

Test area $A_T = P_T \cdot d^2\ = 100 \cdot d$

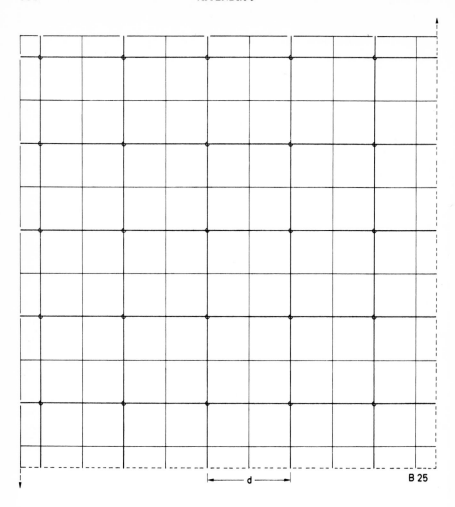

Double square lattice test system B 25

Lattice ratio $q^2 = 4$

Calibration $d =$ distance of coarse lines

Frame width $= 5d$

Test point number:	Coarse points	$P_T = P'_T/q^2$	$= 25$
	All points	$P'_T = P_T \cdot q^2$	$= 100$
Test line length:	Coarse lines	$L_T = P_T \cdot 2 \cdot d$	$= 50 \cdot d$
	All lines	$L'_T = P_T \cdot 2 \cdot q \cdot d = 100 \cdot d$	
	Coarse lines horizontal	$L''_T = P_T \cdot d$	$= 25 \cdot d$
	All horizontal lines	$L'''_T = P_T \cdot q \cdot d$	$= 50 \cdot d$
Test area:		$A_T = P_T \cdot d^2$	$= 25 \cdot d^2$

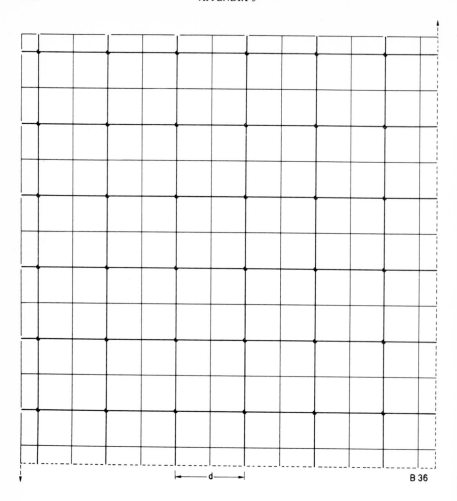

B 36

Double square lattice test system B 36

Lattice ratio $q^2 = 4$

Calibration $d =$ distance of coarse lines

Frame width $= 6d$

Test point number:	Coarse points	$P_T = P'_T/q^2$	$= 36$
	All points	$P'_T = P_T \cdot q^2$	$= 144$
Test line length:	Coarse lines	$L_T = P_T \cdot 2 \cdot d$	$= 72 \cdot d$
	All lines	$L'_T = P_T \cdot 2 \cdot q \cdot d = 144 \cdot d$	
	Coarse lines horizontal	$L''_T = P_T \cdot d$	$= 36 \cdot d$
	All horizontal lines	$L'''_T = P_T \cdot q \cdot d$	$= 72 \cdot d$
Test area:		$A_T = P_T \cdot d^2$	$= 36 \cdot d^2$

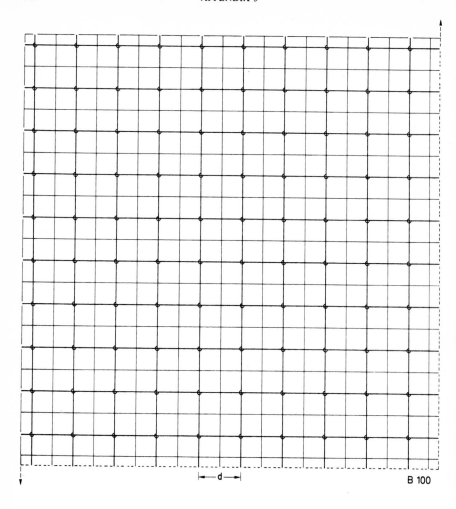

B 100

Double square lattice test system B 100

Lattice ratio $q^2 = 4$

Calibration d = distance of coarse lines

Frame width = $10d$

B 100

Test point number:	Coarse points	$P_T = P'_T/q^2$	$= 100$
	All points	$P'_T = P_T \cdot q^2$	$= 400$
Test line length:	Coarse lines	$L_T = P_T \cdot 2 \cdot d$	$= 200 \cdot d$
	All lines	$L'_T = P_T \cdot 2 \cdot q \cdot d = 400 \cdot d$	
	Coarse lines horizontal	$L''_T = P_T \cdot d$	$= 100 \cdot d$
	All horizontal lines	$L'''_T = P_T \cdot q \cdot d$	$= 200 \cdot d$
Test area:		$A_T = P_T \cdot d^2$	$= 100 \cdot d^2$

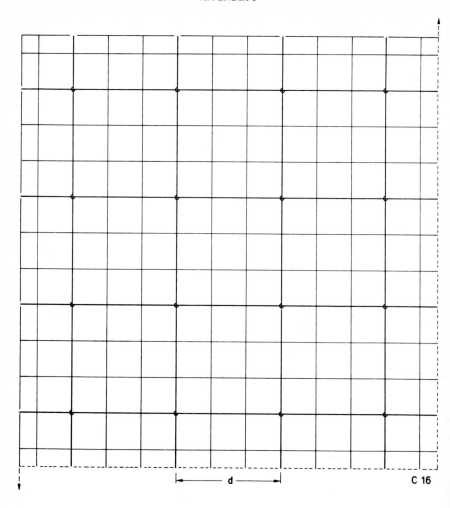

Double square lattice test system C 16

Lattice ratio $q^2 = 9$

Calibration d = distance of coarse lines

Frame width $= 4d$

C 16

Test point number:	Coarse points	$P_T = P'_T/q^2$	=	16
	All points	$P'_T = P_T \cdot q^2$	=	144
Test line length:	Coarse lines	$L_T = P_T \cdot 2 \cdot d$	=	$32 \cdot d$
	All lines	$L'_T = P_T \cdot 2 \cdot q \cdot d$	=	$96 \cdot d$
	Coarse lines horizontal	$L''_T = P_T \cdot d$	=	$16 \cdot d$
	All horizontal lines	$L'''_T = P_T \cdot q \cdot d$	=	$48 \cdot d$
Test area:		$A_T = P_T \cdot d^2$	=	$16 \cdot d^2$

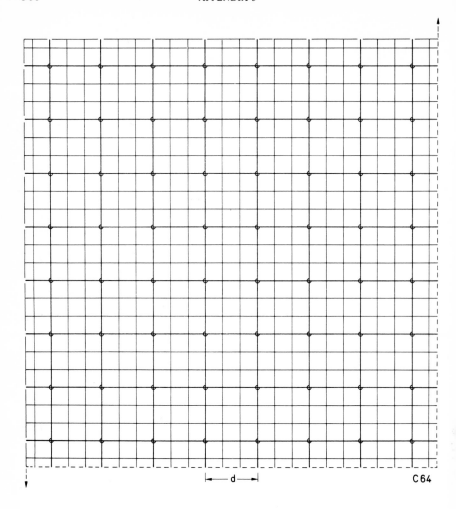

C 64

Double square lattice test system C 64

Lattice ratio $q^2 = 9$

Calibration d = distance of coarse lines

Frame width $= 8d$

C 64

Test point number:	Coarse points	$P_T = P'_T/q^2$	$= 64$
	All points	$P'_T = P_T \cdot q^2$	$= 576$
Test line length:	Coarse lines	$L_T = P_T \cdot 2 \cdot d$	$= 128 \cdot d$
	All lines	$L'_T = P_T \cdot 2 \cdot q \cdot d = 384 \cdot d$	
	Coarse lines horizontal	$L''_T = P_T \cdot d$	$= 64 \cdot d$
	All horizontal lines	$L'''_T = P_T \cdot q \cdot d$	$= 192 \cdot d$
Test area:		$A_T = P_T \cdot d^2$	$= 64 \cdot d^2$

N

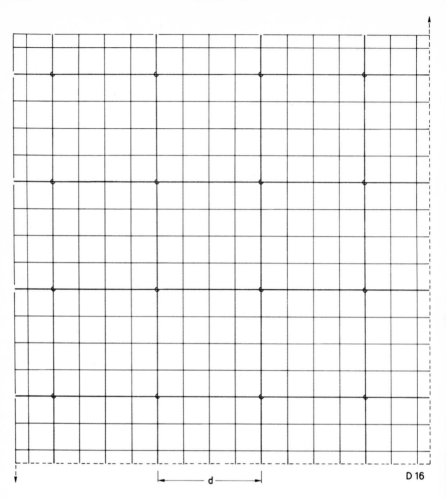

D 16

Double square lattice test system D 16

Lattice ratio $q^2 = 16$

Calibration $d =$ distance of coarse lines

Frame width $= 4d$

D 16

Test point number:	Coarse points	$P_T = P'_T/q^2$	$= 16$
	All points	$P'_T = P_T \cdot q^2$	$= 256$
Test line length:	Coarse lines	$L_T = P_T \cdot 2 \cdot d$	$= 32 \cdot d$
	All lines	$L'_T = P_T \cdot 2 \cdot q \cdot d = 128 \cdot d$	
	Coarse lines horizontal	$L''_T = P_T \cdot d$	$= 16 \cdot d$
	All horizontal lines	$L'''_T = P_T \cdot q \cdot d$	$= 64 \cdot d$
Test area:		$A_T = P_T \cdot d^2$	$= 16 \cdot d^2$

N*

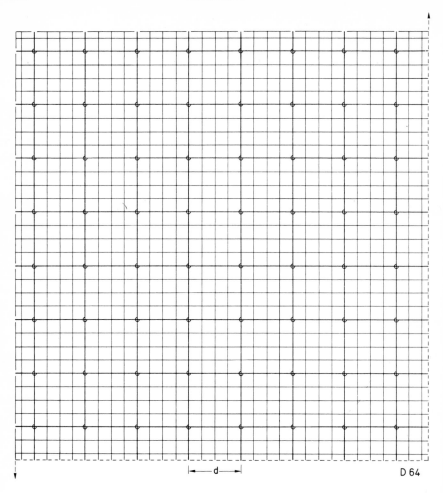

|←— d —→| D 64

Double square lattice test system D 64

Lattice ratio $q^2 = 16$

Calibration $d =$ distance of coarse lines

Frame width $= 8d$

D 64

Test point number:	Coarse points	$P_T = P'_T/q^2$	$= 64$
	All points	$P'_T = P_T \cdot q^2$	$= 1024$
Test line length:	Coarse lines	$L_T = P_T \cdot 2 \cdot d$	$= 128 \cdot d$
	All lines	$L'_T = P_T \cdot 2 \cdot q \cdot d = 512 \cdot d$
	Coarse lines horizontal	$L''_T = P_T \cdot d$	$= 64 \cdot d$
	All horizontal lines	$L'''_T = P_T \cdot q \cdot d$	$= 256 \cdot d$
Test area:		$A_T = P_T \cdot d^2$	$= 64 \cdot d^2$

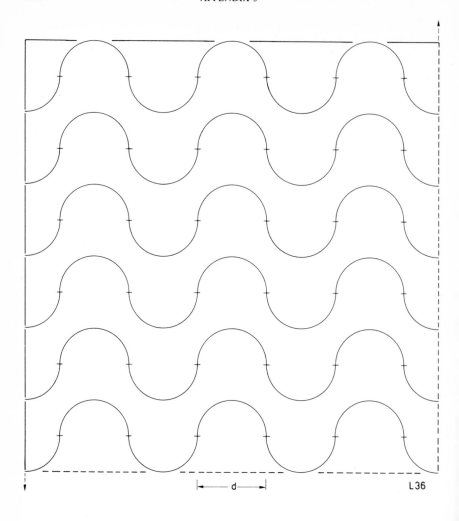

Isotropic test system (after Merz) L 36

Calibration $d =$ distance of test points

Frame width $= 6d$

L 36

Test point number	P_T	= 36
Test line length	$L_T = P_T \cdot (\pi/2)d = 56.55 \cdot d$	
Test area	$A_T = P_T \cdot d^2 = 36 \cdot d^2$	

Isotropic test system (after Merz) L 100

Calibration d = distance of test points

Frame width = $10d$

L100

Test point number	P_T	$= 100$
Test line length	$L_T = P_T \cdot (\pi/2)d = 157 \cdot d$	
Test area	$A_T = P_T \cdot d^2 = 100 \cdot d^2$	

Multipurpose test system M 42

Calibration d = length of short line segment

Horizontal frame width = $6d$

Vertical frame width   = $6.06d$

M 42

Test point number	P_T	$= 42$
Test line length	$L_T = P_T \cdot d/2$	$= 21 \cdot d$
Test area	$A_T = P_T \cdot [\sqrt{(3)}/2]d^2 = 36.37 \cdot d^2$	

M 168

Multipurpose test system M 168

Calibration $d =$ length of short line segment

Horizontal frame width $= 12d$

Vertical frame width $\quad = 12.12d$

M 168

Test point number	P_T	$= 168$
Test line length	$L_T = P_T \cdot d/2$	$= 84 \cdot d$
Test area	$A_T = P_T \cdot [\sqrt{(3)}/2] \cdot d^2 = 145.5 \cdot d^2$

Bibliography

The bibliography is given in three parts. The first part lists a number of references to general works on stereology and related sciences, such as books and proceedings of conferences and symposia; the second part is a compilation of the individual references quoted in the text; the third part is a list of journals where papers on stereology appear regularly.

I. General references to stereology and related sciences

Stereology and geometric probability

De Hoff, R. T. and Rhines, F. N. (1968). "Quantitative Microscopy". McGraw-Hill, New York.

Elias, H. (1967). "Stereology". Proceedings of the Second International Congress for Stereology. Springer Verlag, New York.

Eränkö, O. (1955). "Quantitative Methods in Histology and Microscopic Histochemistry". Karger, Basel–New York.

Hadwiger, H. (1957). "Vorlesungen über Inhalt, Oberfläche und Isoperimetrie". Springer Verlag, Berlin–Göttingen–Heidelberg.

Haug, H. (1963). "Proceedings First International Congress for Stereology Vienna". Congressprint, Wien.

Kendall, M. G. and Moran, P. A. P. (1963). "Geometrical Probability". Charles Griffin, London.

Mardia, K. V. (1972). "Statistics of Directional Data". Academic Press, London and New York.

Matheron, G. (1975). "Random Sets and Integral Geometry". Wiley and Sons, New York.

Miles, R. E. and Serra, J. (1978). "Geometrical Probability and Biological Structures: Buffon's 200th Anniversary". Vol. 23: Lecture Notes in Biomathematics. Springer Verlag, Berlin–Heidelberg–New York.

Nicholson, W. L. (1972). "Proceedings of the Symposium on Statistical and Probabilistic Problems in Metallurgy". Special Supplement to Advances in Applied Probability. Applied Probability Trust, Seattle, Washington.

380

Saltykov, S. A. (1958). "Stereometric Metallography". 2nd edition (3rd edition 1970). State Publishing House for Metals and Sciences, Moscow. (German translation of 3rd edition: VEB Deutscher Verlag für Grundstoffindustrie, Leipzig, 1974.)

Santaló, L. A. (1976). "Integral Geometry and Geometric Probability". Vol. 1 of Encyclopedia of Mathematics and its Applications. Addison-Wesley, Reading, Mass.

Underwood, E. E. (1970). "Quantitative Stereology". Addison-Wesley, Reading, Mass.

Underwood, E. E., de Wit, R. and Moore, G. A. (1976). "Proceedings of the Fourth International Congress for Stereology". National Bureau of Standards, Special Publication Nr. 431, U.S. Government Printing Office, Washington.

Weibel, E. R. (1963). "Morphometry of the Human Lung". Springer Verlag, Berlin, and Academic Press, New York.

Weibel, E. R. and Elias, H. (1967). "Quantitative Methods in Morphology". Springer Verlag, Berlin–Heidelberg–New York.

Weibel, E. R., Meek, G., Ralph, B., Echlin, P. and Ross, R. (1972). "Stereology 3. Proceedings of the Third International Congress for Stereology". Blackwell Scientific Publications, Oxford–London–Edinburgh–Melbourne. (Reprinted from *J. Microscopy*, vol. 95, parts 1 and 2.)

Statistics and sampling techniques
Afifi, A. N. and Azen, S. P. (1972). "Statistical Analysis: A Computer Oriented Approach". Academic Press, New York–San Francisco–London.

Bailey, N. T. J. (1959). "Statistical Methods in Biology". Hodder and Stoughton, London.

Cochran, W. C. (1977). "Sampling Techniques". (3rd edition). Wiley and Sons, New York.

Dixon, W. J. and Massey, F. J. (1969). "Introduction to Statistical Analysis". (3rd edition). McGraw-Hill, New York.

Sachs, L. (1972). "Statistische Auswertungsmethoden". (3rd edition). Springer Verlag, Berlin–Heidelberg–New York.

Snedecor, G. W. (1956). "Statistical Methods". (5th edition). Iowa State University Press, Ames, Iowa.

II. Individual references

Abercrombie, M. (1946). Estimation of nuclear population from microtomic sections. *Anat. Rec.* **94**, 239.

Aherne, W. (1967). Methods of counting discrete tissue components in microscopical sections. *J. R. Microsc. Soc.* **87**, 493.

Aherne, W. (1975). Some morphometric methods for the central nervous system. *J. Neurol. Sci.* **26**, 623.

Altman, P. L. and Dittmer, D. S. (1971). "Biological Handbooks". Federation of American Societies for Experimental Biology. Bethesda, Maryland.

Amstutz, G. C. and Giger, H. (1972). Stereological methods applied to mineralogy, petrology, mineral deposits and ceramics. *J. Microsc.* **95**, 145.

Anderson, T. W. (1960). "An Introduction to Multivariate Analysis". John Wiley, New York.

Arborgh, B., Bell, P., Brunk, U. and Collins, V. P. (1976). The osmotic effect of glutaraldehyde during fixation. A transmission electron microscopy, scanning electron microscopy and cytochemical study. *J. Ultrastruct. Res.* **56**, 339.

Bach, G. (1959). Ueber die Grössenverteilung von Kugelschnitten in durchsichtigen Schnitten endlicher Dicke. *Z. wiss. Mikrosk.* **64,** 265.

Bach, G. (1967). Kugelgrössenverteilung und Verteilung der Schnittkreise; ihre wechselseitigen Beziehungen und Verfahren zur Bestimmung der einen aus der andern. *In:* "Quantitative Methods in Morphology". (E. R. Weibel and H. Elias, eds.) p. 23. Springer Verlag, Berlin–Heidelberg–New York.

Bachmann, L., Salpeter, M. M. and Salpeter, E. E. (1968). Das Auflösungsvermögen elektronenmikroskopischer Autoradiographien. *Histochemie* **15,** 234.

Bachofen, H., Gehr, P. and Weibel, E. R. (1978). Alterations of mechanical properties and morphology in excised rabbit lungs rinsed with a detergent. *J. Appl. Physiol.* (in press).

Barnard, T. (1976). An empirical relationship for the formulation of glutaraldehyde-based fixatives, based on measurements of cell volume change. *J. Ultrastruct. Res.* **54,** 478.

Bartels, P. H. and Wied, G. L. (1977). Computer analysis and biochemical interpretation of microscopic images: current problems and future directions. *Proc. I.E.E.E.* **65,** 252.

Baudhuin, P. (1968). "L'Analyse Morphologique Quantitative de Fractions Subcellulaires". Thèse d'agrégation. Université Catholique, Louvain.

Baudhuin, P. and Berthet, J. (1967). Electron microscopic examination of subcellular fractions. II. Quantitative analysis of the mitochondrial population isolated from rat liver. *J. Cell Biol.* **35,** 631.

Bernimoulin, J. P. and Schroeder, H. E. (1977). Quantitative electron microscopic analysis of the epithelium of normal human alveolar mucosa. *Cell Tiss. Res.* **180,** 383.

Bernimoulin, J. P. and Schroeder, H. E. (1979). Changes of differentiation pattern in oral mucosal epithelium following heterotopic connective tissue transplantation in man. Pathology: Research and Practice (in press).

Bernroider, G. (1976). Recognition and classification of structure by means of stereological methods in neurobiology. *J. Microsc.* **107,** 287.

Bernroider, G. (1978). The foundation of computational geometry: theory and application of the point-lattice-concept within modern structure analysis. *In:* "Geometric Probability and Biological Structures" (R. E. Miles and J. Serra, eds.) Vol. 23 of Lecture Notes in Biomathematics, p. 153. Springer Verlag, Berlin–Heidelberg–New York.

Berry, M., Anderson, E. M., Hollingworth, T. and Flinn, R. M. (1971). A computer technique for the estimation of the absolute three-dimensional array of basal dendritic fields using data from projected histological sections. *J. Microsc.* **95,** 257.

Berry, M., Hollingworth, T., Flinn, R. M. and Anderson, E. M. (1972). Dendritic field analysis—a reappraisal. T.-I.-T. *J. Life Sci.*, **2,** 129.

Bishop, G. H. and Smith, J. M. (1964). The sizes of nerve fibres supplying cerebral cortex. *Exp. Neurol.* **9,** 483.

Blouin, A., Bolender, R. P. and Weibel, E. R. (1977). Distribution of organelles and membranes between hepatocytes and non-hepatocytes in the rat liver parenchyma. *J. Cell. Biol.* **72,** 441.

Böhringer, S. and Hecker, H. (1974). Quantitative ultrastructural differences between strains of the *Trypanosoma brucei* subgroup during transformation in blood. *J. Protozool.* **21,** 694.

Böhringer, S. and Hecker, H. (1975). Quantitative ultrastructural investigations of the life cycle of *Trypanosoma brucei:* a morphometric analysis. *J. Protozool.* **22,** 463.

Bok, S. T. (1936). The branching of the dendrites in the cerebral cortex. *Proc. Roy. Acad.* (Amsterdam) **39,** 1209.

Bok, S. T. (1960). Quantitative analysis of the morphological elements of the cerebral cortex. *In:* "Structure and Function of the Cerebral Cortex". p. 7, Elsevier Publ. Comp., Amsterdam.

Bolender, R. P. (1974). Stereological analysis of the guinea pig pancreas. I. Analytical model and quantitative description of nonstimulated pancreatic exocrine cells. *J. Cell Biol.* **61**, 269.

Bolender, R. P. (1978). Correlation of morphometry and stereology with biochemical analysis of fractions. *Int. Rev. Cytol.* **55**, 247.

Bolender, R. P. and Weibel, E. R. (1973). A morphometric study of the removal of phenobarbital induced membranes from hepatocytes after cessation of treatment. *J. Cell Biol.* **56**, 746.

Bolender, R. P., Paumgartner, D., Losa, G., Muellener, D. and Weibel, E. R. (1978). Integrated stereological and biochemical studies on hepatocytic membranes. I. Membrane recoveries in subcellular fractions. *J. Cell Biol.* **77**, 565.

Bordier, P. J. and Tun-Chot, S. (1972). Quantitative histology of metabolic bone disease. *J. Clin. Endocrinol. Metab.* **1**, 197.

Brody, S. (1968). "Bioenergetics and Growth". Hafner Publ. Comp., New York.

Buffon, G. L. L. (1777). Essai d'arithmétique morale. Suppl. à l'Histoire Naturelle (Paris) **4**.

Burri, P. H. (1974). The postnatal growth of the rat lung. *Anat. Rec.* **180**, 77.

Burri, P. H. and Weibel, E. R. (1971). Morphometric estimation of pulmonary diffusion capacity. II. Effect of environmental PO_2 on the growing lung. *Respir. Physiol.* **11**, 247.

Burri, P. H. and Weibel, E. R. (1977). Ultrastructure and morphometry of the developing lung. *In:* "The Development of the Lung". (W. A. Hodson, ed.) p. 215. Marcel Dekker Inc., New York and Basel.

Burstone, M. S. (1962). "Enzyme Histochemistry". Academic Press, New York and London.

Cahn, J. W. (1967). The significance of average mean curvature and its determination by quantitative metallography. *Trans. AIME* **239**, 610.

Cahn, J. W. and Nutting, J. (1959). Transmission quantitative metallography. *Trans. Am. Inst. Min. metall. Engrs.* **215**, 526.

Caro, L. B. (1962). High resolution autoradiography. II. The problem of resolution. *J. Cell Biol.* **15**, 189.

Caro, L. B. and Palade, G. E. (1964). Protein synthesis, storage and discharge in the pancreatic exocrine cell; an autoradiographic study. *J. Cell Biol.* **20**, 473.

Casley-Smith, J. R. and Crocker, K. W. J. (1975). Quantitative electron microscopy, with especial references to the estimation of section thickness, directly in the microscope and from plates. *J. Microsc.* **103**, 351.

Chalkley, H. W. (1943). Methods for quantitative morphological analysis of tissue. *J. Nat. Cancer Inst.* **4**, 47.

Chalkley, H. W., Cornfield, J. and Park, H. (1949). A method for estimating volume–surface ratios. *Science* **110**, 295.

Cheng, G. C. (1976). What can pattern recognition do for stereology? *In:* "Proceedings Fourth International Congress for Stereology". (E. E. Underwood, R. de Wit, and G. A. Moore, eds.) p. 107, National Bureau of Standards, Special Publication 431, U.S. Government Printing Office, Washington.

Christophers, R. (1960). *"Aedes aegypti:* The yellow fever mosquito: Its life history, bionomics and structure". Cambridge University Press, Cambridge.

Cochran, W. C. (1953). "Sampling Techniques". Wiley and Sons, New York.

Corrsin, S. (1954). A measure of the area of a homogeneous random surface in space. *Quart. Appl. Math.* **12**, 404.

Crofton, M. W. (1885). Probability. *In:* "Encyclopedia Britannica". 9th ed., vol. 19, p. 768.

Cruz-Orive, L. -M. (1976a). Correction of stereological parameters from biased samples on nucleated particle phases. I. Nuclear volume fractions. *J. Microsc.* **106**, 1.

Cruz-Orive, L. -M. (1976b). Correction of stereological parameters from biased samples on nucleated particle phases. II. Specific surface area. *J. Microsc.* **106**, 19.

Cruz-Orive, L. -M. (1976c). Particle size-shape distributions: the general spheroid problem. I. Mathematical model. *J. Microsc.* **107**, 235.

Cruz-Orive, L. -M. (1978). Particle size-shape distributions: the general spheroid problem. II. Stochastic model and practical guide. *J. Microsc.* **112**, 153.

Cullen, M. J. and Fulthorpe, J. J. (1975). Stage in fibre breakdown in Duchenne muscular dystrophy—an electron microscopic study. *J. Neurol. Sci.* **24**, 179.

Cullen, M. J. and Weightman, D. (1975). The ultrastructure of normal human muscle in relation to fibre type. *J. Neurol. Sci.* **25**, 43.

De Hoff, R. T. (1964). The determination of the geometric properties of aggregates of constant-size particles from counting measurements made on random plane sections. *Trans AIME* **230**, 764.

De Hoff, R. T. (1967). The quantitative estimation of mean surface curvature. *Trans AIME* **239**, 617.

De Hoff, R. T. and Bousquet, P. (1970). Estimation of the size distribution of triaxial ellipsoidal particles from the distribution of linear intercepts. *J. Microsc.* **92**, 119.

De Hoff, R. T. and Rhines, F. N. (1968). "Quantitative Microscopy". McGraw-Hill, New York.

De Hoff, R. T. and Rhines, F. N. (1961). Determination of the number of particles per unit volume from measurements made on random plane sections: the general cylinder and the ellipsoid. *Trans. AIME* **221**, 975.

Delesse, M. A. (1847). Procédé mécanique pour déterminer la composition des roches. *C. R. Acad. Sci.* (Paris) **25**, 544.

Dixon, W. J. and Massey, F. J. (1957). "Introduction to Statistical Analysis". McGraw-Hill, New York.

Donovan, A. (1967). The nerve fibre composition of the cat optic nerve. *J. Anat.* **101**, 1.

Duffin, R. J., Meussner, R. A. and Rhines, F. N. (1953). Statistics of particle measurement and of particle growth. Carnegie Inst. Technol., Pittsburgh, Technical Report No. 32, C.I.T. AF8A–TR32.

Dulhunty, A. F. and Franzini-Armstrong, C. (1975). The relative contributions of the folds and caveolae to the surface membrane of frog skeletal muscle fibres at different sarcomere lengths. *J. Physiol.* **250**, 513.

Dunn, R. F., O'Leary, D. P. and Kumley, W. E. (1975). Quantitative analysis of micrographs by computer graphics. *J. Microsc.* **105**, 205.

Ebbeson, S. O. E. and Tang, D. (1965). A method for estimating the number of cells in histological sections. *J. Microsc.* **84**, 449.

Ebbeson, S. O. E. and Tang, D. B. (1967). A comparison of sampling procedures in a structured cell population. *In:* "Stereology". Proceedings 2nd International Congress for Stereology (H. Elias, ed.), p. 131. Springer Verlag, Berlin–Heidelberg–New York.

Eisenberg, B. R. and Kuda, A. M. (1975). Stereological analysis of mammalian skeletal muscle. II. White vastus muscle of the adult guinea pig. *J. Ultrastruct. Res.* **51**, 176.

Eisenberg, B. R. and Kuda, A. M. (1976). Discrimination between fiber populations in mammalian skeletal muscle by using ultrastructural parameters. *J. Ultrastruct. Res.* **54**, 76.

Eisenberg, B. R., Kuda, A. M. and Peter, J. B. (1974). Stereological analysis of mammalian skeletal muscle. I. Soleus muscle of the adult guinea pig. *J. Cell Biol.* **60**, 732.

Elias, H. (1963). Address of the President. *In:* "Proceedings First International Congress for Stereology". (H. Haug, ed.), p. 2. Congressprint, Wien.

Elias, H. (1972). Identification of structure by the common sense approach. *J. Microsc.* **95**, 59.

Elias, H. and Botz, E. (1976). Simple devices for stereology and morphometry. *In:* "Proceedings Fourth International Congress for Stereology". (E. E. Underwood, R. de Wit and G. A. Moore, eds.) p. 431. National Bureau of Standards, Special Publication Nr. 431, U.S. Government Printing Office, Washington.

Elias, H. and Schwartz, D. (1971). Cerebro-cortical surface areas, volumes, lengths of gyri and their interdependence in mammals, including man. *Z. Säugetierkd.* **36**, 147.

Elias, H., Hennig, A. and Elias, P. (1961). Contributions to the geometry of sectioning. V. Some methods for the study of kidney structure. *Z. Wiss. Mikrosk.* **65**, 70.

Elias, H., Kolodny, St. and Schwartz, D. (1967). Surface area and length of convolutions of the cerebral cortex. *In:* "Stereology" Proceedings Second International Congress for Stereology. (H. Elias, ed.) p. 77. Springer Verlag, New York.

Endter, F. and Gebauer, H. (1956). Ein einfaches Gerät zur statistischen Auswertung von mikroskopischen beziehungsweise elektronenmikroskopischen Aufnahmen. *Optik* **13**, 97.

Feret, L. R. (1931). La grosseur des grains. Assoc. Internat. Essais Math. 2D, Zürich.

Fischmeister, H. F. (1968). Scanning methods in quantitative metallography. *In:* "Quantitative Microscopy" (R. T. De Hoff and F. N. Rhines, eds.) p. 336. McGraw-Hill Comp., New York.

Fischmeister, H. F. (1972). Applications of quantitative microscopy in materials engineering. *J. Microsc.* **95**, 119.

Floderus, S. (1944). Untersuchungen über den Bau der menschlichen Hypophyse mit besonderer Berücksichtigung der quantitativen mikromorphologischen Verhältnisse. *Acta path. microbiol. scand. Suppl.* **53**, 1.

Foh, E., Haug, H., König, M. and Rast, A. (1973). Quantitative Bestimmung zum feineren Aufbau der Sehrinde der Katze, zugleich ein methodischer Beitrag zur Messung des Neuropils. *Microsc. Acta.* **75**, 148.

Forssmann, W. G., Siegrist, G., Orci, L., Girardier, L., Pictet, R. and Rouiller, C. (1967). Fixation par perfusion pour la microscopie électronique. Essai de généralisation. *J. Microscopie* **6**, 279.

Fritsch, R. S. (1971). Stereometrische Studie zur Oberflächendichte (s/v-Relation) von Zellen. *Cytobiologie* **3**, 331.

Frost, H. M. (1963). "Bone remodelling dynamics". Thomas, Springfield, Illinois.

Gahm, J. (1971). Current capabilities and limitations of available stereological techniques. I. Stereological measurements with lines, circles and structural standards. *J. Microscopy* **95**, 368.

Geelhaar, A. and Weibel, E. R. (1971). Morphometric estimation of pulmonary diffusion capacity. III. The effect of increased oxygen consumption in Japanese waltzing mice. *Respir. Physiol.* **11**, 354.

Gehr, P. and Weibel, E. R. (1974). Morphometric estimation of regional differences in the dog lung. *J. Appl. Physiol.* **37**, 648.

Gehr, P., Bachofen, M. and Weibel, E. R. (1978). The normal human lung. Ultrastructure and morphometric estimation of diffusion capacity. *Respir. Physiol.* **32**, 121.

Giger, H. and Riedwyl, H. (1970). Bestimmung der Grössenverteilung von Kugeln aus Schnittkreisradien. *Biometr. Zschr.* **12**, 156.

Gil, J. and Weibel, E. R. (1969). Improvements in demonstration of lining layer of lung alveoli by electron microscopy. *Respir. Physiol.* **8**, 13.

Gil, J. and Weibel, E. R. (1972). Morphological study of pressure-volume hysteresis in rat lungs fixed by vascular perfusion. *Respir. Physiol.* **15,** 190.

Gil, J., Bachofen, H., Gehr, P. and Weibel, E. R. (1979). The alveolar volume to surface area relationship in air and saline filled lungs fixed by vascular perfusion. *J. Appl. Physiol.* (in press).

Gillis, J. M. and Wibo, M. (1971). Accurate measurement of the thickness of ultrathin sections by interference microscopy. *J. Cell. Biol.* **49,** 947.

Glagolev, A. A. (1933). On the geometrical methods of quantitative mineralogic analysis of rocks. *Trans. Inst. Econ. Min.,* Moscow, **59,** 1.

Glazier, J. B., Hughes, J. M. B., Maloney, J. E. and West, J. B. (1967). Vertical gradient of alveolar size in lungs of dogs frozen intact. *J. Appl. Physiol.* **23,** 694.

Gnanadesikan, R. (1977). "Methods for Statistical Data Analysis of Multivariate Observations". Wiley–Interscience, Somerset, N.J.

Gooding, R. H. (1972). Digestive processes of haematophagous insects. I. A literature review. *Quaestiones entomol.* **8,** 5.

Gooding, R. H. (1973). Digestive processes of haematophagous insects. IV. Secretion of trypsin by *Aedes aegypti. Can. Entomol.* **105,** 599.

Gundersen, H. J. G. (1977). Notes on the estimation of the numerical density of arbitrary profiles: the edge effect. *J. Microscop.* **111,** 219.

Gundersen, H. J. G., Jensen, T. B. and Østerby, R. (1978). Distribution of membrane thickness determined by lineal analysis. *J. Microscop.* **113,** 27.

Hadwiger, H. (1957). "Vorlesungen über Inhalt, Oberfläche und Isoperimetrie". Springer Verlag, Berlin–Göttingen–Heidelberg.

Hammer, B. and Schroeder, H. E. (1977). Stereologic system and on/off-line computer program for analysing oral epithelia, based on a model of stratification. *Archs. oral Biol.* **22,** 337.

Haug, H. (1955). Die Treffermethode, ein Verfahren zur quantitativen Analyse im histologischen Schnitt. *Z. Anat. Entwickl. Gesch.* **118,** 302.

Haug, H. (1963). "Proceedings First International Congress for Stereology Vienna". (1963). Congressprint, Wien.

Haug, H. (1967a). Ueber die exakte Feststellung der Anzahl Nervenzellen pro Volumeneinheit des Cortex cerebri, zugleich ein Beispiel für die Durchführung genauer Zählungen. *Acta Anat.* **67,** 53.

Haug, H. (1967b). Morphometrie der feinen Markfasern in der Sehrinde der Katze. *Z. Zellforsch.* **77,** 416.

Haug, H. (1967c). Probleme und Methoden der Strukturzählung im Schnittpräparat. *In:* "Quantitative Methods in Morphology". (E. R. Weibel and H. Elias, eds.) p. 58. Springer Verlag, Berlin–Heidelberg–New York.

Haug, H. (1970). Der makroskopische Aufbau des Grosshirns. *Ergebn. Anat. Entwickl. Gesch.* **43,** Heft 4.

Haug, H. (1972). Stereological methods in the analysis of neuronal parameters in the central nervous system. *J. Microsc.* **95,** 165.

Haug, H. (1976). Experiences with optomanual automated evaluation systems in biological research, especially in neuromorphology. *In:* "Proceedings Fourth International Congress for Stereology". (E. E. Underwood, R. de Wit and G. A. Moore, eds.), p. 167. National Bureau of Standards, Special Publication 431, U.S. Government Printing Office, Washington.

Haug, H. (1978). Interest of stereology on central nervous system investigations. *In:* "Quantitative Analysis of Microstructures in Material Science, Biology and Medicine". (J. L. Chermant, ed.): Proceedings of the Symposium "Quantitative Analysis of Microstructures", p. 299. Dr. Riederer Verlag GmbH, Stuttgart.

Haug, H. and Rast, A. (1972). Die Messung der Längen von Fasern in teilorientierten

Strukturen. Untersuchungen des Nervus trigeminus als Beispiel. *Microsc. Acta* **72**, 136.

Haug, H., Kebbel, J. and Wiedemeyer, G.-L. (1971). Die Messung der mittleren Zelldichte und ihre Verteilung in Geweben mit erheblichen Zelldichteunterschieden. Auswertung am Cortex cerebri als Beispiel. *Microsc. Acta* **71**, 121.

Haug, H., Kölln, M. and Rast, A. (1976). The postnatal development of the myelinated fibres in cat's visual cortex. A stereological and electron-microscopical investigation. *Cell Tissue Res.* **167**, 265.

Hecker, H. (1978). Intracellular distribution of ribosomes in midgut cells of the malaria mosquito, *Anopheles stephensi*, in response to feeding. *Int. J. Insect Morphol. Embryol.* **7**, 267.

Hecker, H., Burri, P. H., Steiger, R. and Geigy, R. (1972). Morphometric data on the ultrastructure of the pleomorphic blood-forms of *Trypanosoma brucei*. *Acta Trop.* **29**, 182.

Hecker, H., Burri, P. H. and Böhringer, S. (1973). Quantitative ultrastructural differences in the mitochondrium of pleomorphic bloodforms of *Trypanosoma brucei*. *Experientia* **29**, 901.

Hecker, H., Brun, R., Reinhard, C. and Burri, P. H. (1974). Morphometric analysis of the midgut of female *Aedes aegypti* L. under various physiological conditions. *Cell Tiss. Res.* **152**, 31.

Hennig, A. (1956a). Diskussion der Fehler bei der Volumenbestimmung mikroskopisch kleiner Körper oder Hohlräume aus den Schnitt projektionen. *Z. wiss. Mikrosk.* **63**, 67.

Hennig, A. (1956b). Bestimmung der Oberfläche beliebig geformter Körper mit besonderer Anwendung auf Körperhaufen im mikroskopischen Bereich. *Mikroskopie* **11**, 1.

Hennig, A. (1957). Zur Geometrie von Schnitten. *Z. wiss. Mikrosk.* **63**, 362.

Hennig, A. (1958). Kritische Betrachtungen zur Volumen- und Oberflächenmessung in der Mikroskopie. *Zeiss-Werkzschr.* **30**, 3.

Hennig, A. (1963a). Länge eines räumlichen Linienzuges. *Z. wiss. Mikrosk.* **65**, 193.

Hennig, A. (1963b). Length of a three-dimensional linear tract. In: "Proceedings First International Congress for Stereology". (H. Haug, ed.), p. 44. Congressprint, Wien.

Hennig, A. (1967). Fehlerbetrachtungen zur Volumenbestimmung aus der Integration ebener Schnitte. In: "Quantitative Methods in Morphology". (E. R. Weibel and H. Elias, eds.) p. 99. Springer Verlag, Berlin–Heidelberg–New York.

Hennig, A. (1969). Fehler in der Volumenermittlung aus der Flächenrelation in dicken Schnitten (Holmes-Effekt). *Mikroskopie* **25**, 154.

Hennig, A. and Elias, H. (1970). A rapid method for the visual determination of size distribution of spheres from the size distribution of their sections. *J. Microsc.* **93**, 101.

Hess, F. A., Preisig, R. and Weibel, E. R. (1973). The validity of needle biopsies for liver cytomorphometry. In: "The Liver. Quantitative Aspects of Structure and Function". (G. Paumgartner and R. Preisig, eds.), p. 58. Karger, Basel.

Hilliard, J. E. (1967a). The calculation of the mean caliper diameter of a body for use in the analysis of the number of particles per unit volume. In: "Stereology. Proceedings Second International Congress for Stereology". (H. Elias, ed.) p. 211. Springer Verlag, Berlin–Heidelberg–New York.

Hilliard, J. E. (1967b). Determination of structural anisotropy. In: "Stereology. Proceedings Second International Congress for Stereology". (H. Elias, ed.), p. 219. Springer Verlag, Berlin–Heidelberg–New York.

Hilliard, J. E. (1968). Direct determination of the moments of the size distribution of particles in an opaque sample. *Trans AIME* **242**, 1373.

Hilliard, J. E. (1972). Quantitative analysis of scanning electron micrographs. *J.*

O

Microsc. **95**, 45.

Hilliard, J. E. (1976). Assessment of sampling error in stereological analysis. *In:* "Proceedings Fourth International Congress for Stereology". (E. E. Underwood, R. de Wit and G. A. Moore, eds.), p. 59. National Bureau of Standards, Special Publication 431, U.S. Government Printing Office, Washington.

Hilliard, J. E. and Cahn, J. W. (1961). An evaluation of procedure in quantitative metallography for volume fraction analysis. *Trans. Am. Inst. Min. Engrs.* **221**, 344.

Hokin, L. E. (1968). Dynamic aspects of phospholipids during secretion. *Int. Rev. Cytol.* **23**, 187.

Holländer, H., Wickelmaier, M. and Pastor, W. (1976). Ein Koordinaten und Flächen registrierendes Mikroskop zur topographischen Analyse von Teilchengrössenverteilung. *Microsc. Acta* **78**, 118.

Holmes, A. (1927). "Petrographic Methods and Calculations". Murby, London.

Holtrop, M. E., Raisz, L. G. and Simmons, H. A. (1974). The effects of parathyroid hormone, colchicine, and calcitonin on the ultrastructure and the activity of osteoclasts in organ culture. *J. Cell Biol.* **60**, 346.

Hoppeler, H., Lüthi, P., Claassen, H., Weibel, E. R. and Howald, H. (1973). The ultrastructure of the normal human skeletal muscle. A morphometric analysis on untrained men, women and well-trained orienteers. *Pflügers Arch.* **344**, 217.

Horikawa, E. (1954). On a new method of representation of the mixture of several austhenite grain sizes. *Tetsu to Hagane* **40**, 991.

Horsfield, K. and Cumming, G. (1968). Morphology of the bronchial tree in man. *J. Appl. Physiol.* **24**, 373.

Hougardy, H. P. (1976). Automatic image analysing instruments today. *In:* "Proceedings Fourth International Congress for Stereology". (E. E. Underwood, R. de Wit and G. A. Moore, eds.), p. 141. National Bureau of Standards, Special Publication 431, U.S. Government Printing Office, Washington.

Hucher, M. and Grolier, J. (1977). Fifty-nine tetrakaidecahedra as grain models. *J. Microsc.* **111**, 329.

Hugonnaud, C., Gehr, P., Burri, P. H. and Weibel, E. R. (1977). Adaptation of the growing lung to increased oxygen consumption. II. Morphometric analysis. *Respir. Physiol.* **29**, 1.

Idelman, S. (1970). Ultrastructure of the mammalian adrenal cortex. *Int. Rev. Cytol.* **27**, 181.

Jamieson, J. D. and Palade, G. E. (1967a). Intracellular transport of secretory proteins in the pancreatic exocrine cell. I. Role of the peripheral elements of the Golgi complex. *J. Cell Biol.* **34**, 577.

Jamieson, J. D. and Palade, G. E. (1967b). Intracellular transport of secretory proteins in the pancreatic exocrine cell. II. Transport to condensing vacuoles and zymogen granules. *J. Cell. Biol.* **34**, 597.

Jamieson, J. D. and Palade, G. E. (1971). Synthesis, intracellular transport, and discharge of secretory proteins in stimulated pancreatic exocrine cells. *J. Cell Biol.* **50**, 135.

Jerusalem, F., Engel, A. G. and Peterson, H. A. (1975). Human muscle fiber fine structure: Morphometric data on controls. *Neurology* **25**, 127.

Keller, H. J., Friedli, H. P., Gehr, P., Bachofen, M. and Weibel, E. R. (1976). The effects of resolution on estimating stereological parameters. *In:* "Proceedings Fourth International Congress for Stereology". (E. E. Underwood, R. de Wit and G. A. Moore, eds.), p. 409. National Bureau of Standards, Special Publication 431, U.S. Government Printing Office, Washington.

Kendall, M. G. and Moran, P. A. P. (1963). "Geometrical Probability". Charles Griffin, London.

Kiessling, K. H., Piehl, K. and Lundquist, C. G. (1971). Effect of physical training on ultrastructural features in human skeletal muscle. *In:* "Muscle Metabolism During Exercise". (B. Pernow and B. Saltin, eds.), p. 97. Plenum Press, New York.

Kiessling, K. H., Pilström, L., Karlsson, J. and Piehl, K. (1973). Mitochondrial volume in skeletal muscle from young and old physically untrained and trained healthy men and from alcoholics. *Clin. Sci.* **44**, 547.

Kiessling, K. H. Pilström, L., Bylund, A.Ch., Saltin, B. and Piehl, K. (1974). Enzyme activities and morphometry in skeletal muscle of middle-aged men after training. *Scand. J. Clin. Lab. Invest.* **33**, 63.

Kilarski, W. (1973). Cytomorphometry of sarcoplasmic reticulum in extrinsic eye muscles of the teleost. *Z. Zellforsch.* **136**, 535.

Kleiber, M. (1961). "The Fire of Life". John Wiley, New York.

Klein-Szanto, A. J. P. (1977). Stereologic baseline data of normal human epidermis. *J. Investig. Derm.* **68**, 73.

Klein-Szanto, A. J. P. and Schroeder, H. E. (1977). Architecture and density of the connective tissue papillae of the human oral mucosa. *J. Anat.* (London) **123**, 93.

Klein-Szanto, A. J. P., Andersen, L. and Schroeder, H. E. (1976a). Epithelial differentiation patterns in buccal mucosa affected by lichen planus. *Virchows Arch. B. Cell Path.* **22**, 245.

Klein-Szanto, A. J. P., Banoczy, J. and Schroeder, H. E. (1976b). Metaplastic conversion of the differentiation pattern in oral epithelia affected by leukoplakia simplex. A stereologic study. *Path. Europ.* **11**, 189.

Konigsmark, B. W. (1970). Methods for the counting of neurons. *In:* "Contemporary Research Methods in Neuroanatomy". (W. J. H. Nauta and S. O. E. Ebbeson, eds.), p. 315. Springer Verlag, Berlin–Heidelberg–New York.

Konwinski, M. and Kozlowski, T. (1972). Morphometric study of normal and phytohemagglutinin-stimulated lymphocytes. *Z. Zellforsch.* **129**, 500.

Kraehenbühl, J. P. and Jamieson, J. D. (1972). Enzyme labeled antibody markers for electron microscopy. *In:* "Methods in Immunology and Immunocytochemistry". (C. A. Williams and M. W. Chase, eds.), p. 482. Academic Press, New York.

Krutsay, M. (1963). Methode zur Darstellung einiger Kalziumverbindungen in histologischen Schnitten. *Acta histochem.* **15**, 192.

Landay, M. A. and Schroeder, H. E. (1977). Quantitative electron microscopic analysis of the stratified epithelium of normal human buccal mucosa. *Cell Tiss. Res.* **177**, 383.

Lord, G. W. and Willis, T. F. (1951). Calculation of air bubble size distribution from results of a Rosiwal traverse of aerated concrete. *A.S.T.M. Bull.* **56**, 177.

Losa, G., Weibel, E. R. and Bolender, R. P. (1978). Integrated stereological and biochemical studies on hepatocytic membranes. III. Relative surface of endoplasmic reticulum membranes in microsomal fractions estimated on freeze-fracture preparations. *J. Cell Biol.* **78**, 289.

Loud, A. V. (1968). A quantitative stereological description of the ultrastructure of normal rat liver parenchymal cells. *J. Cell Biol.* **37**, 27.

Loud, A. V., Barany, W. C. and Pack, B. A. (1965). Quantitative evaluation of cytoplasmic structures in electron micrographs. *Lab. Invest.* **14**, 996.

Mandelbrot, B. B. (1967). How long is the coast of Britain? Statistical self-similarity and fractional dimension. *Science* **155**, 636.

Mandelbrot, B. B. (1977). "Fractals. Form, Chance, and Dimension". Freeman and Company, San Francisco.

Mathieu, O., Claassen, H. and Weibel, E. R. (1978). Differential effect of glutaraldehyde and buffer osmolarity on cell dimensions. A study on lung tissue. *J. Ultrastruct. Res.*

63, 20.

Mayhew, T. M. and Cruz, L. -M. (1973). Stereological correction procedures for estimating true volume proportions from biased samples. *J. Microsc.* **99**, 287.

Mayhew, T. M. and Cruz-Orive, L. -M. (1974). Caveat on the use of the Delesse principle of areal analysis for estimating component volume densities. *J. Microsc.* **102**, 195.

Mayhew, T. M. and Cruz-Orive, L. -M. (1975). Some stereological correction formulae with particular applications in quantitative neurohistology. *J. neurol. Sci.* **26**, 503.

Mayhew, T. M. and Williams, M. A. (1971). A comparison of two sampling procedures for stereological analysis of cell pellets. *J. Microsc.* **94**, 195.

Mayhew, T. M. and Williams, M. A. (1974a). A quantitative morphological analysis of macrophage stimulation. I. A study of subcellular compartments and of the cell surface. *Z. Zellforsch. mikrosk. Anat.* **147**, 567.

Mayhew, T. M. and Williams, M. A. (1974b). A quantitative morphological analysis of macrophage stimulation. II. Changes in granule number, size and size distribution. *Cell Tiss. Res.* **150**, 529.

McNutt, N. S. and Weinstein, R. S. (1973). Membrane ultrastructure and mammalian intercellular junction. *Progr. Biophys. Mol. Biol.* **26**, 45.

Meldolesi, J. (1974). Dynamics of cytoplasmic membranes in guinea pig pancreatic acinar cells. *J. Cell Biol.* **61**, 1.

Merz, W. A. (1967). Die Streckenmessung an gerichteten Strukturen im Mikroskop und ihre Anwendung zur Bestimmung von Oberflächen-Volumen-Relationen im Knochengewebe. *Mikroskopie* **22**, 132.

Merz, W. A. and Schenk, R. K. (1970a). Quantitative structural analysis of human cancellous bone. *Acta anat.* **75**, 54.

Merz, W. A. and Schenk, R. K. (1970b). A quantitative histological study on bone formation in human cancellous bone. *Acta anat.* **76**, 1.

Meyer, M. and Schroeder, H. E. (1975). A quantitative electron microscopic analysis of the keratinizing epithelium of normal human hard palate. *Cell Tiss. Res.* **158**, 177.

Miles, R. E. (1972). Multidimensional perspectives on stereology. *J. Microsc.* **95**, 181.

Miles, R. E. (1976). Estimating aggregate and overall characteristics from thick sections by transmission microscopy. *J. Microsc.* **107**, 227.

Miles, R. E. (1978a). The sampling, by quadrats, of planar aggregates. *J. Microsc.* **113**, 257.

Miles, R. E. (1978b). The importance of proper model specification in stereology. *In:* "Geometric Probability and Biological Structures: Buffon's 200th Anniversary". (R. E. Miles and J. Serra, eds.), Vol. 23 of Lecture Notes in Biomathematics. p. 115. Springer Verlag, Berlin–Heidelberg–New York.

Miles, R. E. and Davy, P. J. (1976). Precise and general conditions for the validity of a comprehensive set of stereological fundamental formulae. *J. Microsc.* **107**, 211.

Miles, R. E. and Davy, P. J. (1977). On the choice of quadrats in stereology. *J. Microsc.* **110**, 27.

Miles, R. E. and Serra, J. (1978). "Geometrical Probability and Biological Structures: Buffon's 200th Anniversary". Vol. 23 of Lecture Notes in Biomathematics. Springer Verlag, Berlin–Heidelberg–New York.

Mobley, B. A. and Page, E. (1972). The surface area of sheep cardiac Purkinje fibres. *J. Physiol.* (London) **220**, 547.

Mobley, B. A. and Eisenberg, B. R. (1975). Sizes of components in frog skeletal muscle by methods of stereology. *J. Gen. Physiol.* **66**, 31.

Moor, H. (1969). Freeze-etching. *Int. Rev. Cytol.* **25**, 391.

Moore, G. A. (1972). Recent progress in automatic image analysis. *J. Microsc.* **95**, 105.

Morré, J. D., Keenan, T. W. and Huang, C. M. (1974). Membrane flow and differentiation: Origin of Golgi apparatus membranes from endoplasmic reticulum. *In:* "Advances in Cytopharmacology". (B. Ceccarelli, J. Meldolesi and F. Clementi, eds.), Vol. 2, p. 107. Raven Press, New York.

Morrison, D. F. (1967). "Multivariate Statistical Methods". McGraw-Hill, New York.

Mosteller, F. and Rourke, R. E. K. (1976). "Sturdy Statistic Non-parametrics and Order Statistics". Addison-Wesley, Reading, Mass.

Mulligan, H. W. (1970). "The African Trypanosomiasis". Allen and Unwin Ltd., London.

Needham, J. (1968). "Order and Life". MIT Press, Cambridge Mass. and London, England. (Reprint of original edition by Yale University Press, 1936).

Nicholson, W. L. (1972). "Proceedings of the Symposium on Statistical and Probabilistic Problems in Metallurgy". Special Supplement to Advances in Applied Probability. Applied Probability Trust, Seattle, Washington.

Nicholson, W. L. (1978). Application of statistical methods in quantitative microscopy. *J. Microscopy* **113**, 223.

Olah, A. J. (1974). Histomorphometrie des Knochens. *Verh. Dtsch. Ges. Path.* **58**, 104.

Page, E. (1978). Quantitative ultrastructural analysis in cardiac membrane physiology. *Am. J. Physiol.: Cell Physiol.* **4**, C147.

Page, E. and Oparil, S. (1978). Effect of peripheral sympathectomy on left ventricular ultrastructure in young spontaneously hypertensive rats. *J. Molec. Cell. Cardiol.* **10**, 301.

Page, E. and Upshaw-Earley, J. (1977). Volume changes in sarcoplasmic reticulum of rat hearts perfused with hypertonic solutions. *Circulation Res.* **40**, 355.

Page, E., McCallister, L. P. and Power, B. (1971). Stereological measurements of cardiac ultrastructures implicated in excitation–contraction coupling. *Proc. Nat. Acad. Sci. (USA)* **68**, 1465.

Page, E., Polimeni, P. I., Zak, R., Earley, J. and Johnson, M. (1972). Myofibrillar mass in rat and rabbit heart muscle. Correlation of micromechanical and stereological measurements in normal and hypertrophic hearts. *Circulation Res.* **30**, 430.

Page, E., Earley, J., McCallister, L. P. and Boyd, C. (1974a). Copper content and exchange in mammalian hearts. *Circulation Res.* **35**, 67.

Page, E., Earley, J. and Power, B. (1974b). Normal growth of cardiac ultrastructures in rat left ventricular myocardial cells. *Circulation Res.,* **34** and **35**, Suppl. 2, 12.

Palade, G. E. (1959). Functional changes in the structure of cell components. *In:* "Subcellular Particles". (T. Hayashi, ed.), p. 64. Ronald Press, New York.

Papermaster, D. S., Schneider, B. G., Zorn, M. A. and Kraehenbühl, J. P. (1978). Immunocytochemical localization of opsin in outer segments and Golgi zones of frog photoreceptor cells. An electron microscope analysis of cross-linked albumin-embedded retinas. *J. Cell Biol.* **77**, 196.

Paumgartner, D., Losa, G. and Weibel, E. R. (1979). The effect of resolution on the estimation of stereological parameters: organelles and membranes of liver cells studied by electron microscopy. (in preparation).

Pearse, A. G. E. (1960). "Histochemistry". (2nd edition) Churchill, London.

Petrzilka, G. E. and Schroeder, H. E. (1976). Application of stereologic methodology to inflammatory infiltrates residing in oral mucous membrane lesions. Structural and morphometric comparison of benign mucous membrane pemphigoid and reticular lichen planus. *Beitr. Path.* **159**, 351.

Pluta, M. (1969). A double refracting interference microscope with variable image duplication and half-shade eyepiece. *J. scient. Instrum.* **2**, 685.

Pluta, M. (1971). On the accuracy of microinterferometric measurements of optical path differences by means of the half-shade methods. *J. Microscopy* **93**, 83.

Polimeni, P. I. (1974). Extracellular space and ionic distribution in rat ventricle. *Am. J. Physiol.* **227**, 676.

Pugh, A. (1976). "Polyhedra, a Visual Approach". University of California Press, Berkley and Los Angeles, Calif.

Rao, C. R. (1968). "Linear Statistical Inference". John Wiley, New York.

Rätz, H. U., Gnägi, H. R. and Weibel, E. R. (1974). An on-line computer system for point counting stereology. *J. Microscopy* **101**, 267.

Reith, A. (1976). Is there an unrecognised systematic error in the estimation of surface density of biomembranes? *In:* "Proceedings Fourth International Congress for Stereology" (E. E. Underwood, R. de Wit and G. A. Moore, eds.), p. 427. National Bureau of Standards, Special Publication Nr. 431, U.S. Government Printing Office, Washington.

Rhodin, J. A. G. (1971). The ultrastructure of the adrenal cortex of the rat under normal and experimental conditions. *J. Ultrastruct. Res.* **34**, 23.

Rink, M. (1976). A computerized quantitative image analysis procedure for investigating features and an adapted image process. *J. Microscopy* **107**, 267.

Rohr, H. P. (1976). A new opto-manual semi-automatic evaluation system. *In:* "Proceedings Fourth International Congress for Stereology". (E. E. Underwood, R. de Wit and G. A. Moore, eds.), p. 193. National Bureau of Standards, Special Publication Nr. 431, U.S. Government Printing Office, Washington.

Rohr, H. P., Bartsch, G., Eichenberger, P., Rasser, Y., Kaiser, Ch., and Keller, M. (1976). Ultrastructural morphometric analysis of the unstimulated adrenal cortex of rats. *J. Ultrastruct. Res.* **54**, 11.

Rosiwal, A. (1898). Ueber geometrische Gesteinsanalysen. *Verh. K. K. Geol. Reichsanst. Wien,* 143.

Ross, R. and Benditt, E. P. (1962). Wound healing and collagen formation. III. A quantitative radioautographic study of the utilization of proline-H^3 in wounds from normal and scorbutic guinea pigs. *J. Cell Biol.* **15**, 99.

Salpeter, M. M. and Bachmann, L. (1964). Autoradiography with the electron microscope. A procedure for improving resolution, sensitivity and contrast. *J. Cell Biol.* **22**, 469.

Salpeter, M. M. and Bachmann, L. (1972). Radioautography. *In:* "Principles and Techniques of Electron Microscopy: Biological Applications". (M. A. Hayat, ed.), Vol. 2, p. 221. Van Nostrand Reinhold Company, New York–London.

Saltykov, S. A. (1946). *Zavodskaja laboratorija* **12**, 816.

Saltykov, S. A. (1958). "Stereometric Metallography". 2nd edition (3rd edition 1970). State Publishing House for Metals and Sciences, Moscow. (German translation of 3rd edition: VEB Deutscher Verlag fur Grundstoff industrie, Leipzig, 1974).

Samorajski, T. and Friede, R. L. (1968). A quantitative electron microscopic study of myelination in the pyramidal tract of rat. *J. Comp. Neur.* **134**, 323.

Santaló, L. A. (1976). "Integral Geometry and Geometric Probability". Vol. 1 of Encyclopedia of Mathematics and its Applications. Addison-Wesley, Reading, Mass.

Schadé, J. P. and Meeter, K. (1963). Neuronal and dendritic pattern in the uncinate area of the human hippocampus. *Progr. Brain Res.* **3**, 89.

Schenk, R. K. (1965). Zur histologischen Verarbeitung von unentkalktem Knochen. *Acta anat.* **60**, 3.

Schenk, R. K., Merz, W. A. and Müller, J. (1969). A quantitative histological study on bone resorption in human cancellous bone. *Acta anat.* **74**, 44.

Schenk, R. K., Olah, A. J. and Merz, W. A. (1973). Bone cell counts. *In:* "Clinical Aspects of Metabolic Bone Disease". (B. Frame, A. M. Parfitt, H. Duncan, eds.),

p. 103. Excerpta Medica, Amsterdam.

Scherle, W. (1970). A simple method for volumetry of organs in quantitative stereology. *Mikroskopie* **26**, 57.

Schlenska, G. (1969). Messung der Oberfläche und der Volumenanteile des Gehirnes menschlicher Erwachsener mit neuen Methoden. *Z. Anat. Entwickl. Gesch.* **128**, 47.

Schroeder, H. E. and Graf-de Beer, M. (1976). Stereologic analysis of chronic lymphoid cell infiltrates in human gingiva. *Archs. oral Biol.* **21**, 527.

Schroeder, H. E. and Münzel-Pedrazzoli, S. (1970a). Application of stereologic methods to stratified gingival epithelia. *J. Microscopy* **92**, 179.

Schroeder, H. E. and Münzel-Pedrazzoli, S. (1970b). Morphometric analysis comparing junctional and oral epithelium of normal human gingiva. *Helv. odont. Acta* **14**, 53.

Schroeder, H. E. and Münzel-Pedrazzoli, S. (1973). Correlated morphometric and biochemical analysis of gingival tissue. Morphometric model, tissue sampling and test of stereologic procedures. *J. Microscopy* **99**, 301.

Schroeder, H. E., Münzel-Pedrazzoli, S. and Page, R. (1973). Correlated morphometric and biochemical analysis of gingival tissue in early chronic gingivitis in man. *Archs. oral Biol.* **18**, 899.

Schroeder, H. E., Graf-de Beer, M. and Attström, R. (1975). Initial gingivitis in dogs. *J. periodont. Res.* **10**, 128.

Serra, J. (1972). Stereology and structuring elements. *J. Microscopy* **95**, 93.

Serra, J. (1978). One, two, three, . . . infinity. *In*: "Geometric Probability and Biological Structures: Buffon's 200th Anniversary". (R. E. Miles and J. Serra, eds.), vol. 23: Lecture Notes in Biomathematics. p. 137, Springer Verlag, Berlin–Heidelberg–New York.

Silverman, L., Schreiner, B. and Glick, D. (1969). Measurement of thickness within sections by quantitative electron microscopy. *J. Cell Biol.* **40**, 768.

Simar, L. (1973). "L'Ultrastructure des Ganglions Lymphatiques au Cours des Ré, actions Immunitaires". Thèse d'agrègation de l'enseignement supérieur, Université de Liège.

Simar, L. J. (1975). Cinétique de la réaction ganglionnaire au cours d'une immunisation humorale. Etude autoradiographique ultrastructurale. *Path. Biol.* **23**, 440.

Sitte, H. (1967). Morphometrische Untersuchungen an Zellen. *In*: "Quantitative Methods in Morphology". (E. R. Weibel and H. Elias, eds.), p. 167. Springer Verlag, New York.

Small, J. V. (1968). Measurements of section thickness. *In*: "Proceedings 4th European Congress on Electron Microscopy", (D. S. Bocciarelli, ed.), Vol. 1, p. 609. Tipografia Poliglotta Vaticana, Roma.

Smit, G. J. and Colon, E. J. (1969). Quantitative analysis of the cerebral cortex. I. A selectivity of the Golgi-Cox staining technique. *Brain Res.* **13**, 485.

Smith, C. S. and Guttman, L. (1953). Measurement of internal boundaries in three-dimensional structures by random sectioning. *Trans AIME* **197**, 81.

Smith, H. E. and Page, E. (1976). Morphometry of rat heart mitochondrial sub-compartments and membranes: Application to myocardial cell atrophy after hypophysectomy. *J. Ultrastruct. Res.* **55**, 31.

Smith, H. E. and Page, E. (1977). Ultrastructural changes in rabbit heart mitochondria during the perinatal period: Neonatal transition to aerobic metabolism. *Develop. Biol.* **57**, 109.

Snedecor, G. W. (1956). "Statistical Methods Applied to Experiments in Agriculture and Biology". The Iowa State University Press, Ames, Iowa.

Stahl, W. R. (1967). Scaling of respiratory variables in mammals. *J. Appl. Physiol.* **22**, 453.

Stäubli, W., Hess, R. and Weibel, E. R. (1969). Correlated morphometric and biochemical studies on the liver cell. II. Effects of phenobarbital on rat hepatocytes. *J. Cell Biol.* **42**, 92.

Stäubli, W., Schweizer, W., Suter, J. and Weibel, E. R. (1977). The proliferative response of hepatic peroxisomes of neonatal rats to treatment with SU-13 437 (Nafenopin). *J. Cell Biol.* **74**, 665.

Steiger, R. F. (1973). On the ultrastructure of *Trypanosoma brucei* in the course of its life-cycle and some related aspects. *Acta Trop.* **30**, 64.

Stewart, J. and Page, E. (1978). Improved stereological techniques for studying myocardial cell growth: Application to external sarcolemma, T-system and intercalated disks of rabbit and rat hearts. *J. Ultrastruct. Res.* **65**, 119.

Thomson, E. (1930). Quantitative microscopic analysis. *J. Geol.* **38**, 193.

Thompson, D'Arcy W. (1942). "Growth and Form". Second Edition, Cambridge University Press, Cambridge, England. (Second printing, 1956).

Thurlbeck, W. M. (1967). The internal surface area of non-emphysematous lungs. *Am. Rev. Resp. Dis.* **95**, 765.

Tomkeieff, S. I. (1945). Linear intercepts, areas and volumes. *Nature* **155**, 24.

Treff, W. M., Meyer-König, E. and Schlote, W. (1971). Morphometric analysis of a fibre system in the central nervous system. *J. Microscopy* **95**, 337.

Underwood, E. E. (1968). Particle size distribution. *In*: "Quantitative Microscopy". (R. T. De Hoff and F. N. Rhines, eds.), p. 149. McGraw-Hill, New York.

Underwood, E. E. (1970). "Quantitative Stereology". Addison–Wesley, Reading, Mass.

Uylings, H. B. M., Smit, G. J. and Veltman, W. A. M. (1975). Ordering methods in quantitative analysis of branching structures of dendritic trees. *Adv. Neurol.* **12**, 247.

Vickerman, K. (1965). Polymorphism and mitochondrial activity in sleeping sickness trypanosomes. *Nature* **208**, 762.

Walton, W. H. (1948). Feret's statistical diameter. *Nature* **162**, 329.

Wann, D. F., Woolsey, T. A., Dierker, M. L. and Cowan, W. M. (1973). An on-line digital-computer system for the semiautomatic analysis of Golgi-impregnated neurons. *IEEE Trans. Bio-Med. Eng.* **BME 20**, 233.

Weibel, E. R. (1963). "Morphometry of the Human Lung". Springer Verlag, Berlin, and Academic Press, New York.

Weibel, E. R. (1967a). Postnatal growth of the lung and pulmonary gas-exchange capacity. *In*: "Ciba Foundation Symposium: Development of the Lung". (A.V.S. de Reuck and R. Porter, eds.), p. 131. Churchill, London.

Weibel, E. R. (1967b). Ordnung und Leben. *Schweiz. Med. Wschr.* **97**, 629.

Weibel, E. R. (1967c). Structure in space and its appearance on sections. *In*: "Stereology". (H. Elias, ed.) p. 15. Springer Verlag, New York.

Weibel, E. R. (1969). Stereological principles for morphometry in electron microscopic cytology. *Int. Rev. Cytol.* **26**, 235.

Weibel, E. R. (1970a). Morphometric estimation of pulmonary diffusion capacity. I. Model and method. *Resp. Physiol.* **11**, 54.

Weibel, E. R. (1970b). An automatic sampling stage microscope for stereology. *J. Microscopy* **91**, 1.

Weibel, E. R. (1972). A stereological method for estimating volume and surface of sarcoplasmic reticulum. *J. Microscopy* **95**, 229.

Weibel, E. R. (1973a). A simplified morphometric method for estimating diffusing capacity in normal and emphysematous human lungs. *Am. Rev. Respir. Dis.* **107**, 579.

Weibel, E. R. (1973b). The morphological basis of alveolar-capillary gas exchange. *Physiol. Rev.* **53**, 419.

Weibel, E. R. (1974). Selection of the best method in stereology. *J. Microscopy* **100**, 261.

Weibel, E. R. (1978). The non-statistical nature of biological structure and its implications on sampling for stereology. *In*: "Geometric Probability and Biological Structures: Buffon's 200th Anniversary". (R. E. Miles and J. Serra, eds.), Vol. 23: Lecture Notes in Biomathematics, p. 171. Springer Verlag, Berlin-Heidelberg-New York.

Weibel, E. R. and Bolender, R. P. (1973). Stereological techniques for electron microscopic morphometry. *In*: "Principles and Techniques of Electron Microscopy". (M. A. Hayat, ed.), Vol. 3, p. 237. Van Nostrand Reinhold Company, New York.

Weibel, E. R. and Gil, J. (1977). Structure-function relationships at the alveolar level. *In*: "Bioengineering Aspects of Lung Biology. (J. West, ed.), Vol. 3, p. 1. Marcel Dekker, New York–Basel.

Weibel, E. R. and Gomez, D. M. (1962). A principle for counting tissue structures on random sections. *J. Appl. Physiol.* **17**, 343.

Weibel, E. R. and Knight, B. W. (1964). A morphometric study on the thickness of the pulmonary air-blood barrier. *J. Cell Biol.* **21**, 367.

Weibel, E. R. and Paumgartner, D. (1978). Integrated stereological and biochemical studies on hepatocytic membranes. II. Correction of section thickness effect on volume and surface density estimates. *J. Cell Biol.* **77**, 584.

Weibel, E. R., Kistler, G. S. and Scherle, W. F. (1966). Practical stereological methods for morphometric cytology. *J. Cell Biol.* **30**, 23.

Weibel, E. R., Stäubli, W., Gnägi, H. R. and Hess, F. A. (1969). Correlated morphometric and biochemical studies on the liver cell. I. Morphometric model, stereologic methods and normal morphometric data for rat liver. *J. Cell Biol.* **42**, 68.

Weibel, E. R., Untersee, P., Gil, J. and Zulauf, M. (1973). Morphometric estimation of pulmonary diffusion capacity. VI. Effect of varying positive pressure inflation of air spaces. *Respir. Physiol.* **18**, 285.

Weibel, E. R., Losa, G. and Bolender, R. P. (1976). Stereological methods for estimating relative membrane surface area in freeze-fracture preparations of subcellular fractions. *J. Microscopy* **107**, 255.

Whitehouse, W. J. (1974a). The quantitative morphology of anisotropic trabecular bone. *J. Microscopy* **101**, 153.

Whitehouse, W. J. (1974b). A stereological method for calculating internal surface areas in structures which have become anisotropic as the result of linear expansions or contractions. *J. Microscopy* **101**, 169.

Whur, P., Herscovics, A. and Leblond, C. P. (1969). Radioautographic visualisation of the incorporation of galactose-^{3}H and mannose-^{3}H by rat thyroids *in vitro* in relation to the stages of thyroglobulin synthesis. *J. Cell Biol.* **43**, 289.

Wicksell, S. D. (1925). The corpuscle problem I. *Biometrica* **17**, 84.

Wicksell, S. D. (1926). The corpuscle problem II. *Biometrica* **18**, 152.

Williams, M. A. and Mayhew, T. M. (1973). Quantitative microscopical studies of the mouse peritoneal macrophage following stimulation *in vivo*. *Z. Zellforsch. mikrosk. Anat.* **140**, 187.

Winkelmann, E., Kunz, G., Wenzel, J. und Kirsche, W. (1974). Licht- und elektronenmikroskopische Untersuchungen an synaptischen Formationen im Neocortex (Area 17) der Albinoratte. *Z. mikrosk.-anat. Forsch.* **88**, 148.

Wisse, E. (1969). An electron microscopic study of the fenestrated endothelial lining of rat liver sinusoids. *J. Ultrastruct. Res.* **31**, 125.

Wyss, U. R. (1971). Analysis of dendrite patterns by use of an adaptive scan system. *J. Microscopy* **95**, 269.

Yang, G. C. H. and Shea, S. M. (1975). The precise measurement of the thickness of ultrathin sections by a "re-sectioned section" technique. *J. Microscopy* **103**, 385.

Zilles, K., Schleicher, A. and Kretschmann, H.-J. (1976). Semi-automatic morphometric analysis of the nucleolar development in the nucl. N. oculomotorii of Tupaia
belangeri during ontogenesis. *Anat. Embryol.* **149**, 15.

III.   List of journals publishing articles on stereology and its application on a regular basis

Journal of Microscopy *(J. Microsc.)* (Blackwell's, Oxford). Official journal of the
International Society for Stereology, published with Royal Microscopical Society,
London/Oxford; formerly (until 1968) Journal of the Royal Microscopical Society
(J. Roy. Microsc. Soc.).
Microscopica Acta *(Microsc. Acta)* (Stuttgart), formerly (until 1971) Zeitschrift für
wissenschaftliche Mikroskopie und mikroskopische Technik *(Z. wiss. Mikrosk.)*.
Transactions of the American Institute of Mining, Metallurgical (and Petroleum)
Engineers *(Trans. AIME)* (New York), particularly branch of Metals Society.
Biometrika *(Biometrika)* (Cambridge).
Biometrische Zeitschrift *(Biometr. Zschr.)* (Berlin, DDR).
Journal of Applied Probability *(J. appl. Prob.)* (East Lansing, Mich.).
Journal of Cell Biology *(J. Cell Biol.)* (New York).
Journal of Ultrastructure Research *(J. Ultrastruct. Res.)* (New York).
Laboratory Investigation *(Lab. Invest.)* (New York).
Micron: Quarterly journal of electron microscopy *(Micron)* (West Watford, Herts.).
Mikroskopie *(Mikroskopie)* (Wien).

Author Index

Subject Index

403